Combinatorial
Materials Synthesis

Combinatorial Materials Synthesis

edited by
Xiao-Dong Xiang

Intematix Corporation
Moraga, California, U.S.A.

Ichiro Takeuchi

University of Maryland
College Park, Maryland, U.S.A.

CRC Press
Taylor & Francis Group
Boca Raton London New York

CRC Press is an imprint of the
Taylor & Francis Group, an **informa** business

CRC Press
Taylor & Francis Group
6000 Broken Sound Parkway NW, Suite 300
Boca Raton, FL 33487-2742

First issued in paperback 2019

ISBN-13: 978-0-8247-4119-8 (hbk)
ISBN-13: 978-0-367-39508-7 (pbk)

Library of Congress Cataloging-in-Publication Data
A catalog record for this book is available from the Library of Congress.

Visit the Taylor & Francis Web site at
http://www.taylorandfrancis.com

and the CRC Press Web site at
http://www.crcpress.com

Foreword

Combinatorial materials synthesis and characterization is a new technique, borrowed from the pharmaceutical industry, and adapted and improved by materials scientists, that has now achieved major status in the materials profession, all in the space of some seven years. It has already acquired that ultimate feature of successful innovations, a popular nickname: combi. The book assembled here by Drs. Xiang and Takeuchi offers an effective overview of this new technique of synthesizing and comparatively assessing large ''libraries'' of novel materials in a rapid and efficacious manner. Synthesis and assessment by rapid techniques are both equally important, and they receive balanced attention in this book.

Some chapters focus on techniques of synthesis; two of them center on the ''continuous composition spread'' approach (a.k.a. ''continuous phase diagrams''), which is at the heart of combi. The idea of creating a film of (say) a ternary system with continuously and systematically varying composition is decades old, but the methods of achieving it had to await the advent of sophisticated computer techniques. Other techniques generate large numbers of tiny, separate deposits, for example, by molecular beam epitaxy, presented here by a lively team in Japan. Yet other chapters focus on general techniques of characterization, specialized to determine properties from tiny areas of film or powder. A chapter on the use of x-ray diffraction in combi offers a particularly detailed and subtle analysis of the best design of a diffractometer for this purpose, while another chapter describes in great detail a microwave approach for determining electrical impedance of items in a combi library. Other chapters exemplify actual combi exercises, for instance, the combinatorial synthesis and assessment of large libraries of phosphors (luminescent materials) and the synthesis and measurement at a range of temperatures of materials for gas sensors. This last chapter makes use of MEMS techniques for shaping tiny, thermally insulated freestanding platforms—a proper exemplification of a "nano technique."

The preceding paragraph is by no means an exhaustive list of the valuable features of this book. It would not have been possible, within the covers of a single volume, to include all aspects of combinatorial technique; thus, polymers and structural alloys, both of which groups have benefited from combi approaches, are not featured. The reader is offered a unique, authoritative overview of a major new field of materials science and engineering. I salute the editors' and contributors' achievement.

Robert W. Cahn, FRS
University of Cambridge
Cambridge, England

Preface

The year 2003 marks the tenth anniversary of the first combinatorial materials science experiment performed at Lawrence Berkeley National Laboratory, Berkeley, California. There have been substantial developments in high-throughput experimentation in materials science in the past decade. It has been truly remarkable to watch the basic idea of combinatorial strategy unfold in so many different ways in a variety of materials applications. From integrated materials chips and thin-film composition spreads to novel sensor arrays for catalysis and scanning microanalysis techniques, the combinatorial approach is having a major influence in advancing our materials exploration capabilities. It is important to remember that the basic premise of the approach was a simple one: to enhance the throughput of one's experimentation.

The combinatorial strategy is increasingly applied to new types of studies. It is now much more than a way to look for new materials. It is used to address materials issues at different levels in a wide range of topics ranging from rapid optimization of process and synthesis conditions to phase diagram mapping of complex functional materials. These studies are significantly broadening the scope of combinatorial experimentation. In many applications such as catalysis and polymers, the strategy is well on its way to becoming a mainstream method, just as it became a commonplace technique in the area of pharmaceutical research. At the same time, there are many materials systems that are yet to be explored in a high-throughout way, and many challenges remain in the areas of effective rapid characterization techniques and data management.

This book captures the general state of the field of combinatorial materials science as it stands today. Based on the progress made thus far, we are hopeful that researchers around the world will continue to find ways to apply the strategy for accelerated innovation with large and immediate impacts on various aspects of materials science.

We would like to take this opportunity to acknowledge the support and advice of Peter G. Schultz. We are deeply indebted to him for the vision and encouragement that started us on this journey.

Xiao-Dong Xiang
Ichiro Takeuchi

Contents

Contributors

Rodion V. Belosludov, Ph.D. Department of Materials Chemistry, Graduate School of Engineering, Tohoku University, Sendai, Japan

Hauyee Chang, Ph.D. Department of Chemistry, University of California, Berkeley, Berkeley, California, U.S.A.

William Chang, Ph.D. Instrument and Components Department, Advanced Research and Applications Corporation, Sunnyvale, California, U.S.A.

Chang-Ming Cheng, Ph.D. Shanghai Institute of Nuclear Research, Chinese Academy of Sciences, Shanghai, China

Joseph J. Hanak, Ph.D. Consultant, Ames, Iowa, U.S.A.

Masashi Kawasaki, Ph.D. Institute for Materials Research, Tohoku University, Sendai, Japan

Hideomi Koinuma, Ph.D. Frontier Collaborative Research Center and Materials and Structures Laboratory, Tokyo Institute of Technology, Yokohama, Japan

Momoji Kubo, Ph.D. Department of Materials Chemistry, Graduate School of Engineering, Tohoku University, Sendai, Japan

Min-Qian Li, Ph.D.   Shanghai Institute of Nuclear Research, Chinese Academy of Sciences, Shanghai, China

Mikk Lippmaa, D.Tech.   Institute for Solid State Physics, University of Tokyo, Kashiwa, Japan

Xin-Quan Liu, Ph.D.   National Laboratory for Infrared Physics, Shanghai Institute of Nuclear Research, Chinese Academy of Sciences, Shanghai, China

Akira Miyamoto, Ph.D.   Department of Materials Chemistry, Graduate School of Engineering, Tohoku University, Sendai, Japan

Krishna Rajan, Sc.D.   Combinatorial Materials Science and Informatics Laboratory, Department of Materials Science and Engineering and Faculty of Information Technology, Rensselaer Polytechnic Institute, Troy, New York, U.S.A.

Lynn F. Schneemeyer, Ph.D.   Bell Laboratories, Lucent Technologies, Murray Hill, New Jersey, U.S.A.

Steve Semancik, Ph.D.   Chemical Science and Technology Laboratory, National Institute of Standards and Technology, Gaithersburg, Maryland, U.S.A.

Ted X. Sun, Ph.D.   Corporate Research and Development, General Electric Company, Schenectady, New York, U.S.A.

Seiichi Takami, Ph.D.   Department of Materials Chemistry, Graduate School of Engineering, Tohoku University, Sendai, Japan

Ichiro Takeuchi, Ph.D.   Department of Materials Science and Engineering, and Center for Superconductivity Research, University of Maryland, College Park, Maryland, U.S.A.

R. Bruce van Dover, Ph.D.*   Bell Laboratories, Lucent Technologies, Murray Hill, New Jersey, U.S.A.

Gang Wang, Ph.D.   Intematix Corporation, Moraga, California, U.S.A.

Xiao-Dong Xiang, Ph.D.   Intematix Corporation, Moraga, California, U.S.A.

Young K. Yoo, Ph.D.   Research & Development, Intematix Corporation, Moraga, California, U.S.A.

---

* *Current affiliation*: Department of Materials Science and Engineering, Cornell University, Ithaca, New York, U.S.A.

Combinatorial
Materials Synthesis

1
Introduction

Xiao-Dong Xiang
Intematix Corporation, Moraga, California, U.S.A.

Ichiro Takeuchi
University of Maryland, College Park, Maryland, U.S.A.

From Stone Age to Information Age, the development of functional materials has always played a key role in bringing advances to our society. In particular, applications of electronic, photonic, and magnetic materials have led to revolutions in computer and communication technology and in the electronics industry over the past several decades. This trend is expected to continue far into the future.

The periodic table forms the basis of modern chemistry and materials science. During the last century, materials scientists and engineers have focused mostly on understanding and applications of simple materials that are formed by one or two elements from the periodic table. However, serendipitous discoveries of complex materials with fascinating properties, such as high-temperature superconductivity, have begun to draw attention to multicomponent compounds with complicated structures. The number of known complex functional materials is miniscule compared to the vast phase space of all possible multicomponent compounds that can be formed by combining different elements from the periodic table. At the end of the 1980s, J. C. Phillips estimated that approximately 24,000 inorganic phases were known to mankind. Among them, 16,000 are binary and pseudobinary compounds and only 8,000 are ternary and pseudoternary compounds (1). If we choose about 60 elements from the periodic table to form ternary compounds, then there are $\sim$34,000 possible ternary systems. A large

1

fraction of them has yet to be explored. Within each ternary system, different stoichiometries can naturally give rise to myriad different structures and properties. The most comprehensive way to study a ternary system is to map its compositional phase diagram. To map the structure–composition–physical property relationship of ternary systems with the conventional method of making and testing one composition at a time with small compositional increments would require an enormous number of experiments. Given that theoretical techniques such as the density-functional approach and first-principles calculations cannot actually predict properties of complex structures, experimental investigation will continue to play a central role in materials research. It is our view that to systematically explore the inexhaustibly large phase space of unknown complex systems, the conventional one-by-one trial-and-error method has to be replaced with more time-efficient and cost-effective methods.

There have always been demands to increase the rate and efficiency of the materials exploration and development process. One traditional approach, of mapping a binary phase diagram, is to use a diffusion couple (2). At the interface of a diffusion couple, different phases are formed, depending on the local stoichiometry and the annealing condition. By using chemical and structural analysis techniques based on scanning electron microscopy, a phase diagram can be constructed. This approach was extended to ternary and higher-order systems in the 1970s (2). One drawback of this approach is the small size of useful diffused regions, which is typically limited to tens of microns. Most physical property characterization tools cannot be applied for detailed quantitative mapping at this scale.

Another approach researchers have been employing since the 1960s is the composition-spread technique, which makes use of much lower-spatial-resolution tools for both structural and physical property mapping. Kennedy et al. published a paper in 1965 that described this method for determination of a ternary-alloy phase diagram using e-beam coevaporation (3). The authors proposed to replace the conventional metallurgical technique with this method in order to perform quick mapping of ternary phase diagrams. However, due to nonuniform composition profiles across wafers, the resulting spread did not provide good one-to-one correspondence between compositional distribution on the wafers and regions in the phase diagram. The authors were able to demonstrate a qualitative agreement between the result of their phase diagram mapping and a reported phase diagram determined by the conventional technique, but there were some discrepancies due to a limitation in the accuracy of the determined stoichiometry and structure. In 1967, Miller and Shirn published a paper on a Au-SiO_2 composition spread fabricated by cosputtering (4). In this technique, the geometric arrangement of targets and the substrate allows a continuously varying composition ratio of Au/SiO_2 to be deposited on the substrate. Electrical resistivity was measured as a changing function of the weight percentage of Au. A similar method was used

by Hanak et al. to study the effect of grain size in transition metal alloy superconductors (5) and by Sawatzky and Kay to study the $Gd_3Fe_5O_{12}$ system (6) in 1969.
In 1970, Hanak described an approach using a one-cathode, multicomponent (two
or three) spliced target to cosputter binary or ternary composition spreads (7). In
fact, Hanak had invested a substantial effort in search of superconductors using
this method at the RCA Laboratories. Independently, Berlincourt, who was the
Director of the Physical Science Division at the U.S. Office of Naval Research,
proposed to launch a national effort to conduct a search for high-temperature
superconductors using similar synthesis methods and automated screening techniques in 1973 (8). This proposal was circulated among top materials scientists
and physicists, but it was ultimately rejected. Subsequently, the general high-
throughput approach was abandoned for the most part. One can speculate on the
reasons why the approach did not find widespread acceptance at the time. The lack
of appropriate tools (computers and sophisticated high-resolution characterization
tools) was by far the biggest reason. Compared to presentday technology, the
equipment available at the time was primitive in its measurement capabilities as
well as its data-handling capacities and speed. The lack of support by funding
agencies also contributed to the demise of the approach at the time.

At the height of research activities in high-T_c superconductivity, high-
throughput approaches began to re-emerge. Scientists again started developing
techniques to accelerate materials discovery processes. For instance, in Japan a
robotic approach was used to emulate and replace human activities of bulk materials synthesis (9). A solution synthesis technique was developed for the rapid
investigation of inorganic compounds in the United Kingdom in an effort to
quickly search new compositions of high-T_c superconductors (10).

In the 1980s and '90s, the field of biochemistry and the pharmaceutical
industry underwent a revolution with the development of methodology for speed-
ing up the drug discovery process. A variety of high-throughput synthesis and
rapid screening schemes were invented to dramatically increase the efficiency of
the discovery process for organic compounds. These methods are collectively
known as *combinatorial chemistry* (11–17). Among these techniques, spatially
addressable libraries and gene chips had a significant impact and directly inspired
our work on the combinatorial approach to materials science. We began our effort
to develop a methodology of systematic synthesis and high-throughput screening
of inorganic materials at Lawrence Berkeley National Laboratory in 1993
(18–20). Combinatorial materials libraries can be constructed to contain diverse
discrete compositions aimed at exploring a large segment of a particular compositional landscape. They may also be designed to contain continuously varying
compositions (as in ternary phase diagrams) aimed at identifying phase boundaries
and the detection of important narrow-phase regions of interest. Mathematical
strategies are sometimes incorporated to optimize the efficiency of the library
design, i.e., to maximize the number of different combinations of elements created

in individual libraries. Many examples of these experiments are described in this book.

The timeliness of this work was evident in the fact that over the last few years, this approach has been widely embraced and further developed by scientists around the world in both academic and industrial research labs. It has been used in a wide range of electronic, magnetic, and photonic applications as well as in battery, fuel cell, and catalysis applications. Nowadays, materials experimentations involving any type of multisample-at-a-time, high-throughout procedures are often grouped together under the name of *combinatorial materials science*.

''A blind shotgun technique'' and ''Edisonian science'' are some of the phrases opponents as well as proponents have used to describe the combinatorial approach. Such characterizations miss the fact that the approach is merely a reflection of advances in experimental techniques and tools, which now enable thousands of experiments to be carried out within the time frame of one traditional experiment. The notion that good experiments require good insight and design has not changed. The approach simply represents an innovation in the materials exploration process.

One can now begin to think about systematically studying and classifying materials systems. For instance, one can create catalogues of ternary phase diagrams (including quaternary oxides, nitrides, and hydrides) and their important physical properties. Building such comprehensive databases on complex materials systems will deepen and expand our understanding of materials properties. Such an exercise still represents a daunting task, even with the aid of state-of-the-art technology. Yet, at least in principle, it is technically feasible now. In many ways, this is analogous to genome mapping. The benefits one can reap from such databases are enormous.

Experimental tasks involved in combinatorial materials research can be divided into three major activities that are closely interconnected: synthesis, materials diagnostics, and rapid characterization of physical properties. Synthesis is naturally an essential part of the experiment. It is desirable to have versatile fabrication tools, which can be used to create a variety of combinatorial libraries and composition spread samples. Our main emphasis has been thin-film experiments, anticipating and focusing on immediate applications to the electronics industry. In addition to compositional variables, materials processing parameters such as temperature profiles and atmosphere in heat treatment can also be varied in a combinatorial way.

One's ability to successfully perform a combinatorial investigation critically hinges on having the right characterization tools, and their availability largely determines what materials systems can be tackled. Thus, it is desirable to have a tool that can be used to measure multiple functionalities and multiple figures of merit for different applications. Generally, screening techniques involved in the combinatorial materials science are more challenging and sophisticated than those of combinatorial chemistry in the life sciences, where the luminescence-tagging technique is predominantly used. Properties ranging from superconductivity, electrical and dielectric properties, magnetic properties, mechanical properties, thermal properties, and optical properties, with each containing a large group of parameters, have to be characterized. Fortunately, many high-throughput screening tools have been made available by physicists and engineers. Scanning probe microscopes represent ideal screening tools for the combinatorial approach since they provide a noncontact means of spatially resolved physical property mapping with a relatively quick turnaround time. There are always developments in measurement instrumentation that will continue to bring about advances in novel mapping/imaging techniques.

An integral part of any materials synthesis effort is materials diagnostics: There is a need to check and confirm the formation and presence of intended phases at intended locations on libraries. Obtaining accurate composition and phase mapping is of paramount importance in establishing the composition–structure–property relationship of materials. The significance of this aspect of research is often overlooked. The necessity to confirm the compositional and phase distribution across libraries is perennial and common to all fields of combinatorial research, including the pharmaceuticals industry. In combinatorial materials science, a variety of analytical characterization tools are employed for this purpose. They include x-ray microdiffraction, Rutherford backscattering, and electron probe techniques.

In addition to time and cost efficiencies, the environmental benefits of the combinatorial approach are also large. For instance, assuming that a typical bulk sample size is 0.1 cm^3 in volume, if one were to synthesize 100–$10,000$ bulk samples to map a ternary phase diagram, the total amount of materials needed is 10–1000 cm^3. Compare this to the amount of materials used to make a thin-film library on a 1-cm^2 chip. If the library consists of 100-nm-hick films, the volume of all materials on the library is only 10^{-5} cm^3. This represents a reduction in materials resources used, by more than six orders of magnitude.

This book contains detailed accounts of many aspects of combinatorial materials science, with examples from the synthesis of composition spreads to rapid characterization using novel scanning probe microscopes. The scope of the approach is far-reaching, but the topics covered in this book mostly reflect our main interest and background in thin-film materials and devices. This is also where we believe some of the most exciting developments in the field are taking place right now. We have also included chapters on some recent developments in the "nonexperimental" aspects of the field, including materials informatics and the computational approach. Although this book is not intended to be comprehensive, we hope it provides a good flavor and "snapshot" of this rapidly emerging and changing field. By exploring topics that are both application specific and fundamental to the approach, we hope it will lead to new ideas and applications in the future.

REFERENCES

1. Phillips, J.C. Physics of High-T_c Superconductors; Academic Press: Washington DC, 1989.
2. Hasebe, M.; Nishizawa, T. Application of Phase Diagrams in Metallurgy and Ceramics: Washington DC, 1977, 2, NBS special pub. 496, p 911.
3. Kennedy, K.; Stefansky, T.; Davy, G.; Zacky, V. F.; Parker, E. R. J. Appl. Physics. 1965, 36, 3808.
4. Miller, N.C.; Shirn, G.A. Applied Physics Letters. 1967, 10, 86.
5. Hanak, J.J.; Gittleman, J. I.; Pellicane, J. P.; Bozowski, S. Physics Letters. 1969, 30A, 201.
6. Sawatzky, E.; Kay, E. IBM J. Res. Develop. Nov. 1969, 696.
7. Hanak, J.J. J. Mat. Sci. 1970, 5, 964.
8. Berlincourt, T. private communication.
9. Yamauchi et al., H. ISTEC J. (in Japanese). 1992, 5, 25.
10. Hall, S. R.; Harrison, M. R. Chemistry Britain. Sept. 1994, 739.
11. Tonegowa, S. Nature. 1983, 302, 575.
12. Scott, J. K.; Smith, G. P. Science. 1990, 249, 386.
13. Fodor, S. P. A.; Read, J. L.; Pirrung, C.; Stryer, L.; Lu, et al., A. T. Science. 1991, 251, 767.
14. Tuerk, C.; Gold, L. Science. 1990, 249, 505.
15. Lam, K. S.; Salmon, S. E.; Hersh, E. M.; Hruby, V.; Kazmierski, et al., M. Nature. 1992, 354, 82.
16. Needels et al., M. C. Proc. Natl. Acad. Sci. USA. 1993, 90, 10700.
17. Bunin, B. A.; Plunkett, M. J.; Ellman, J. A. Proc. Natl. Acad. Sci. USA. 1994, 91, 4708.
18. Xiang, X.-D. Annu. Rev. Mater. Sci. 1999, 29, 149.
19. Xiang, X.-D.; Shultz, P. G. Physica C. 1997, 282–287, 428.
20. Xiang, X.-D.; Sun, X.; Briceño, G.; Lou, Y.; Wang, K.-A.; Chang, H.; Wallace-Freedman, W. G.; Chen, S.-W.; Schultz, P. G. Science. 1995, 268, 1738–1740.

2

Multiple-Sample Concept: The Forerunner of Combinatorial Materials Science

Joseph J. Hanak
Consultant, Ames, Iowa, U.S.A.

I. INTRODUCTION

In the middle of the 20th century, RCA, then known as the Radio Corporation of America, became the leader in electronics technology, having invented, produced, and popularized black-and-white and color television. This position led the company to an interest and involvement in a broad spectrum of electronics. One of the consequences was that by the 1960s the RCA Research Center in Princeton, NJ, became the second leading electronics research center in the United States, the first being Bell Laboratories.

The evolution of electronics introduced the requirement of developing a multitude of new, solid-state materials. At that time, the traditional approach to the search for new materials was that of handling one material at a time in the processes of synthesis, chemical analysis, and testing of properties, a time-consuming and expensive process. There were several reasons for searching for a faster process, the main one being that the number of known materials in comparison to the possible ones is dismally small for the combinations of three or more elements and that the situation becomes progressively worse for higher combinations, as shown in Figure 1.

Consequently, in the early days of my career at RCA, while at first developing new superconductors, I sought a way to increase the productivity of the search for new materials and succeeded in introducing a novel approach, named the multiple-sample concept, identified by the *Chemical and Engineering News* (1)

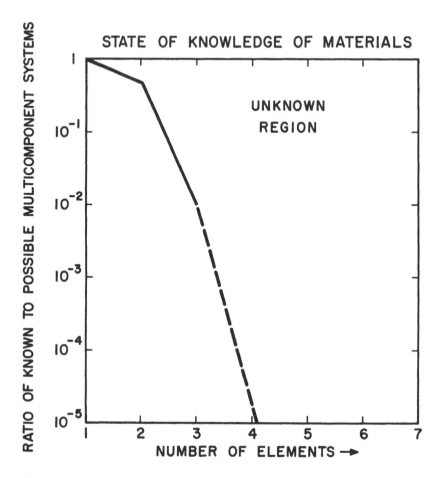

Figure 1   The ratio of known vs. possible materials as a function of the number of elements in the multicomponent system (ca. year 1970).

and two other publications (2,3) as the forerunner of combinatorial materials science. Ron Dagani in *C&EN* stated that:

> Joseph J. Hanak was close to sparking a revolution in materials research some three decades ago, but the world wasn't ready for it yet . . .
>
> Hanak's idea was the forerunner of a concept that is now taking materials science by storm: the application of the combinatorial approach, which the pharmaceutical industry uses for the rapid synthesis and screening of large collections of new drug candidates, to the discovery of useful new materials.

This chapter reviews the multi-sample concept (MSC) approach (4–7) and lists some of its results and useful consequences. Due to the fact that most of the references on MSC are several decades old, all of them are listed, to provide a complete description.

II. OUTLINE OF THE MULTIPLE-SAMPLE CONCEPT

The automated MSC approach (4) involves processing of entire or partial multi-component systems at one time, instead of just one composition. A significant increase in the rate of acquiring new information results from the fact that a given multicomponent system can be processed in almost the same time as a single sample. Implementation of the new approach involved first adopting the existing radiofrequency (rf) cosputtering technique (8,9), capable of synthesizing most and nearly complete binary and/or ternary solid alloy systems, deposited on flat substrates in the form of thin films, having continuously variable composition. The rf cosputtering technique was developed further by using single, composite targets instead of multiple, separately powered targets. Single, planar substrates parallel to the targets were used to synthesize binary systems (10,11) and then ternary solid alloy systems (4). Figures 2 and 3 show schematic arrangements of the targets and substrates for cosputtering films of two- and three-component systems, respectively, having continuously variable composition (4).

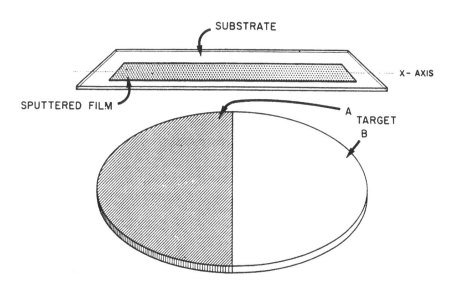

Figure 2 Schematic arrangement for cosputtering two-component systems. (From Ref. 4.)

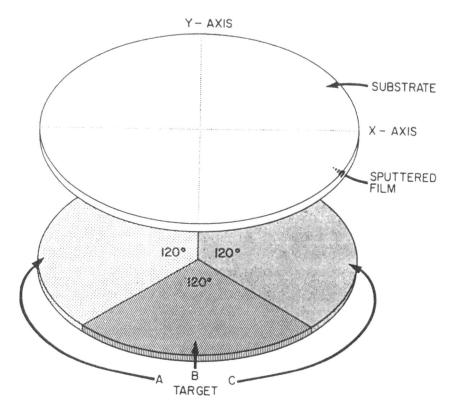

Figure 3   Schematic arrangement for cosputtering three-component systems. (From Ref. 4.)

The second step in implementing the MSC approach was the development of a unique method of compositional determination applicable to all cosputtered films (4). This method is based on the measurement of a simple, extensive property common to all deposited films, namely, film thickness. *In order to obtain analyses for the entire composition range, the only requirements are two thickness measurements for a binary system and three thickness measurements for a ternary system.* This method is based on the superposition principle and on the use of empirical determination of deposition profiles G_j, both in Eq. (1) and both discussed further on. This method was used with success primarily for the study of binary systems, such as superconductors, as discussed later.

The sputtering arrangement in Figure 4, which uses polar coordinates, is helpful in explaining this method (6). On the target, point O is the center of the target, point O' is a reference point on the substrate (here, above the center of the target disk), and P_O is any point on the substrate. The superposition principle, on which the determination of composition is based, states that the total film thickness $T(P_O)$ at P_O consists of the sum n of the thicknesses $T_j (P_O)$ of each constituent j:

$$T(Po) = t \sum_{j=1}^{n} T_j(Po) = t \sum_{j=1}^{n} G_j(Po)R_j(O') \qquad (1)$$

Eq. (1) also shows that each thickness consists of a product of the sputtering time t, a deposition profile $G_j(P_O)$, which is assumed to be dependent only on the geometry of the sputtering arrangement, and a deposition rate $R_j(O')$, measured at some reference point on the substrate.

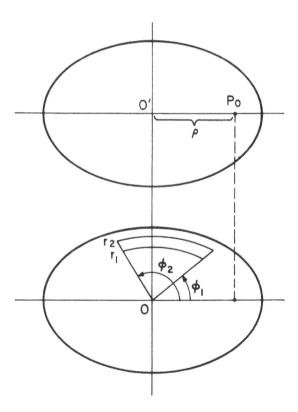

Figure 4 Sputtering arrangement showing polar coordinates appropriate for a segment of the annulus and a disk segment. (From Ref. 6.)

On the right-hand side of Eq. (1), t is known and $G_j(P_O)$ was first determined empirically (4,5) and used primarily for two-component systems.

A much faster and more practical method of calculating the deposition profiles was then developed by Hanak, Lehmann, and Wehner (6), based on expressions derived from Knudsen's cosine law. Expressions have been derived for target segments of an annulus, shown in Figure 4, using polar coordinates, and for a rectangle, using Cartesian coordinates, which together are applicable for calculating deposition profiles of most practical composite target shapes, such as those shown in Figure 5 (6). This method became very useful for ternary systems, shown in the examples that follow. An example of a deposition profile for a binary target, using two half-disk targets, is shown in Figure 6. A set of profiles for a ternary target having a shape of a 120° segment of a disk is shown in Figure 7 (6).

The unknown deposition rates $R_j(O')$ can be determined in several ways. One of them is by measuring the total thickness $T(P_O)$ at n or more locations on the substrate, substituting them in Eq. (1), and solving the resulting simultaneous equations (4,5). For low concentrations, such as doping, where small targets placed on large targets are used, one method consists of measuring the target emission rates by weight loss or sputter-etch depths and the sticking coefficients and then calculating $R_j(O')$ (7). Another method consists of using actual chemical or instrumental analysis and thickness measurements on one or more points and calculating $R_j(O')$ (7).

Once the rates $R_j(O')$ are known, the thickness at any point of the substrate can be calculated, as implied by Eq. (1). Composition, in volume percent, can be obtained from Eq. (2),

$$\text{Vol.}\%\, j = 100[T_j(Po)/T(Po)] \tag{2}$$

which uses the calculated thicknesses. Composition, in terms of mole %, can be obtained with the knowledge of the density and atomic weight of each of the components.

Although approximate, the first method for determining $R_j(O')$, which is applicable to concentrations of ~15–85%, has been shown to yield accuracy (rms) of better than 10% of the amount of each constituent (6). The second method, applicable from large concentrations down to about 0.1%, yielded accuracy of 25% or better. The third method, remaining to be tested, should be applicable to concentrations in the ppm level.

An example of calculated and observed composition of a binary Cu-Ni system as a function of position is shown in Figure 8, for the first and fourth runs, for which the difference will be discussed later. For each set, the maximum deviation between the calculated and observed is 5.3 volume % while the average deviation is 2.3 volume %.

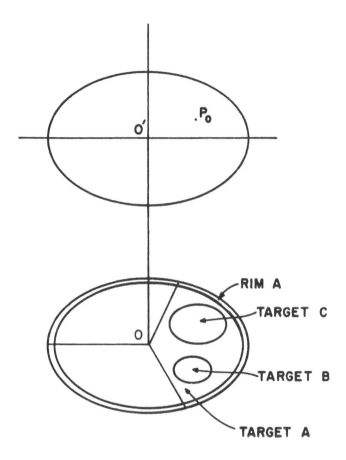

Figure 5 Sputtering arrangement with rim segments. (From Ref. 6.)

The third step for implementing the MSC involved the development of apparatus and methods, which facilitated measurement of materials properties by scanning methods (4,13,15,18). The final step involved collection of the large amounts of data by means of a data-acquisition system and by computer processing, both of which were in their infancy. (The lack of computers in the early 1970s is believed to be responsible for the non use of the MSC outside of RCA at that time.) An interim summary of the MSC automation process appeared in Ref. 18.

Although the reliability of cosputtering and the method of compositional calculations have withstood numerous tests, there have been reports of unexpected phenomena, which posed serious challenges. Among them are cross-contamina-

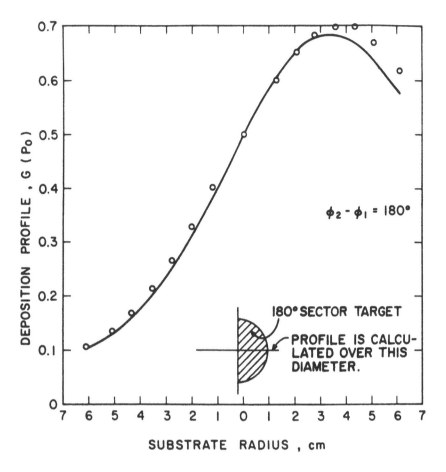

Figure 6 Experimental (○) and theoretical (—) deposition profile for a half-disk. (From Ref. 6.)

tion of targets, substrate bombardment by focused secondary electrons, and (then) newly observed etching of the substrate by anions (19). These problems have been investigated at RCA both experimentally and by modeling (16,17). A simplified model has been developed for cross-contamination of sputtering targets occurring during sputtering (16). Approximate, position-dependent expressions have been derived for the relative target-surface composition, target etch rates, and the time dependence of surface composition. Parameters that were found to affect cross-contamination include (a) the re-emission (resputtering) coefficient, (b) the deposition profile, (c) the ratio of emission rates, and (d) surface concentra-

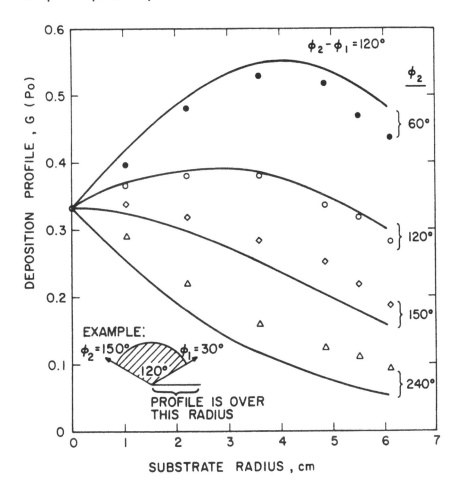

Figure 7 Experimental and theoretical profiles for a 120° disk segment. (From Ref. 6.)

tion. Taking into account the target cross-contamination, modifications were suggested for existing methods of the compositional calculations of cosputtered films.

Comparison of experimental data with the derived expressions was made and qualitative agreement obtained. The experiments involved cosputtering of nickel and copper, in which case the copper becomes extensively contaminated with nickel while the nickel target remains essentially uncontaminated. The sputtering rate of the copper target is thereby reduced nonuniformly, which affects

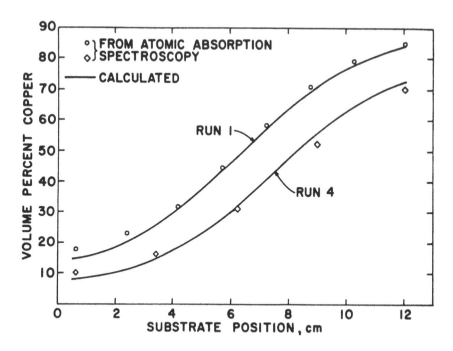

Figure 8   Composition of cosputtered Cu-Ni films. (From Ref. 6.)

the calculated thickness and composition profiles of the films, as shown in Figure 8.

An example of a problem caused by substrate bombardment by focused secondary electrons is a cosputtered system, Al_2O_3–In_2O_3 (19). In this case Al_2O_3 has sizable secondary electron emission characteristics, which produce a negative bias on the substrate and thereby reduce its sputtering rate. The secondary electrons were found to depress deposition rates over the Al_2O_3 target and to cause changes in the composition of the films. Consequently, thickness measurements as a function of position on the substrate revealed a discontinuous decrease in the region above the Al_2O_3 target. Fitting the data in the usual way, over the combined targets, produced a nonfitting curve. A practical solution to this problem was found by fitting the data using calculated deposition profiles over each partial target separately, which produced two separate, fitting curves, discontinuous over the center of the target. This method also applies to the calculation of the film composition.

The observed etching of the substrate by anions, in this case halide ions, including fluoride and chloride, was found not only to depress the deposition rates to zero but also to etch the substrate (19). In this experiment ZnS was cosputtered from a large, 6-inch-diameter target with TbF_3 or $TbCl_3$ using 1-inch diameter targets placed at the center. In both cases, vertically above the halide targets, the glass substrates were etched, while around this etched zone, film deposition with nearly normal thickness took place, although with vastly different concentration profiles of Tb and the halides. No plausible remedy has been offered for circumventing the nonideal concentration distribution due to the focused negative ions.

III. ELECTRONIC MATERIALS STUDIES

The MSC process has been applied to the study of electronic materials of interest to RCA. They included materials with superconducting, magnetic, optical, electroluminescent, photoconducting, and photovoltaic properties. Superconducting materials became important when materials maintaining superconductivity in a high magnetic field, such as Nb_3Sn, were discovered and the first high-field solenoid built; new magnetic materials were sought for constructing magnetic heads for high-speed recording; optical and electroluminescent materials were studied for possible applications in television displays; photoconducting materials were investigated for computer application and photography; photovoltaic materials were synthesized and tested for developing efficient and low-cost solar arrays.

Several of these materials were *cermets*, which are mixtures of finely divided metal particles and dielectric particles, formed preferably by rf cosputtering. When the volume fraction of the metal component is large, the metallic particles touch and the film exhibits metallic behavior. As the volume fraction of the metal decreases below approximately 0.6, the metal particles become smaller and isolated and the properties become nonmetallic. Electrical resistivity increases rapidly with decreasing metal content, exhibiting a negative temperature coefficient. The ordering temperature for both ferromagnetism and superconductivity decreases rapidly and the films become transparent in the infrared.

A. Granular and Compound Superconductors

RCA was very successful in developing a chemical vapor deposition method for the Nb_3Sn coating of steel ribbons to build high-field solenoids. Up to 14-tesla magnets, 15 inches long and 6 inches in diameter, have been built. The transition-temperature Nb_3Sn was 18 K. Consequently, to lower the cost of operating superconducting devices, materials with increased transition temperature (T_c) were sought, if possible at least over the boiling point of nitrogen (77 K).

One set of materials studied included binary mixtures of several metals and silica. They were cosputtered in the search for a high superconducting (sc) transition temperature (T_c) (11,13,14,18). The films were deposited onto 12.5-cm-long glass or ceramic alumina substrates, as shown in Figure 9. The substrates were precoated with 50 narrow gold strips to serve as electrical contacts for the sputtered layers. The rf targets used consisted of two half-disks. For depositing cermets, the silica targets were coated on the bottom side with a conducting metal layer, such as silver, to increase its electrical contact with the underlying disk electrode. Several metals have shown an increase in T_c, such as Mo-SiO$_2$, for which T_c increased from 0.9 K to 6.3 K, while other metals, such as La-SiO$_2$, have shown a decrease, as shown in Figure 10.

Mixtures of immiscible metals also show typical granular metal behavior, as shown in Figure 11. When both metals show enhancement in T_c of a cermet, one maximum in T_c occurs; when one metal enhances and the other does not, a maximum and a minimum in T_c are observed. Finally, when one metal enhances and the other is not a superconductor, a maximum in T_c is observed and the curve extrapolates to zero at some intermediate composition.

Other studies established a theoretical upper limit for the normal resistivity of granular superconductors at which superconductivity vanishes. Good agree-

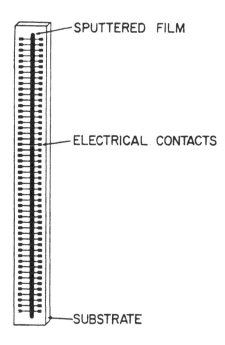

Figure 9   Substrate for the deposition of superconducting materials. (From Ref. 4.)

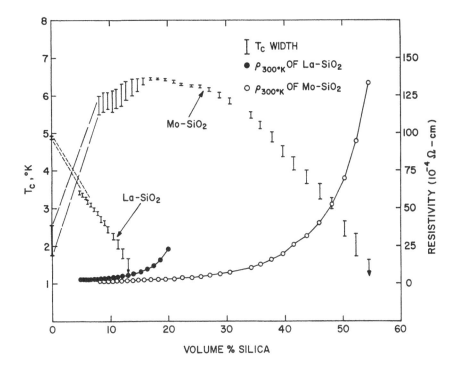

Figure 10 Superconducting transition temperature vs. composition of cermets. (From Ref. 12.)

ment has been found with the experimental value of 10^{-2} ohm-cm for binary cermets consisting of Al, Mo, W, and Be in the form of 30-Å grains embedded in SiO_2 (17). Fluctuation rounding of the superconducting transition in the three-dimensional regime also has been investigated. Excess conductivity due to superconducting fluctuations has been observed (16).

In the category of the A15 compounds, work has been done (10,13,18) both to reproduce T_c values of bulk materials in films and also to obtain metastable compositions with T_c higher than obtainable in the bulk. For example, the compound Nb_3Ge, when prepared by sintering, is deficient in Ge and its T_c is low (6 K), while when cosputtered onto frigid alumina substrates, followed by careful anneal, its T_c increases up to 17 K.

Although the progress on increasing T_c was promising, work on superconductivity was discontinued after RCA sold this business to a GE subsidiary.

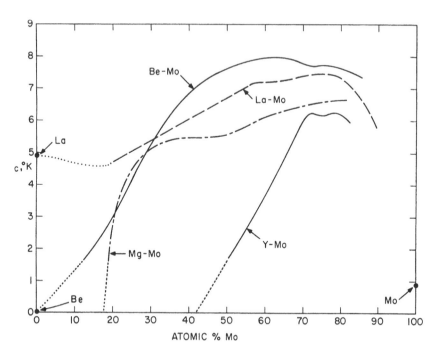

Figure 11 Superconducting transition temperature vs. composition for binary granular film mixtures of immiscible metals. (From Ref. 18.)

B. Granular Ferromagnets

Following the analogy of granular superconductors, Gittleman et al. (22) performed a systematic study of the electrical and magnetic properties of a cosputtered Ni-SiO$_2$ system as a function of composition, including resistivity, magnetoresistance, and magnetization, the latter by the use of the magneto-optic Kerr effect. In another study, also at RCA, Rayl et al. (23) have shown a decrease of Ni grain size and in Curie temperature with decreasing Ni content for similar samples. Noting a wide range of permeabilities available in the Fe-SiO$_2$ and 80 permalloy-SiO$_2$ cermets, Hanak and Gittleman (15) cosputtered the ternary system (Ni$_y$Fe$_{1-y}$)$_{1-x}$(SiO$_2$)$_x$ and measured its permeability and resistivity. In this case the rf-sputtering target consisted of three disk segments of Ni, Fe, and SiO$_2$ placed on the disk target electrode. The data shown in Figure 12 indicate that the material (Ni$_{.7}$Fe$_{.3}$)$_{.55}$(SiO$_2$)$_{.45}$ (by volume) had a particularly prominent permeability peak of $\mu \simeq 170$. A rather high electrical resistivity of 10^{-1} ohm-cm, shown in the Figure 13, was observed at the same composition. This material is applicable for high-frequency applications, such as magnetic recording heads and inductor cores. This composition can be reproduced in thick continuous films by using either sintered powder targets or a disk target of Fe overlayed with small islands of Ni and SiO_2.

It is important to note that the distribution of the data, as shown in Figures 12 and 13, follows the computed composition of the three-component system but not the geometry of the square substrate. The distribution of the composition as a function of position on a similar substrate for another three-component system is shown in Figure 15.

C. Optical Properties of Cermets

The first systematic investigation of the optical properties of cosputtered $Ag-SiO_2$ and $Au-SiO_2$ films, varied over nearly the entire range, was made by Cohen et

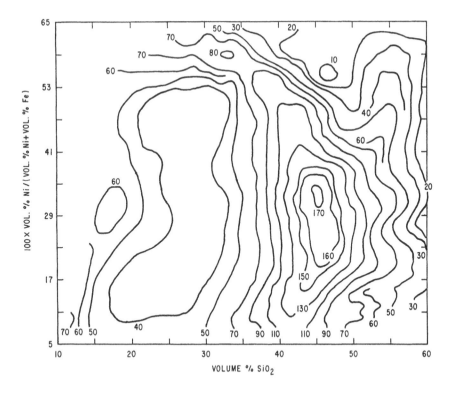

Figure 12 Contours of initial permeability as a function of composition for the ternary $(Ni_yFe_{1-y})_{1-x}(SiO_2)_x$ system. (From Ref. 15.)

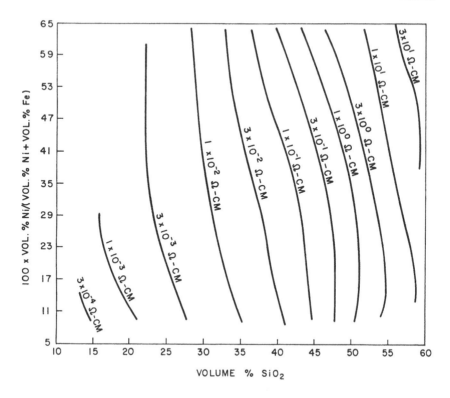

Figure 13 Contours of electrical resistivity as a function of composition for the $(Ni_y\text{-}Fe_{1-y})_{1-x}(SiO_2)_x$ system. (From Ref. 15.)

al. (24), also at RCA. They observed a gradual evolution of the characteristic absorption peaks in the visible and have studied the transition from metallic to dielectric behavior in the near infrared. The results were interpreted in terms of the theory of Maxwell–Garnett, which takes into account the modification of the applied electric field at any point within the medium by the dipole fields of the surrounding metal particles.

   One practical application of an optical property of cermets (e.g., Ni–SiO₂) involved their use in developing an improved intensity-correction filter for adjusting the light intensity over the field of the cathode ray tube (CRT) (25).

D.   Electroluminescence

The system $ZnS:Mn_x:Cu_y$ is well known for its yellow dc electroluminescence. However, electroluminescent cells that reached efficiency and life practical for

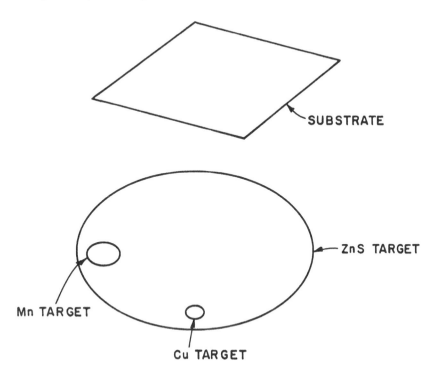

Figure 14 Schematic of a cosputtering setup of a ternary mixture of electroluminescents of $ZnS:Mn_x:Cu_y$. (From Ref. 18.)

displays were made in the past only from powder phosphors. In this work (26) a thorough study was made of the dc electroluminescent properties of this system on cosputtered films on glass and sandwiched between transparent conducting SnO_2 or In_2O_3 and a current-limiting cosputtered layer of $Ni–SiO_2$ cermet. A square array of electroluminescent cells was then formed, using a cosputtering target consisting of a large ZnS disk and two small disks of Mn and Cu, shown in Figure 14.

The resulting concentrations of the Mn and Cu dopants in ZnS, determined by the MSC method, are shown in Figure 15. To test the brightness, voltage, and current, a square array of 34×34 cells was defined on the sample, similar to that shown in Figure 17 for photovoltaic cells. Then tests were carried out at several levels of voltage. The test data of the EL performance in Figure 16 show that the optimum dopant concentration was Cu $\sim 0.3\%$ and Mn $\sim 0.7\%$. Mean brightness of up to 770 fL and 20 fL with pulsed dc excitation at 12.5% and 0.1% duty cycle, respectively, were achieved. A bright clock display was then made, using a target consisting of a large ZnS disk covered with numerous small,

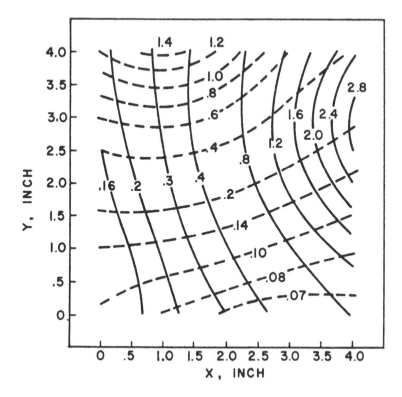

Figure 15   Concentration of Mn (—) and Cu (--) in mole % in $ZnS:Mn_x:Cu_y$, as a function of substrate position (x,y). (From Ref. 18.)

uniform pieces of Mn and Cu, yielding a uniform composition for the display. A detailed description of this work was reported in the final government contract report (27).

Additional systems were successfully studied to determine their potential applications for electroluminescent TV screens. They included green-emitting $ZnS:Tb^{+3}$ and a new, red-emitting, ternary, all-oxide electroluminescent film $[(Y_2O_3)_{y-1}(In_2O_3)_y]_{1-x} (Eu_2O_3)_x$ (28,29).

E.   Photovoltaics

The development of hydrogenated amorphous silicon (a-Si:H) photovoltaic (PV) cells was initiated first at RCA by Carlson and Wronski (30). The cells are multilayer thin-film devices, which were subject to the optimization of numerous

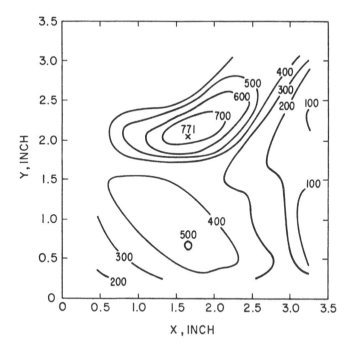

Figure 16 Contours of electroluminescence vs. composition of $ZnS:Mn_x:Cu_y$ in fL, showing the optimum dopant concentration as Cu $\sim$ 0.3% and Mn $\sim$ 0.7%. (From Ref. 18.)

of its parameters before becoming applicable for commercial use. These parameters include the deposition temperature, thickness, composition, deposition rate, gas pressure, deposition methods, deposition geometry, cell structures, and front and rear contact layers, among others.

In our approach to optimization (31,32) of a-Si:H photovoltaics (PV) cells we used the MSC to generate square, planar samples, in most cases having continuously graded pairs of some of these parameters at right angles to each other along the surface. This approach was used either on single layers having variables of deposition temperature and film thickness, which can be varied uniformly. In most cases, however, this approach was used for pairs of sequentially deposited layers, for which the thickness was varied in the direction perpendicular to each other. Cosputtered layers of a variety of cermets were used as contact layers to enhance the open-circuit voltage of the cells.

The a-Si:H films were deposited by the decomposition of silane (SiH_4), using the rf-capacitive, glow-discharge technique. The cell structure consisted of

Figure 17   Rear view of a solar cell sample having graded variables in the x- and y-directions and a matrix of 25 × 25 metal electrodes to yield numerous sets of performance data. (From Ref. 31.)

a glass substrate coated with a transparent conducting layer of In_2O_3, a thin cermet layer (Pt-Y_2O_3), p-i-n layers of a-Si:H, and a film of rear metal electrode (Cr/Al). Pure SiH_4 was used to form undoped i-layers. Mixtures of PH_3 or B_2H_6 in hydrogen and silane to form n- and p-doped layers of a-Si:H. Of particular interest for this chapter was the method of deposition of the semiconducting films having continuous variation of the substrate deposition temperature (T) and film thickness (D) to be applicable to MSC.

Variation of T was accomplished by mounting the glass substrate, in close contact, onto the bottom side of a graphite plate, heated by a cylindrical quartz lamp inserted in a cylindrical hole on one side of the plate. Typically, a gradient of up to a 220°C was generated along the plate. Variation of film thickness D along the transverse direction was achieved by using a motor-driven square mask covering the substrate surface.

A photograph of the rear side of test sample having continuously variable parameters in the x- and y-directions is shown in Figure 17 (31). This ''device library'' consists of a matrix of 25 × 25 metal electrodes deposited on a continuous PV cell structure, and it yields numerous sets of data upon testing. Tested variables, depending on the continuously built-in twin sets of variables, included photovoltage (V_{oc}), photocurrent (J_{sc}), fill factor (FF), cell efficiency (η), and optical stability. Examples of such data appear in Figures 18, 19, and 20, which show three different properties of the sample (31).

Steady improvements in the performance of a-Si:H PV cells were achieved, which then called for the development of series-interconnected cells to form simplified PV panels having the desired voltage to power devices. A "monolithic solar cell panel of amorphous silicon" was developed (33), starting with a large, continuous cell, then separated by laser scribing in one direction into parallel, 0.7-cm-wide cells, and then interconnected by sputtered or evaporated metal film, followed by further laser scribing. This method was successful and remains in use in commercial PV arrays to date.

Another type of device that was developed is that of multiple-junction, or "stacked," solar cells, because this concept held the promise of improved effi-

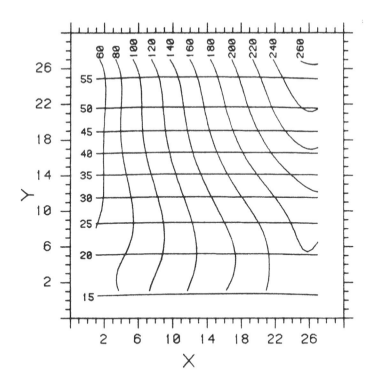

Figure 18 Contours of composition (vol. % Pt along y) and thickness (Å along x) of the transparent Pt-SiO$_2$ cermet contact layer as a function of substrate position. (From Ref. 31.)

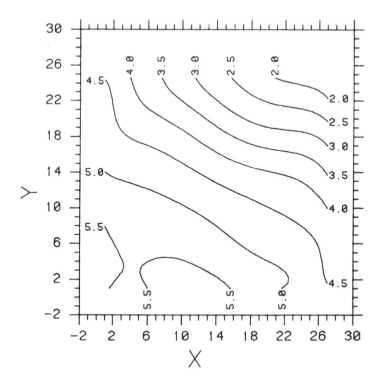

Figure 19 Dependence of current density (J_{sc}) on the thickness (x) and Pt content (y) of the cermet layer shown as a function of substrate position. (From Ref. 31.)

ciency and stability (34,35). The stacked cell thickness design was as follows. The front cell had its i-layer graded from 20 to 350 nm, while for the second cell it was graded in the transverse direction from 120 to 800 nm. The optimum cell performance occurred at an i-layer thickness of 38 nm for the first cell and 600 nm for the second cell. Later on, a-Si$_x$Ge$_{(1-x)}$H cells were used as the second or third cell in the stacked cells to increase the efficiency.

A description of additional photovoltaic systems or their applications using MSC research follows. Study of the optical stability of a-Si:H cells revealed a strong inverse dependence of optical stability on the i-layer thickness (36). For a particular i-layer the cell efficiency (η) remained stable up to a thickness of 200 nm. At higher thicknesses, η decreased because of a large increase in internal series resistance, an indication of the Staebler–Wronski effect. As a consequence of the stability of the thin films, a good stability was found in stacked cells.

Detailed quantum efficiency measurements performed on a-Si:H p-n diodes have shown that p layers are photovoltaically active (37). A 10-nm-thick p layer

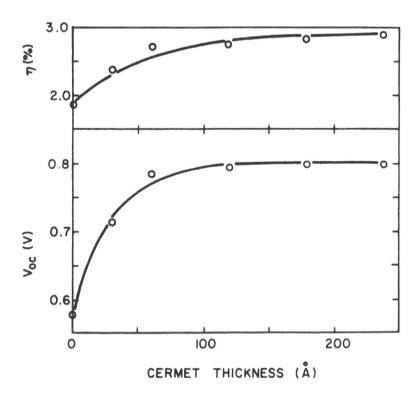

Figure 20 Dependence of PV cell efficiency and V_{oc} of the Pt-SiO$_2$ cermet, 12 vol. % Pt. (From Ref. 31.)

typically used in p-i-n cells could collect up to 1 mA/cm^2 if the electric field was greater than 10^5 V/cm.

A process has been developed at Sovonics Solar Systems, in Troy, MI, by J.J. Hanak (38) for fabrication of roll-up, monolithic photovoltaic modules made of a-Si:H alloys. They consist of tandem-junction solar cells deposited by a continuous roll-to-roll process onto thin foil substrates of bare metal or on high-temperature resin coated with metal film or metal coated with insulators. The modules have high power density and stability; they are portable, stowable, deployable, retractable, and tolerant to projectile impact.

The ultralight a-Si:H module was further developed into a deployable aerospace photovoltaic array for possible aerospace applications (39). The array design utilized 20 large cells, each 1566 cm^2 in area, connected in series to provide power of 207 W at AM0 illumination. The weight of the array was only 800 g, which resulted in a power density of 258 W/kg, about one order of magnitude higher than existing space arrays.

The multiple-sample concept, now referred to as combinatorial materials science, was also used by J.J. Hanak et al. (40) at First Solar, LLC, in rapid optimization of $CdTe_{(1-x)}S_x$ absorber films and photovoltaic devices. Vapor transport deposition was used to prepare the films. These films were shown to be effective as a barrier against interdiffusion between the semiconductor films, thereby facilitating the use of thin CdS ''window'' film and leading to enhanced photocurrent and efficiency of the photovoltaic cells. This method was also used to improve the cells by rapid optimization of other film parameters, including thickness, deposition temperature, energy gap, and surface texture, and by making the process suitable for high-throughput manufacturing environments.

At the end of the list of References at chapter's end is a list of 12 U.S. Patents, [P-1 to P-12], which describe the use of the MSC for the development of electroluminescent displays, photovoltaic cells, magnetic recording heads, liquid crystal displays, and optical shadow masks for kinescopes.

IV. CONCLUSIONS

At RCA, MSC was applied successfully in demonstrating quantum increases in the rate of synthesis, characterization, optimization, and application of materials. For two-component and three-component systems, the increases were, respectively, 50 and 1000 times greater than in the traditional synthesis involving just a single composition. Additionally, in each system studied, the composition of maximum performance was observed and, in some cases, the composition of a combination of desired properties.

MSC also was useful in generating 12 U.S. Patents at RCA in the areas of cermets, including CRT applications, superconductors, magnetic recording, and for solar cell contact films; electroluminescent films emitting yellow, green, and red light; liquid crystal devices having diode characteristics; and a variety of photovoltaic cell applications.

The technology that benefited most from MSC was thin-film a-Si:H photovoltaics, in which RCA made significant progress. Nevertheless, after making estimates of the costs for its commercialization, RCA sold the technology to Solarex, now owned by BP Solar, which became the largest a-Si:H PV array manufacturer. The next significant manufacturer of a-Si:H PV arrays in Troy, MI, is Baekart-USSC, which produces *flexible* arrays. Various other companies in the United States and worldwide are also producing a-Si:H arrays.

The technology that was likely to benefit RCA from MSC was superconductivity. RCA pioneered the production of high-field superconducting magnets using superconducting metal ribbons coated via a chemical vapor deposition

(CVD) process with Nb_3Sn (41). In the search for much better materials, up to nine times greater increases in T_c in binary granular metal mixtures have been discovered using MSC. However, in 1970 RCA sold the superconducting technology to a General Electric (GE) spinoff, Intermagnetics General (IG), which is now one of the leading companies in superconducting magnets, partially based on RCA's innovations. Since then, the high-field superconducting magnet technology has progressed substantially. The high-superconducting-T_c technology also progressed rapidly, and the new high-T_c materials have been confirmed by using the new combinatorial methodology (1).

Later, in 1985, RCA, numbering about 120,000 employees, was sold to GE, which now participates actively in the application and further development of combinatorial materials science.

ACKNOWLEDGMENTS

Acknowledgments are due to the personnel at the former RCA David Sarnoff Research Center, in Princeton, NJ, including Fred Rosi, a VP of happy memory; George Cody and David Carlson, group leaders in superconductivity and photovoltaics; John I. Gittleman, physicist; Victor Korsun, Joseph P Pellicane, and numerous other collaborators, most of whom are listed as coauthors in the References.

Special acknowledgments are due to Hans W. Lehmann and Roland K. Wehner (6), of the former RCA Labs in Zurich, Switzerland, for their major contribution in developing the method of calculating the deposition profiles based on Knudsen's cosine law.

REFERENCES

1. Dagani, R. A faster route to new materials. Chem. Eng. News. March 8, 1999, 77(10), 51–60.
2. Bruendli, C. Mit High Speed zu neuen Materialen, wachsende Interesse an kombinatorischer Chemie. Neuer Zuricher Zeitung. March 14, 2001, p 81.
3. Cahn, R. W. Combinatorial materials synthesis and screening. In: The coming of Materials Science; Elsevier Science: London, 2001, 444–446.
4. Hanak, J.J. The ''multiple-sample concept'' in materials research: synthesis, compositional analysis and testing of entire multicomponent systems. J. Materials Sci. 1970, 5, 964–971.
5. Hanak, J.J. Compositional determination of RF cosputtered multicomponent systems. J. Vacuum Sci. Technol. 1971, 8(1), 172–175.
6. Hanak, J.J.; Lehmann, H.W.; Wehner, R.K. Calculation of deposition profiles and compositional analysis of cosputtered films. J. Appl. Phys. 1972, 43, 1666–1673.

7. Hanak, J.J.; Bolker, B.F.K. Calculation of composition of dilute cosputtered multi-component films. J. Appl. Phys. 1973, 44(11), 5142–5147.
8. Miller, N.C.; Schirn, G.A. Appl. Phys. Letters. 1967, 10, 86.
9. Schwartz, G.C.; Jones, R.E.; Maissel, L.E. J. Vac. Sci. Technol. 1969, 6, 359.
10. Hanak, J.J. RF cosputtering of multicomponent materials. Proceedings of Conference and School on the Elements, Techniques and Applications of Sputtering: Brighton: England; October 20–22, 1969.
11. Hanak, J.J.; Gittleman, J.I.; Pellicane, J.P.; Bozowski, S. RF sputtering of supercon-ducting materials. Proceedings of Conference and School on the Elements, Tech-niques and Applications of Sputtering: Brighton: England; October 20–22, 1969.
12. Hanak, J.J.; Gittleman, J.I.; Pellicane, J.P.; Bozowski, S. The effect of grain size on the superconducting transition temperature of the transition metals. Phys. Letters. 1969, 30A, 201–202.
13. Hanak, J.J.; Gittleman, J.I.; Pellicane, J.P.; Bozowski, S. RF-sputtered films of beta-tungsten structure compounds. J. Appl. Phys. 1970, 41, 4958.
14. Hanak, J.J.; Gittleman, J.I. The effect of grain size on the superconducting transition temperature of transition metals. Physica. 1971, 55, 555–561.
15. Hanak, J.J.; Gittleman, J.I. Iron-nickel-silica ferromagnetic cermets. In: Magnetism and Magnetic Materials—1972; 18th Annual Conference, Denver; Graham, C.D. Jr., Rhyne, J.J., eds.; AIP Conference Proceedings, Part 2(10): New York, 1973, 961–965.
16. Abeles, B.; Hanak, J.J. Superconducting and semiconducting phases of granular films. Phys. Letters. 1971, 34A, 165–166.
17. Gittleman, J.I.; Cohen, R.W.; Hanak, J.J. Fluctuation rounding of the superconducting transition in the three-dimensional regime. Phys. Letters. 1969, 29A, 56–57.
18. Hanak, J.J. Automation of the search for electronic materials by means of co-sputter-ing. Proc 1er colloque international de pulverization cathodique et ses applications: Les Couches Minces; 1973, 177.
19. Hanak, J.J.; Pellicane, J.P. The effect of secondary electrons and negative ions on sputtering of films. J. Vac. Sci. Technol. 1976, 13(2), 406–409.
20. Hanak, J.J. Cosputtering—its limitations and possibilities. Le Vide. January/Febru-ary 1975, 175, 11.
21. Hanak, J.J.; Klopfenstein, R.W. Model of target cross-contamination during cosput-tering. RCA Rev. 1976, 37(2), 220–233.
22. Gittleman, J.I.; Goldstein, Y.; Bozowski, S. Phys. Rev. 1972, B1(5), 3609.
23. Rayl, M.; Wojtowicz, P.J.; Abrahams, M.S.; Harvey, R.L.; Buiocchi, C.J. Magnetism and Magnetic Materials—1972; Graham, C. D. Jr., Rhyne, J.J., eds.; AIP Conference Proceedings Part 2(5): New York, 1972, 472.
24. Cohen, R.W.; Cody, G.D.; Coutts, M.D.; Abeles, B. Phys. Rev. 1973, B8(8), 3689.
25. Hanak, J.J. Photodeposition of CRT screen structures using cermet IC filter. U.S. Patent No. 4157215, June 5, 1979.
26. Hanak, J.J. Electroluminescence in $ZnS:Mn_x:Cu_y$ RF-sputtered films. Japan J. Appl. Phys. 1974(Suppl 2, Part 1), 809–812.
27. Hanak, J.J.; Yocom, P.N. DC-Electroluminescent flat-panel display. R&D Technical Final Report, ECOM-0290-F, U.S. Army Electronics Command, Fort Monmouth: NJ, August 1973.

28. Pankove, J.I.; Lampert, M.A.; Hanak, J.J.; Berkeyheiser, JE Dependence of DC electroluminescence and host-excited photoluminescence on Tb^{3+} concentration in sputtered $ZnS:Tb^{3+}$ films. J. Luminescence. 1977, 15, 349–352.

29. Hanak, J.J. Electroluminescent device comprising electroluminescent layer of indium oxide and/or tin oxide. U.S. Patent No. 4027192, May 31, 1975.

30. Carlson, D.E.; Wronski, C.R. Appl. Phys. Letters. 1976, 28(11), 671.

31. Hanak, J.J.; Korsun, V.; Pellicane, J.P. Optimization studies in hydrogenated amorphous silicon solar cells. Proceedings of the 2nd European Communities Photovoltaic Solar Energy Conference, Berlin (West); D Reidel Publishing Co.: Dordrecht: The Netherlands, 1979, 220.

32. Hanak, J.J.; Korsun, V.; Pellicane, J.P. Optimization studies of materials in hydrogenated amorphous silicon solar cells, II. Proceedings of the 15th IEEE Photovoltaic Specialists Conference; 1981, 697–684.

33. Hanak, J.J.; Faughnan, B.W.; Korsun, V.; Pellicane, J.P. Development of amorphous silicon stacked cells. Proceedings of the 14th IEEE Photovoltaic Specialists Conference: San Diego: CA; 1980, 1209–1213.

34. Hanak, J.J. Stacked solar cells of amorphous silicon. J. Non-Crystalline Solids. 1980, 3536, 755–759.

35. Hanak, J.J. Monolithic solar cell panel of amorphous silicon. Solar Energy. 1979, 23, 145–147.

36. Hanak, J.J.; Korsun, V. Optical stability studies of a-Si:H solar cells. Proceedings of the 16th IEEE Photovoltaic Specialists Conference; 1982, 1381–1383.

37. Faughnan, B.W.; Hanak, J.J. Photovoltaicaly active P-layers of amorphous silicon. Appl. Phys. Lett. April 1983, 42(8), 722–724.

38. Hanak, J.J. Ultralight monolithic photovoltaic modules of amorphous silicon alloys. Proceedings of the 18th IEEE Photovoltaic Specialists Conference: Las Vegas: NV; 1985, 89–94.

39. Hanak, J.J.; Walter, L.; Dobias, D.; Flaisher, H. Deployable aerospace PV array based on amorphous silicon alloys. Proceedings of 9th Space Photovoltaic Research and Technology Conference; NASA Lewis Research Center: Cleveland: OH, April 19–21, 1988.

40. Hanak, J.J.; Bykov, E.; Elgamel, H.; Grecu, D.; Putz, J.; Reiter, N.; Shvydka, D.; Powell, R.C. Rapid optimization of $CdTe_{(1-x)}S_x$ absorber films and PV devices by means of the combinatorial method. Proceedings of the 26th IEEE Specialists Conference. September 2000, 495–498.

41. Synthesis, characterization, and application of superconducting niobium stannide Nb_3Sn. RCA Rev. XXV (3) (Special Issue). September 1964.

U.S. Patents

P-1. Hanak, J.J. Electroluminescent film and method for preparing same. U.S. Patent No. 3803438, April 9, 1974.

P-2. Gittleman, J.I.; Hanak, J.J. Sputtered granular ferromagnetic iron-nickel-silica films. U.S. Patent No. 843420, October 22, 1974.

P-3. Hanak, J.J. Electroluminescent cell with a current limiting layer of high resistivity. U.S. Patent No. 3919589, November 11, 1975.

P-4. Hanak, J.J. Electroluminescent device comprising electroluminescent layer containing indium oxide and/or tin oxide U.S. Patent No. 4027192, May 31, 1975.

P-5. Hanak, J.J.; Friel, R.N.; Goodman, L.A. Liquid crystal devices having diode characteristics. U.S. Patent No. 4042293, August 16, 1977.

P-6. Hanak, J.J. Photodeposition of CRT screen structures using cermet IC filter. U.S. Patent No. 4157215, June 5, 1979.

P-7. Hanak, J.J. Inverted amorphous silicon solar cell utilizing cermet layers. U.S. Patent No. 4162505, July 24, 1979.

P-8. Hanak, J.J. Cermet layers for amorphous silicon solar cell. U.S. Patent No. 4167015, September 4, 1975.

P-9. Hanak, J.J. Tandem junction amorphous silicon solar cell. U.S. Patent No. 4272641, June 9, 1981.

P-10. Hanak, J.J. Laser processing technique for fabricating series-connected and tandem junction series-connected solar cells into a solar battery. U.S. Patent No. 4292092, September 29, 1981.

P-11. Hanak, J.J. High voltage series-connected tandem junction solar battery.U.S. Patent No. 4316049, February 16, 1982.

P-12. Hanak, J.J. Method of depositing a semiconductor layer from a glow discharge. U.S. Patent No. 4,481,230. Nov 6, 1984.

3
The Continuous Composition Spread Approach

R. Bruce van Dover* and Lynn F. Schneemeyer
Lucent Technologies, Murray Hill, New Jersey, U.S.A.

I. INTRODUCTION

The continuous composition-spread (CCS) approach is based on a synthetic technique in which material is deposited on a substrate simultaneously from two or more sources that are spatially separated and chemically distinct, producing a thin film with an inherent composition gradient. With three sources, an entire ternary phase diagram may be produced in a single experiment, at least in principle. When this synthetic technique is combined with rapid, automated evaluation, the overall effect is a dramatic increase in the throughput of materials evaluation, compared to conventional single-sample studies. To date, CCS synthesis has been implemented using physical vapor deposition techniques, specifically evaporation and sputtering. It has been used to create both alloys and compounds in chemical systems, including intermetallics, nitrides, and oxides.

Initial work establishing the CCS approach dates back to the mid 1960s, so composition-spread approaches unquestionably antedate all other high-throughput discovery/evaluation methods. In 1965, Kennedy and coworkers published the first description of evaporated composition-spread films for rapid determination of an alloy phase diagram (1). Sawatzky and Kay (2) published a description of the CCS synthetic technique using cosputtering with independent sources. In 1970, Hanak (3) published a prescient declaration of the potential of high-throughput synthesis/evaluation techniques for materials investigations and provided details and examples for the particular case of multitarget sputtering. He

---

* *Current affiliation:* Cornell University, Ithaca, New York, U.S.A.

discussed the possibilities of one-dimensional and two-dimensional composition spreads using this technique, recognized the persistent problem of determining film compositions efficiently, and identified the need for rapid automated testing to complement the parallel synthetic technique.

A. Systematic Exploration of Materials Systems—One-Dimensional Spreads

Although Hanak (3) elucidated many of the issues involved in applying the CCS technique to two-dimensional composition spreads in his initial paper, for many years the only cases in which this technique was used were one-dimensional spreads. Typically, these were prepared for studies where the goal was to investigate properties as a function of composition in a known system rather than to discover new materials. In situations where run-to-run variations could confound systematic trends, the ability to create a set of samples prepared under identical conditions could be a critical improvement in experimental technique. For example, Hanak (3) reported on the composition dependence of the superconducting transition temperature of Mo-Mg alloys, and Sawatzky and Kay used the technique to study the effect of cation deficiency in a magnetic insulator "in an expeditious manner" (2).

For composition-spread synthesis, coevaporation is a much more challenging technique to implement than cosputtering, since evaporation rates are sensitive to the condition of the evaporant source and depend exponentially on the input power, thus making it very difficult to maintain constant rates during the run. Materials with relatively low evaporation temperatures can be deposited using temperature-stabilized crucibles in a UHV environment (i.e., MBE growth), but for refractory materials, electron beam evaporation is necessary, along with careful feedback control. Hammond and coworkers developed a scheme for this feedback that employed choppered ion gauge rate monitors and a dual time-constant feedback algorithm. This provided routine control to less than 1 at. % for each constituent and allowed Hammond and coworkers to synthesize spreads comprising refractory metals, such as V-Al and Nb-Al-Ge (4).

The one-dimensional composition-spread technique was also used by van Dover et al. (5,6) to elucidate the properties of amorphous Tb-Fe films as a function of composition—the first instance of cosputtering with independent magnetron guns known to the authors. The Tb-Fe system is the prototype for magneto-optical recording media because the large uniaxial anisotropy permits stable domains with the magnetization perpendicular to the plane of the film. Figure 1 shows the composition dependence of the intrinsic anisotropy as determined by measurements on a composition-spread sample. The origin of the anisotropy had been controversial, and the ability to show that the measured anisotropy

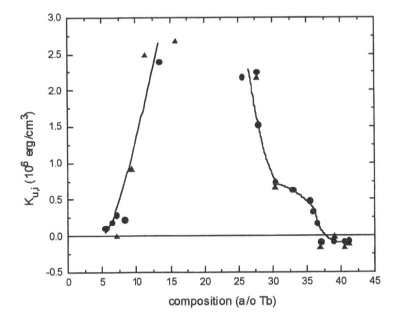

Figure 1 Intrinsic magnetic anisotropy, $K_{u,i}$ (i.e., with shape effects removed by analysis) of Tb-Fe films as a function of composition. Triangles represent data measured with a torque magnetometer, and circles represent data measured with a vibrating sample magnetometer. Data could not be analyzed for $16 < x < 25$ because the sample is too close to magnetic compensation to permit reliable measurements. (From Ref. 5.)

did not correlate with the magnetization allowed the authors to conclude that microstructural shape effects are not the main source of anisotropy.

A closely related technique is that of producing so-called "wedge" structures for investigating the effect of thickness (typically of a single component) on some property or behavior. This technique dates at least as far back as 1961, when Middelhoek (7) used a wedge-geometry thin film of Ni-Fe (Permalloy) to demonstrate the transition in magnetic domain wall structures from Bloch walls (where the film was thick) to Néel walls (where the film was thin). In 1981, the technique was used to investigate extension of the A15 phase boundary in homoepitaxial $Nb_{100-x}Al_x$. (8). In that work, Kwo and Geballe used tunneling measurements to determine that layers with a composition $x = 24$ could be stabilized in the A15 phase if grown on (equilibrium) material with $x = 20$, but only for thicknesses less than 200 Å, neatly resolving a puzzle that had been noted earlier (9). More recently, the technique was used by Ungaris et al. to great advantage in studying the effect of metal interlayer thickness on exchange coupling between magnetic thin films (10,11). Figure 2 presents a SEMPA (scanning electron microscopy with polarization analysis, a technique that is sensitive to the magnetic orientation of the imaged film) image of an Fe-Cr-Fe trilayer with varying Cr thickness, showing the quasiperiodic reversal of the sign of the exchange coupling due to the RKKY interaction. This wedge technique has also now been employed in a two-dimensional geometry, e.g., with variation of Cr thickness along one direction and variation of the Fe overlayer thickness in the orthogonal direction.

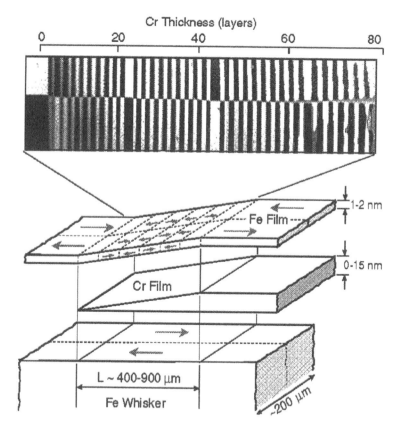

Figure 2 SEMPA (scanning electron microscopy with polarization analysis) image of a "wedge" sample, showing the quasiperiodic oscillations of the exchange coupling between two Fe films separated by a Cr film of 0–150Å. The light and dark contrast represents regions in which the magnetization of the top Fe layer is aligned or antialigned with that of the bottom layer. (From Ref. 11.)

B. Discovery of New Compounds or New Materials—One-Dimensional Spreads

Until recently, the CCS technique has been used primarily for systematic exploration of known materials systems. Nevertheless, the technique has also been used sporadically for the intentional discovery of new materials. Indeed, one of the initial reports of the technique concerned a new (albeit expected from simple materials considerations) compound, A15-structure V_3Al (12). A crucial aspect of this work was the ability to form the compound in situ, since the A15 structure was inferred to be unstable above $\sim 700°C$, well below the temperature required to process these intermetallic compounds by conventional metallurgy ($>1000°C$). Codeposited phase spreads were extensively used by Geballe, Hammond, and coworkers in subsequent work exploring binary and pseudobinary superconducting compounds.

The CCS technique was used by van Dover et al. (13) in 1993 to discover a true ternary nitride compound. In this experiment, a one-dimensional spread was employed, with the addition of a reactive sputtering atmosphere, specifically Ar/N_2. The authors found that extremely fine-grained material was obtained, even at a deposition temperature of 850°C, due to the refractory nature of the nitrides. Electron diffraction measurements on films with a composition $Gd_xCr_{1-x}N_y$, $0.23 < x < 0.4$ showed a powder ring pattern not derived from the endmember NaCl structure. A magnetic transition at about 10 K was associated with this phase. However, the structure of this phase could not be identified due to the extremely small crystallite size. The nominal stoichiometry of the phase was speculated to be $GdCr_2N_3$.

C. Extension to Two-Dimensional Spreads

Deposition of two-dimensional compositional spreads was anticipated by Hanak (3) in 1970, but very few studies have been published until recently. In 1972, Hanak and Gittleman (14) reported on the magnetic properties of $Fe-Ni-SiO_2$, in which the metal–insulator composites (cermets) were prepared by the three-sector on-axis sputtering technique and measured by mounting the sample on a small milling machine with a commercial recording head mounted on the milling head. It is not clear whether this work had any impact on subsequent development of magnetic cermets.

A more successful approach was reported by van Dover, Schneemeyer, and Fleming (15) in 1998. This approach exploited the versatility of off-axis reactive magnetron sputtering and was used in a systematic search for new materials with targeted properties. The off-axis approach allowed a large part of a ternary composition diagram to be synthesized in one run. The use of independent magnetron sputter guns made it possible to alter the region explored simply by varying the power applied to each gun and made it easy to vary the choice of elements deposited. The authors combined this synthetic technique with automated, rapid measurement of critical properties using custom software, and they found that the overall approach allowed complete preliminary evaluation of a ternary phase diagram in 24 hours, from choosing the elements to be incorporated to evaluating a map of the properties as a function of composition. This allowed the high-throughput synthesis/evaluation concept to achieve its full promise, in that over 30 multinary composition spreads with various elements were examined in less than three months.

The authors used the CCS technique to address a specific imminent techno-logical problem: the need for an alternative to SiO_2 for capacitors in integrated circuit (IC) technology. For a variety of reasons, to be discussed later, this problem appeared well matched to the capabilities provided by the CCS technique. Indeed, the authors reported the discovery of a new dielectric material, amorphous $Zr_{.2}Sn_{.2}Ti_{.6}O_2$. This material is superior in many ways to the alternatives and may eventually be incorporated into Si IC technology.

An alternative scheme for the CCS technique was recently reported by Xiang and coworkers (16,17). In this scheme, a traveling shutter is used to create sequential, overlapping wedges (as described earlier) of two or more elements. The multilayers are then processed with a low-temperature anneal that permits diffusion but not crystallization (at least in the cases reported), creating a homoge-neous film with a composition graded in one or two dimensions. The sample can then be given a high-temperature anneal to crystallize it. This technique is notable in that it allows high concentration gradients (10 at. % per lateral millimeter) to be achieved, enabling a complete composition diagram to be synthesized on a small sample. That is advantageous when epitaxial growth is desired and only small seed crystals are available for use as substrates.

D. Comparison with the Discrete Combinatorial Synthesis Technique

The CCS synthetic technique is distinct from the conventional ''combinatorial chemistry'' approach as well as the discrete combinatorial synthesis (DCS) ap-proach pioneered by Xiang and Schultz (18). In the former, sets of discrete chemi-cal entities (distinct molecules) are generated using rapid sequential synthetic techniques (19 –21). The properties of each molecule are fixed by its composition and structure, and the molecule sets (''libraries'') are in fact permutational, be-cause each distinct molecule is created by performing the synthetic routines in a distinct sequence. This approach is suited to the investigation of biologically active molecules, for example, in the effort to discover new pharmaceuticals, but it is less useful for the exploration of inorganic materials. In the DCS approach, mixtures of inorganic components are created by sequential deposition of layers of precursors followed by moderate-or high-temperature diffusion and reaction steps. This procedure generally results not in distinct molecules but, rather, multi-phase mixtures of various compounds.

An important advantage of the DCS technique is that arbitrary compositions with a large number of constituents can be prepared as desired. Thus, for example, an entire four-component phase diagram could be prepared with intervals of 10 at.% (yielding 220 samples) on a single substrate (22). In principle, the CCS technique cannot yield a full phase diagram in a single run for cases of more than three components. In practice, however, the technique of codeposition implies that only a portion of the possible compositions are obtained in a single run, even for cases of two or three components. A two-dimensional spread with four sources would have an even smaller range of possible compositions, though this might nevertheless be useful for many purposes.

An advantage of the CCS technique is the very fine compositional resolution that can be achieved. Fluctuations in the sputter rates probably limit the ultimate resolution to 0.1–1 at.% In many cases, the resolution obtained in an experiment will be limited by the lateral extent of the measurement probe rather than this synthetic technique. The DCS technique can be designed to explore a selected region of a phase diagram with fine resolution, but practical masking schemes do not permit both fine resolution and broad composition coverage. The cosputtering approach also has the advantage that no mask needs to be positioned in close proximity to the substrate. Depending on the deposition conditions, such masks can be a significant source of contamination.

Another advantage of the CCS technique is the fact that intimate mixing is guaranteed by codeposition. Phase segregation can occur in polycrystalline samples. However, even microscopic segregation must be driven by a strong chemical tendency and enabled by a sufficiently high deposition or annealing temperature. The intimate mixing has proven particularly valuable in preparing uniform amorphous materials, such as dielectric oxides. Indeed, one must be careful when preparing materials in a nonequilibrium state, since the particular nonequilibrium state achieved may depend on the details of the synthesis; thus the properties one measures in, for example, off-axis sputtered films may not be the same as found in on-axis sputtered films or chemical-vapor-deposited (CVD) films. We discovered exactly this problem in conventional amorphous dielectrics (SiO_2, Ta_2O_5) prepared by off-axis sputtering and determined that it could be solved only by modifying the off-axis process with the addition of rf bias on the substrate. Similar problems can occur even for crystalline equilibrium materials, if the properties being considered depend significantly on extrinsic factors such as microstructure rather than on intrinsic ones such as crystal chemistry.

II. REQUIREMENTS FOR SUCCESSFUL APPLICATION OF THE TECHNIQUE

The strategy and tactics used in the CCS approach depend on the sort of problem being addressed. Two types of problems are typically discussed: those in which the technique is used to explore the properties of known systems, and those in which the technique is used to explore new systems, that is, to search for candidates with new (or improved) properties. In the latter case, the scenario includes subsequent, focused exploration of the best materials that are uncovered. In the former case the emphasis is on developing an analytic technique that is sophisticated enough to be able to reliably measure the properties in question, while in the latter case the analytic technique need only be sufficient to lead the investigator to the best materials, which then can be explored in greater detail in a more sophisticated manner.

A. Exploration of Known Materials Systems

High-throughput synthesis and evaluation of solid-state materials can allow systematic exploration of materials systems, dramatically improving both the throughput and the reliability of comparisons, due to the reduction of uncontrolled systematic experimental variations. For experiments of this kind, one requirement is that the synthetic technique be capable of producing materials with the desired properties, such as dense microstructure and low level of impurities. That is, it is necessary that materials can be made in which the composition is the dominant variable, because if some other parameter is both variable and important to the properties, it will confound interpretation of the data.

The CCS explorations of properties within known materials systems will be most effective when a high-throughput evaluation is feasible. This may be straightforward when a simple property is all that is to be measured, such as the superconducting transition temperature or the dielectric constant of an insulating film. In other cases, a more sophisticated measurement may be desired, and these cases represent a great challenge to high-throughput techniques. An example might be the measurement of soft magnetic properties. For this there are measurements that are both relevant and amenable to high-throughput data collection, such as measuring Kerr-effect θ_k-H loops or measuring B-H loops with a small Hall-effect sensor. However, the Kerr effect does not give information on the absolute magnetization, and even measurements with a Hall-effect sensor are susceptible to being confounded by the effects of magnetic anisotropy in the film. To date, reliable characterization of soft magnetic films has required breaking up the sample and measuring each piece in a painstaking fashion. With sufficient development a system could certainly be imagined that would be able to acquire the necessary information, though the details of the system would depend on the specific requirements of the study. In general, then, a requirement for the most effective use of high-throughput techniques for exploration of known systems is that a suitable analysis can be executed with minimal human intervention.

B.   Targeted Search for New Materials

Until recently, high-throughput synthesis/evaluation techniques have been employed only sporadically for the systematic search for and discovery of new solid-state materials. Part of the reason is that there is already a long history of materials exploration in many fields, so the prospects of finding important new materials are slight. Only carefully chosen problems are likely to be addressed successfully by this technique. A clear example was the problem of finding a superior dielectric for embedded DRAM, a problem in electronic materials that is discussed more fully later. The significant aspect of this problem was that a material was desired that could be processed below 400°C, a requirement that largely excluded the extensively investigated crystalline dielectric materials. In fact, the possibility of developing an amorphous material was considered an advantage from the start, since such materials would not suffer from possible grain-boundary-based leakage current. Use of the CCS approach to address this problem allowed the authors to explore regions of composition space that might otherwise have been left unexplored, based on the precedent of previous thin-film work. The successful discovery of an amorphous dielectric in just such an unexpected region of composition explicitly established the validity of this motivation for ''shotgun-style'' high-throughput techniques.

The choice of synthetic approach is also important for the search for and discovery of new materials. The most reliable results will be obtained if the synthetic technique is identical to the one ultimately employed to synthesize useful material, but this is not generally possible, since those techniques are often not amenable to direct modification to include a composition spread. Instead, one must carefully consider the properties that are likely to be important and attempt to develop a synthetic technique that is meaningfully related to the desired final form. Furthermore, it is essential that the number of chemical combinations be rationally limited, since even massively parallel synthesis will not allow exhaustive searches of all possible materials.

The requirements for an analytic technique are somewhat relaxed for materials discovery, as compared to the exploration of known systems. This is because it is only necessary to define a generally reliable indicator that will lead the researcher to the best materials. Once a small number of promising candidates is established, it is feasible to return and evaluate them in detail. In order to obtain suitable feedback about the materials choices being pursued, it is desirable that the measurement technique be very rapid. In particular, we have found it helpful to be able to measure a sample set (e.g., a two-dimensional composition spread with >4000 sample positions) in 24 hours or less. It is also desirable to choose an indicator, a figure of merit, that is a scalar. The advantage is that the value can then be displayed simply on a two-dimensional false-color map, allowing trends to be discerned by eye.

Finally, once one or more promising candidates are identified, it is necessary to focus on their synthesis and properties in more detail. For example, it might be necessary to demonstrate that they can be prepared using a technique that permits synthesis of large areas with uniform composition and properties. It might be desirable to vary the processing parameters to obtain optimum properties. Alternatively, it might be necessary to explore other properties not captured by the chosen figure of merit, in order to ascertain the real value of the newly discovered material. It is very challenging to discover a new material that truly outperforms other materials, and the definitive test is to actually use the new material in the intended application.

III. EXPERIMENTAL TECHNIQUES

A. On-Axis Cosputtering System

Our work on composition spreads in metallic and nitride films employed a three-gun cosputtering system in the approximately on-axis geometry. Two planar magnetron guns (US Gun I) and one S gun type (Sputtered Films, Inc.) were arranged to sputter almost confocally, as illustrated schematically in Figure 3. Substrates were generally mounted in a one-dimensional spread 10 cm across, with resulting film compositions typically ranging over a 70 at.% spread. This system included an LN_2-cooled cryoshroud that enveloped the sputter sources as well as the substrate holder/furnace to provide a gettering environment (23). The system was pumped with a Varian VHS-6 diffusion pump, with a typical base pressure of 2 $\times 10^{-8}$ torr, rising to 4×10^{-8} torr when the pump is throttled for sputtering. Inside the cryoshroud the partial pressure of O_2 and related reactive species was estimated to be about 5×10^{-10} torr, based on the measured properties of Nb films deposited very slowly. This allowed the deposition of films with very little oxygen contamination, often a serious concern for deposition of reactive metals, such as Tb, in a non-UHV environment. Depositions were typically carried out at a pressure of 5 mtorr and a flow rate of 50 sccm Ar. When a ferromagnetic material was one of the components (e.g., Fe or Gd), it was always sputtered using the S gun, which provided enough flux density to saturate the magnetic target and enough field above the target to sustain the low-pressure plasma.

Early work established that dc sputtering provided extremely stable rates, so cosputtered alloys could be synthesized using independent guns. Sharp features in the magnetic properties of Tb-Fe-U (pseudobinary) films, for example, demonstrate that the composition variation during the run must be less than 1 at. %.

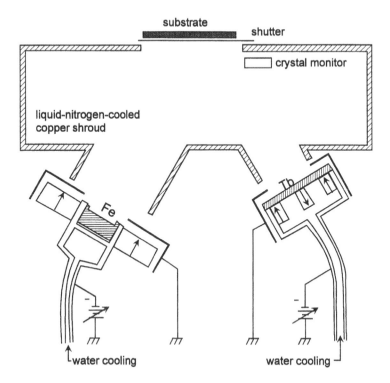

Figure 3 Schematic arrangement of the three-gun system used for metal and nitride depositions, showing the cryoshroud (Meissner trap) that surrounds the sources and substrates. The third sputter gun is not shown, for clarity. (Adapted from Ref. 6.)

Compositions could be reproduced accurately by starting one gun and adjusting the dc power to obtain a desired rate, as measured with a crystal rate monitor adjacent to the substrate position, and then turning on a second gun and adjusting the power until the observed rate was equal to the sum of the desired rates. Film compositions were measured using Rutherford backscattering spectrometry (RBS) at selected spots.

B. Off-Axis Cosputtering System

For one- and two-dimensional spreads of oxides, we developed a three-gun sputtering system using the 90° off-axis configuration. A schematic view of this system is shown in Figure 4. In this system we used US Gun II sources with a 2-mm separation between the substrate holder and the ground shield of the guns.

The guns were powered by independent rf supplies, with powers typically in the range of 20–150 W, chosen to provide a suitable deposition rate for each constituent. The substrate holder could be heated as well as biased with an independent rf source, typically at a level of 10 W. Substrates were roughly 70 × 70-mm rectangles cut from standard wafers that had been coated with 500 Å of TiN in a conventional IC process line.

We used reactive sputtering with metallic targets and a very oxygen-rich sputter ambient (20–50% O_2 in Ar by volume). Under these conditions, we expect the target to achieve coverage by oxide, i.e., operating in the so-called "poisoned" mode. The depositions were performed at 30 mT. At this pressure, the mean free path for the sputtered atoms/molecules is very short, resulting in diffusive trans-

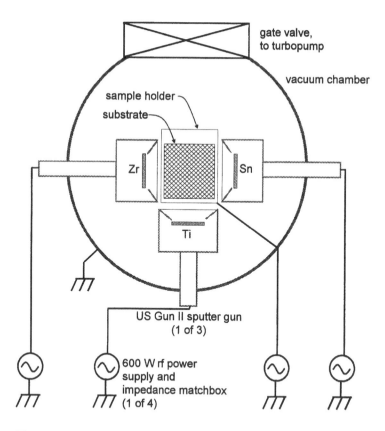

Figure 4   Schematic arrangement of the three-gun system used for oxide depositions. The normal to the gun targets is oriented 90° to the normal of the substrate, hence the designation "90° off-axis-deposition."

port and an exponential decrease of deposition rate with distance from the target. Calibration runs were used to establish the deposition rate and decay constants for each cation species. The decay constants range from 24 mm for TiO_x to 43 mm for TaO_x. The deposition rate immediately adjacent to the gun varies with power; typical rates used were roughly 1 Å/s. Compositions were calculated as a function of position on the substrate using the calibration data and a simple model for the deposition profile. At selected spots the film was also analyzed using RBS, which supported the model calculations. In cases where it was desirable to know the thickness of the film accurately, a surface profilometer (Dektak II) was used in conjunction with a liftoff mask pattern.

C. Magnetic Measurements (M_s, H_c, K_u)

The essential magnetic parameters for the films studied were the saturation magnetization, M_s, the coercive force, H_c, and the uniaxial anisotropy energy, K_u. In order to obtain highly reliable measurements, the samples were sectioned into small pieces that could be measured with a vibrating sample magnetometer (VSM, ~3 % accuracy, calibrated with a Ni standard) and torque magnetometer (~5% accuracy, calibrated with an yttrium iron garnet standard). The VSM was used to take M–H hysteresis loops at up to 15 kOe, yielding values for M_s and H_c. Torque curves of the Tb-Fe films indicated some deviations from a simple uniaxial anisotropy, but the anisotropy constant and magnetization were nevertheless calculated using the approach of Miyajima. (24) The magnetooptic activity (polar Kerr rotation and ellipticity and magneto-optic hysteresis loop) was measured with an apparatus similar to that described by Sato (25), with a spatial resolution of about 50 μm.

D. Electrical Measurements

The electrical properties of the dielectric films were evaluated using either a scanning Hg-probe instrument or an automatic probe station in conjunction with deposited counterelectrodes. The scanning Hg probe utilized a Teflon capillary with an inner diameter of about 0.9 mm, filled with Hg and positioned in light contact on the sample (~17 g). The Hg makes a reliable contact with a surface area of about 4×10^{-3} cm^{-2}. The capillary was then translated across the surface of the sample without lifting. The sample was typically measured at positions forming a rectangular matrix with 1-mm centers (more than 4000 positions in a typical two-dimensional spread sample). The probe-station approach required the additional step of depositing Pt (or other metal thin film) counterelectrodes through a shadow mask by e-beam evaporation, or blanket evaporation and photolithography, followed by probing the individual 0.2-mm-diameter dots with an automatic probing station. The electrical properties observed for both Hg and Pt counterelectrodes were found to be comparable. Contact to the base electrode was made by soldering a wire either to the TiN underlayer, for metal-oxide-metal (MOM) capacitors, or to the Si substrate, for metal-oxide-semiconductor (MOS) capacitors.

The capacitance of the MOM or MOS structures was measured using an HP4274 LCR meter at 10 kHz and 0.1 V_{ac}. MOM structures were measured at 0 V_{dc}, and C-V curves were taken at each point for the MOS capacitors. In the latter case, we tabulated the flatband voltage and the capacitances in accumulation and depletion. For both structures, an I-V curve was taken at each physical location using a Keithley 617 source/electrometer. We used an algorithm in which the voltage was stepped (usually by 0.5 V) and after a 1-second pause the resulting current was measured. When the current rose above 2 μA, the dielectric was assumed to have broken down. Sometimes this corresponded to a ''hard'' breakdown, i.e., an abrupt and irreversible jump in the current, but often this was a somewhat arbitrary choice on a smoothly increasing trend. The complete I-V curve was stored for each capacitor. The I-V curves of MOS capacitors were taken in accumulation, which means the Pt or Hg counterelectrode was biased negatively with respect to the substrate for the case of n-type Si. The shape of the I-V curves in MOM capacitors also depends on the polarity of the voltage due to work-function differences between the electrodes and the oxide as well as other factors; we found that biasing the counterelectrode negatively gave the most reliable and meaningful data, so this is the polarity we used by convention in all the work reported here.

A useful figure of merit (FOM) for capacitor dielectrics can be defined by the product $\varepsilon_r\varepsilon_0 E_{br}$, where ε_r is the relative permittivity of the material, ε_0 is the permittivity of free space, and E_{br} is the breakdown field. This FOM corresponds physically to the maximum charge per unit area that can be stored on the capacitor using a given material. It can be rewritten as $\varepsilon_r\varepsilon_0 E_{br} = CV_{br}/A$, where C is the capacitance, V_{br} is the breakdown voltage, and A is the area of the test capacitor. Thus, this FOM can be trivially evaluated from the electrical measurements. In particular, it is not necessary to measure the film thickness to evaluate this FOM. Of course, the value of this FOM depends on the assumption that ε_r is independent of thickness. While that is inappropriate for some materials, notably including (26) $(Ba,Sr)TiO_3$, we have found that ε_r is essentially independent of thickness for amorphous oxides.

Another useful indicator for capacitor dielectrics is the leakage current that will pass through the material at relatively low fields. In order to compare materials with different values of ε_r, it is appropriate to evaluate the leakage at a specified stored charge density, i.e., at a particular value for $\varepsilon_r\varepsilon_0 E$. In our work we chose the value $\varepsilon_r\varepsilon_0 E = 7$ μC/cm^2, which corresponds to the charge on a 28-fF capacitor operated at 1.25 V, a reasonable target for dynamic random-access memory (DRAM). The leakage current at this charge density can be evaluated automatically at each position on the composition spread using the measured capacitance and I-V curve data. A disadvantage of this indicator is that it requires a specific choice for the target value for $\varepsilon_r \varepsilon_0 E$. It therefore may provide misleading information if the target value is not chosen carefully or, indeed, if the target is subject to change, for example, due to device design changes.

IV. SPECIFIC INVESTIGATIONS

A. Tb-Fe Amorphous Magnetic Films

Amorphous rare earth/transition metal alloy thin films are the active components in magneto-optic disk media. While essentially all of the combinations, such as Tb-Fe, Gd-Fe, and Gd-Co, are magneto-optically active, Tb-Fe is the prototype system. The large uniaxial anisotropy that can be obtained in amorphous Tb-Fe alloys allows small, stable domains to be written with the magnetization perpendicular to the plane of the film. That is, the effective uniaxial anisotropy is positive (defined as $K_{ue} = K_{ui} - 2\pi M_s^2$, where K_{ui} is the intrinsic uniaxial anisotropy and M_s is the saturation magnetization). In 1985, the origin of the intrinsic uniaxial anisotropy in amorphous Tb-Fe had been variously attributed to stress, microstructure, or pair ordering. The systematic data shown in Figure 1, along with supporting microstructural data and data from films deposited at 77 K, allowed van Dover et al. (5) to rule out the first two possibilities and make a plausible argument for the third.

Hellman et al. used the same system in a much more extensive examination of local order in amorphous Tb-Fe (27). The direction of the easy axis and the dependence of the torque on applied field at various fixed angles revealed a transition with changing composition within what is typically considered a featureless amorphous state as judged by x-ray and electron diffraction. These experiments took advantage of the extreme sensitivity of magnetic anisotropy to local structure and demanded continuous composition spreads in order to allow subtle effects to be discerned. Hellman et al. concluded, among other things, that Tb-rich material (i.e., >22 at.% Tb) develops anisotropy from local clusters that are oriented by the film growth process. Fe-rich Tb-Fe, on the other hand, has a more complex behavior that suggests the coexistence of two separate amorphous phases.

Dillon et al. (28) also used the same system to examine the effects of doping the Tb-Fe system with U. In some cases the addition of dopants can enhance magneto-optical effects. For example, it is well known that the addition of Bi^{3+} to iron garnets greatly increases the Faraday rotation. This is due to the large spin-orbit interaction of the heavy ion, which increases the splittings responsible for magneto-optical rotation. It is also known that some crystalline metallic uranium compounds show exceedingly large values for the magneto-optical polar

Kerr effect and that uranium forms binary compounds with $3d$ transition metals that are strongly magnetic. Therefore, the addition of U might be expected to have an effect on the Kerr rotation amorphous of Tb-Fe. The results contradicted this hypothesis, however. The authors concluded first that the U had no net magnetization. Second, the U substitutes on both the Tb and Fe subnetworks randomly. Third, there was no evidence of spectral structure associated with U in the variation of rotation with photon energy in the range 1.3 eV $< h\nu <$ 2.4 eV. Fourth, the main effect of U additions was the prosaic result that the Curie temperature was depressed, especially for Tb-rich compositions. The use of composition-spread samples greatly facilitated the intercomparisons that were necessary to reach these conclusions.

B. Thin-Film Dielectrics for Embedded DRAM

The well-documented trend for integrated circuits to employ smaller feature sizes (29) has driven interest in deposited thin-film insulators with a dielectric constant, ε_r, significantly greater than that of the standard material, amorphous silicon dioxide (a-SiO$_x$). For example, basic circuit considerations in dynamic random-access memory (DRAM) require a minimum capacitance of 20–40 fF/cell. As the area of a cell has shrunk, designers have maintained the cell capacitance by resorting to extremely thin a-SiO$_x$ films, limited by the decreasing reliability of thinner films due to the finite breakdown fields, and exotic trenches, limited by the ability to obtain reliable high-aspect-ratio trenches and contacts (30). A promising geometry is that of the so-called "stack" capacitor, in which the capacitor is fabricated as a dielectric inserted between two metal (typically Al, TiN, or W) layers, as illustrated schematically in Figure 5a.

The limitations of a-SiO$_x$ might be avoided by utilizing a different material with improved properties—a material with a higher dielectric constant or breakdown field. As discussed earlier, a useful figure of merit can be defined by the product $\varepsilon_r\varepsilon_0E_{br}$. As a baseline value, consider that deposited films of a-SiO$_x$N$_y$ have $\varepsilon_r = 4.0$, E$_{br} \sim 10$ MV/cm, for an FOM of 3.5 μC/cm^2. Most attention to date has focused on two alternative dielectric materials with high values of ε_r and FOM: a-Ta$_2$O$_{5-\delta}$ (amorphous tantalum pentoxide, $\varepsilon_r \sim 23$, E$_{br} \sim 4$ MV/cm, FOM ~ 8.1 μC/cm^2) and polycrystalline (31) x-(Ba,Sr)TiO$_3$ ($\varepsilon_r \sim 400$, E$_{br} \sim 0.9$ MV/cm, FOM ~ 41 μC/cm^2). Each offers advantages as well as disadvantages. The high dielectric constant of x-(Ba,Sr)TiO$_3$ is a great advantage, but the breakdown field in polycrystalline material is typically quite low—a problem attributed to defects at the grain boundaries. Furthermore, crystallinity necessitates processing at over 650°C and therefore precludes fabrication of the capacitor structures during "back-end" steps (i.e., fabrication after the critical steps needed to form transistors are completed) and requires exotic metallurgy to make metallic contacts that withstand such high temperatures. On the other hand, a-TaO$_x$ films

(a)

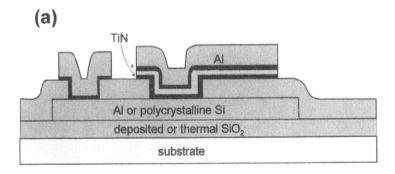

(b)

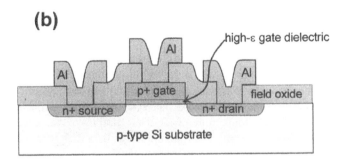

Figure 5 (a) Schematic depiction of a metal-oxide-metal capacitor structure. (b) Schematic depiction of a MOSFET, showing the gate region where SiO_2 will be replaced with a deposited hig-ε_r dielectric.

can be deposited at low temperatures (<450°C), are free of grain boundaries, and can have high breakdown fields and low leakage currents (32). Ta is also generally considered compatible with Si IC fabrication, unlike Ba and Sr, which, among other things, are difficult to etch cleanly. While the breakdown field of a-TaO_x is relatively high, the dielectric constant is low, so the material has a figure of merit only modestly higher than SiO_2, making it probably only a short-term solution. A schematic depiction of a proposed "back-end" capacitor is shown in Figure 5a. Obviously, a transistor can be located directly under the capacitor, saving space and thereby increasing the DRAM density.

The literature does not provide much guidance for choosing alternative amorphous dielectrics. While the dielectric constants of many polycrystalline bulk materials are known, the breakdown fields, even when (rarely) measured, do not correlate well with breakdown fields in thin films of the same composition.

Reliable data for amorphous films are sparse, even for straightforward properties such as the dielectric constant. Therefore, a broad search using a combinatorial-type approach is indicated for this problem. The CCS technique is almost ideally suited for this search, since it is straightforward to obtain intimately mixed, homogeneous amorphous thin films by this technique. Nevertheless, the search for a material that combines both a high dielectric constant and a high breakdown field was not expected to be trivial, because the two parameters tend to correlate inversely.

We constrained our search in a few ways. First, we decided to focus on oxides exclusively, since these would be most readily compatible with existing IC fabrication processes. Oxides also offer many promising systems for finding high values for ε_r and high breakdown strengths. Second, we recognized that known materials with high values for ε_r typically incorporate one or more early transition metals (columns IIIB, IVB, VB, and VIB) and therefore chose to include at least two in most of the ternary systems we examined. Third, we favored as constituents metallic elements already accepted in Si IC fabrication facilities. We ruled out metals that we judged unlikely to be accepted, including all radioactive elements and most of the late transition metals, because they are fast diffusers and form recombination centers in Si and because their tendency to have variable oxidation states is likely to create traps in the oxide itself, leading to excessive leakage current. The established difficulties of incorporating Ba and Sr into the Si process flow biased us against the investigation of phase spreads incorporating alkali and alkaline earth metals as constituents. Finally, we used intuition (and prejudice) to choose the particular combinations that would be explored first.

From the outset we chose to look for materials that could be formed at a temperature that would be compatible with "back-end" Si processing. This conventionally limits process temperatures to below 400°C (since grain growth in Al interconnects begins at this temperature); we therefore chose a fixed deposition temperature of about 200°C. We also anticipated that deposition at a relatively high oxygen partial pressure would lead to the fewest oxygen vacancies in the amorphous network and therefore the lowest leakage current. On this basis we narrowed the range of oxygen/argon ratios to 20–50% (vol. %) in the sputtering ambient.

Improvements in dielectric constant and breakdown strength can depend critically on the processing and microstructure of any material, so it was important for this search that the dielectrics be produced in a manner closely related to that in which they would ultimately be used. It was not a priori obvious that off-axis sputtering at low temperature would produce films comparable in quality to those produced by more conventional means, such as on-axis sputtering or chemical vapor deposition (CVD). Nevertheless test runs of Si-O and Ta-O demonstrated definitively that high-quality oxide thin-films could be grown by off-axis sputtering. We observed that it was crucial to apply an rf bias to the growing film during deposition. This bias causes ion bombardment of the growing film, presumably leading to enhanced surface mobility and a denser microstructure.

1. Zr-Sn-Ti-O

An initial survey of about 30 multinary phase systems led us quickly to the Zr-Ti-Sn-O pseudo-ternary phase system, which we then explored in detail (15). Figure 6 shows the experimentally obtained capacitance values and breakdown voltages, displayed as a function of position. The locations of the guns with respect to the substrate are indicated for the capacitance plot. The capacitance and breakdown voltage were multiplied point by point to obtain the FOM data shown in Figure 7. The leakage current at a fixed value of stored charge $Q = 7 \, \mu C/cm^2$ was evaluated from the I-V curve data and is also plotted, on a logarithmic scale, in Figure 7. With knowledge of the composition, inferred by interpolation of RBS data measured at selected points on the film, we can then map these data onto a ternary phase diagram, as shown in Figure 8. In the raw data, a high-FOM region (black points) can be seen in the data in the region closest to the Ti gun; the corresponding region is also displayed in black on the ternary chart.

Regions of high FOM or low leakage, "sweet spots," can be characterized schematically by plotting, as in Figure 9, the 95th percentile contours calculated from the full data sets. In this figure, we show the 95th percentile FOM regions for thin films of amorphous Zr-Ti-Sn-O. Also shown is the location of the single-

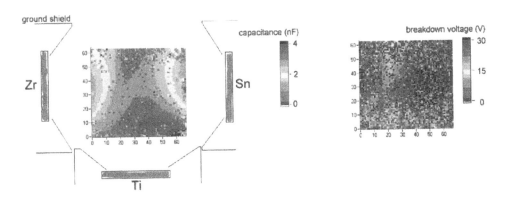

Figure 6 Raw capacitance and breakdown voltage for a Zr-Sn-Ti-O film, plotted as a function of position on the rectangular substrate. The position of the sputter guns during deposition is shown schematically with respect to the capacitance data. The vertical and horizontal scales shown are in millimeters. (From Ref. 15.)

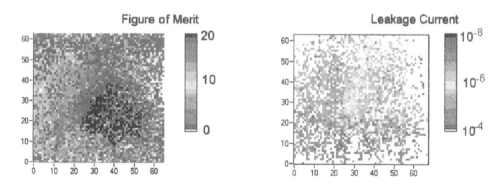

Figure 7 Figure of merit (defined as FOM $= \varepsilon_r\varepsilon_0E_{br} = CV_{br}/A$, measured in $\mu C/cm^2$) and leakage currents (measured at a stored charge of $7\mu C/cm^2$, current density measured in A/cm^2) as a function of position on the substrate, for the same sample as in Figure 6.

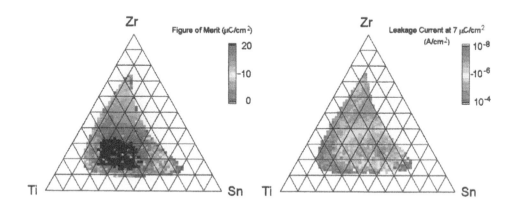

Figure 8 Figure of merit and leakage current data of Figure 8, mapped onto a conventional ternary composition diagram.

phase crystalline ceramic of composition, $Zr_{1-x}Sn_xTiO_4$, which is used for low-loss dielectric filter elements in wireless applications (33,34). The regions of this phase space that were previously explored in thin film deposition experiments are indicated by the black line and the dot, which refer to a film with the composition $Zr_{.8}Sn_{.2}TiO_4$ prepared by sputtering from a single target by Nakagawara, et al. (35) and to several samples with varying Zr/Ti ratios prepared by Ramakrishnan et al. (36). These studies were clearly inspired by the crystalline $Zr_{1-x}Sn_xTiO_4$ material and were aimed at producing films displaying the behavior of the crystalline ceramic, including its near-zero temperature coefficient of resonant frequency. In the case of a-Zr-Ti-Sn-O films, neither the high-FOM region nor the low-leakage region corresponds to the area of phase space occupied by the dielectric ceramic. Thus, one of the promises implied by high-throughput synthesis/evaluation was redeemed: By using a survey technique, we managed to avoid prejudging the most likely composition in the chemical system and focusing on that exclusively.

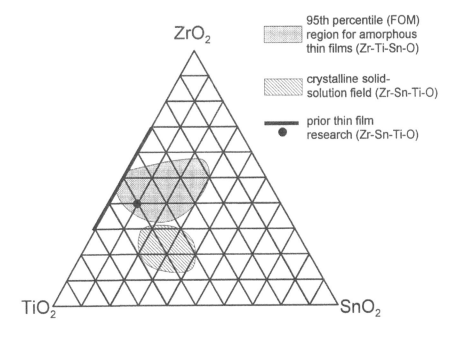

Figure 9 Schematic illustration of the Zr-Sn-Ti-O ternary composition diagram, showing the ternary phase field, the region that yields the best properties in amorphous off-axis sputtered films and the regions that were explored in previous work (Refs. 35 and 36). (Adapted from Ref. 15.)

Further development of this material has demonstrated that $Zr_{.2}Sn_{.2}Ti_{.6}O_2$ can be synthesized by on-axis sputtering. With a modest amount of optimization, dielectric films with a figure of merit $\varepsilon_r\varepsilon_0E_{br} = 25$ $\mu C/cm^2$ were obtained (37). Figure 10 shows the J-E curve of a $Pt/Zr_{.2}Sn_{.2}Ti_{.6}O_2/TiN$ capacitor made on an optimized film, illustrating the very low leakage current that is achievable in this material. Figure 11 replots the same data as well as the J-E curves of the other standard thin-film dielectric materials. Here the abscissa is scaled by the dielectric constant for each material, to provide a meaningful comparison. As indicated, this axis physically measures stored charge. In the desired region for operation for some generic embedded DRAM designs (indicated by the hatched area), SiO_2 is well past breakdown and a-TaO_x is close to hard breakdown (vertical jump at $\varepsilon_rE = 90$ MV/cm). Only the $(Ba,Sr)TiO_3$ film has properties superior to that of the $Zr_{.2}Sn_{.2}Ti_{.6}O_2$, but recall that $(Ba,Sr)TiO_3$ is ruled out for back-end processing. Table 1 summarizes the status of various alternative dielectrics.

Present work is focused on evaluating various practical aspects of processing $Zr_{.2}Sn_{.2}Ti_{.6}O_2$, such as the feasibility of CVD deposition, the etching charac-

Table 1 Various Dielectric Materials for Thin-Film Integrated Circuit Capacitors

Material	ε_r	$E_{br},$ MV/cm	FOM, $\mu C/cm^2$	J_{leak} (E = 1 MV/cm), A/cm^2	Comments
SiO_2	4.0	10	3.5 (typical)	6×10^{-12}	Deposited TEOS (i.e., not thermally grown). ε_r is too low for future ICs
TaO_x	23	4	8.1	6×10^{-10}	Moderate ε_r, good breakdown field
$(Ba,Sr)TiO_3$ BST	400	1	20–41	1×10^{-8}	High ε_r, low breakdown field. difficult to integrate with Si IC processing
Zr-Sn-Ti-O a-ZTT	62	4.4	24–35	2×10^{-9}	Good ε_r, good breakdown field, good compatibility with standard processing
Hf-Sn-Ti-O	58	4	19	1×10^{-6}	Higher T_{dep} than aZTT
Ti-Ln-O	48	3.3	14	3×10^{-8}	Binary, simple

The data for SiO_2 was taken on a deposited film rather than a thermally oxidized film in order to provide a fair comparison among materials. The data for other materials represent the state of the art. The data for TaO_x is typical of the best material prepared in our lab. The data for BST is taken from Park, et al.[31] The data for a-ZTT is taken from Ref. 37, for a HfTT from Ref. 39, and the data for Ti-Ln-O from Ref. 42. FOM refers to the Figure of Merit, FOM$\equiv\varepsilon_r\varepsilon_0E_{br}$.

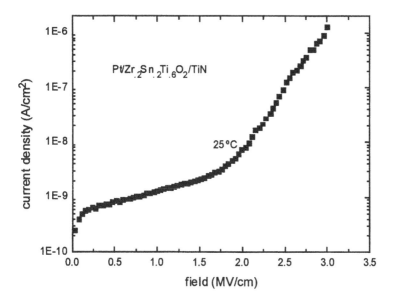

Figure 10   Current density versus field for an on-axis deposited $Zr_{.2}Sn_{.2}Ti_{.6}O_2$ film, showing excellent leakage behavior. The dielectric constant for this film was 62, and the film was about 400 Å thick.

teristics of the film, and the effect of thermal annealing. Despite the daunting challenge associated with introducing a new material into a Si IC processing line, the prospects for a-$Zr_{.2}Sn_{.2}Ti_{.6}O_2$ are bright.

2.   Hf-Sn-Ti-O

The lanthanide contraction leads to essentially identical atomic and ionic radii for hafnium and zirconium. Both are group IV early transition metals, and thus they have nearly identical chemistries. Therefore, having discovered excellent properties in the amorphous Zr-Sn-Ti-O system, we explored the Hf-Sn-Ti-O system using the CCS approach (38,39). We have found that the properties of thin films in the a-Hf-Sn-Ti-O system depend strongly on deposition conditions, particularly substrate temperature. Phase spreads were prepared at three different substrate temperatures, 185°C, 260°C, and 305°C, with all other conditions held constant, including the composition of the gas mix (O_2 40% by volume in Ar), the total pressure (30 mtorr) and the rf power to each sputter gun.

Figure 12 summarizes the results by showing the regions of high FOM, using thresholds as indicated. The FOM of the "best region" is strongly depen-

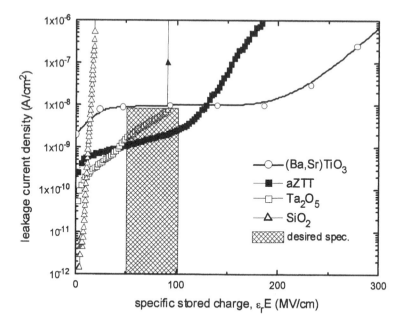

Figure 11   Current density versus stored charge for various capacitor dielectric candidates. The ordinate represents stored charge normalized to ε_0. The hatched box represents approximate specifications for an embedded DRAM capacitor that is to fit into a 0.7 μm cell. Clearly aZTT represents a superior choice for this application, because the leakage current is so low at high stored charge (recalling that $(Ba,Sr)TiO_3$ is ruled out because it required high temperature processing). The a-TaO$_x$ film exhibited hard breakdown at $\varepsilon_r E$ = 90 MV/cm and should not be operated above $\varepsilon_r E \sim 45$ MV/cm.

dent on the deposition temperature—the film deposited at 260°C is definitely superior to the others. This finding contrasts with the behavior of the amorphous Zr-Sn-Ti-O system, where the optimum temperature was found to be 200 ± 25°C, so it may indicate a somewhat greater thermal stability for the Hf system. The location of the high-FOM region is substantially different than that for Zr-Sn-Ti-O, as may be seen by comparison with Figure 9. The shape of the contours for Hf-Sn-Ti-O also change systematically with deposition temperature. For the lowest temperature the region is seen to be more spread out or diffuse, conforming to our expectation that properties will change more gradually with compositions in systems that are more disordered. The highest-temperature deposition yielded high-FOM contours, suggesting three independent composition regions. In the crystalline Hf-Sn-Ti-O equilibrium (high-temperature) phase diagram there is a substituted $HfTiO_4$ phase (40) (analogous although not identical to the $Sn_xZr_{1-x}TiO_4$ phase), which may correspond to the large FOM region obtained at T_{dep} = 305°C. The smaller regions do not corrrespond to any known phase. We speculate that the regions might reflect low-temperature or metastable phases not accessible by conventional ceramic synthetic techniques. Perhaps a systematic effort to synthesize bulk Hf-Sn-Ti-O at these compositions, using low temperature processing, could successfully lead to these hypothetical ternary compounds. We note parenthetically that a phase of composition $Hf_{.75}Sn_{.25}O_2$ with the α-PbO_2 structure (41) has also been identified, although compositions in that region were not explored in the present work.

Capacitors made with the composition $Hf_{.2}Sn_{.05}Ti_{.75}O_2$ at the optimum temperature of 260° show reasonable leakage, near 1×10^{-6} A/cm^2 at E = 1 MV/cm, a dielectric constant of 58 and a breakdown field of 4 MV/cm. Further work on this material might be expected to result in films with reduced leakage currents and still higher breakdown fields. Certainly the most surprising result of this study was that such substantial differences in dielectric properties should be obtained between the Zr and Hf analogs of these amorphous compositions, considering the chemical similarity of the two elements.

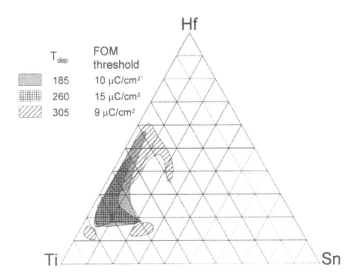

Figure 12 Schematic illustration of the Hf-Sn-Ti-O ternary composition diagram, showing the regions where the $\varepsilon_r\varepsilon_0E_{br}$ is highest, for films deposited at the three indicated substrate temperatures. Compare these regions with the Zr-Sn-Ti-O diagram of Figure 9.

3.   Ta-Ti-Sn-O

Since we have shown that the Zr-Sn-Ti-O and Hf-Sn-Ti-O systems both include some specific compositions/processing conditions that yield excellent thin-film dielectrics, one might ask whether this is a fairly ubiquitous result. The answer, in short, is no. But considering the attention given to amorphous TaO_x as a possible high-ε dielectric for Si IC capacitors, one might speculate that the Ta-Sn-Ti-O system in particular might yield a superior dielectric. Again, however, the answer is no. Figure 13 shows the figure of merit for a Ta-Sn-Ti-O composition spread deposited under standard conditions (40% O_2 in Ar, 30 m torr, rf bias, 200°C substrate temperature) on TiN. The results are interesting, if discouraging for practical applications. An abrupt transition is seen, and only compositions with more than 50 at. % Ta and less than 30 at. % Ti have a high FOM. The underlying cause for this abrupt transition is not known, although it presumably reflects some packing consideration. In any case, even the highest FOM in this system, about 8 $\mu C/cm^2$, is no higher than that obtained for unsubstituted a-TaO_x (see Table 1). Comparably low values for the FOM were obtained for a wide variety of elemental combinations in our (nonexhaustive) search.

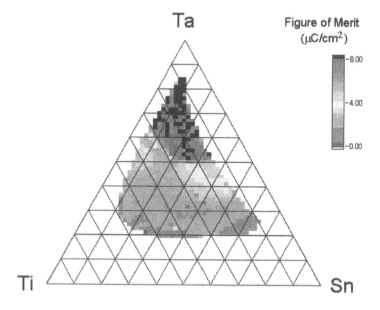

Figure 13   Figure of merit ($\varepsilon_r\varepsilon_0 E_{br}$) for amorphous films in the Ta-Ti-Sn-O composition diagram, showing that the Ti-rich region of the diagram has particularly poor properties. The abrupt increase in FOM for the Ta-rich region is not understood.

4. Ln-Ti-O

Using the CCS approach, van Dover recently showed (42) that lanthanide additions to amorphous titania produce films that have high dielectric constants ($\varepsilon \sim 50$) and greatly improved leakage characteristics compared to undoped aTiO$_x$ films ($J_{leak} \sim 10^{-8}$ A/cm^2 at 1 MV/cm, compared to $\sim 10^{-5}$ A/cm^2 for undoped TiO$_x$). The dielectric constant, breakdown field, and leakage behavior were explored as a function of lanthanide concentration. The particular cases Ln = Nd, Tb, and Dy were examined, of which the Ti$_{1-x}$Dy$_x$O$_y$ films were found to be slightly superior. Figure 14 compares the I-V curves for Ti-O and Ti-Dy-O films deposited in the same sputtering system. Incorporation of the lanthanide, in this case Dy, produces dramatically improved leakage currents. The advantage of Ti$_{1-x}$Ln$_x$O$_y$ films compared to aZTT or (Ba,Sr)TiO$_3$ films is that they have only two cation components, which may simplify process development for integration into a conventional Si IC process flow.

C. Gate Oxides

Another problem facing the Si IC industry is that of the gate oxide in metal-oxide-semiconductor field-effect transistors (MOSFETs), the basis for digital IC technology (see Figure 5b). Thermally grown SiO$_2$ has been the dielectric of choice for over 40 years, but low-voltage operation requires that the SiO$_2$ thick-

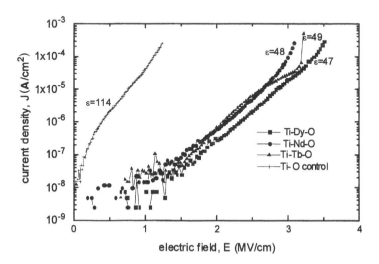

Figure 14 Current density versus electric field plots for lanthanide-doped TiO$_x$ films compared to an undoped TiO$_x$ film. (From Ref. 42.)

ness be less than 13 Å in future generations of the technology; for layers this thin or less, direct tunneling through the SiO_2 will lead to unacceptably high leakage currents. Therefore a dielectric is sought that can provide the same electrical properties (i.e., a specific capacitance of 2.8 $\mu F/cm^2$, or 28 $fF/\mu m^2$) in a thicker film. This can be done by choosing a dielectric that is compatible with Si processing and that has a dielectric constant in the range of $\varepsilon_r = 15-20$. Amorphous TaO_x has been used previously (43) to extend the scaling of SiO_2, but it is found to crystallize during the rapid thermal anneals (RTA) used for dopant activation and interface state passivation in Si. The crystallites have lateral dimensions comparable to the gate length, as shown in Figure 15, which can lead to excessive fluctuations in threshold voltage. This crystallization transition also leads to growth of SiO_2 at the a-TaO_x/Si and the a-TaO_x/poly-Si interface and can lead to deterioration of the gate capacitance.

We have found (44) that the crystallization transition can be suppressed by adding Al or Si (or, with somewhat less effect, Ge) to the a-TaO_x film, as shown in Figure 16. The electrical properties of $Ta_{1-y}Al_yO_x$ for $0.1 < y < 0.4$ are superior to those of undoped TaO_x although the dielectric constant is slightly reduced. It is expected that CVD of Ta-Al-O or Ta-Si-O alloys should be straightforward, since CVD of the individual elements is well established.

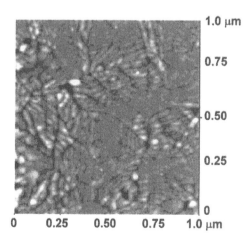

Figure 15   Atomic force microscope image of a 1-μm × 1-μm region of a TaO_x film deposited on Si, after annealing to 850°C in nitrogen. Before annealing, the root mean square (rms) roughness was 3 Å, while after annealing the roughness was 36 Å. The lateral extent of typical features after annealing can be seen to be 1000 Å.

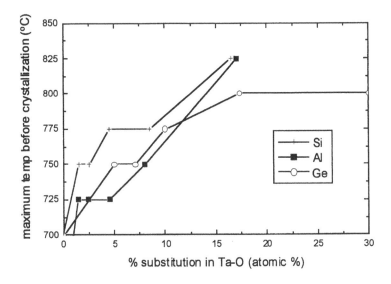

Figure 16 Maximum temperature at which a-Ta-X-O film remains crystalline, as a function of doping concentration, where X is either Al, Si, or Ge.

V. OTHER PROBLEMS FOR WHICH THE CCS APPROACH MIGHT BE WELL SUITED

Many other studies are also well matched to a combinatorial-style approach. We note, however, that these methods, including the CCS approach, are no panacea for materials science. A few problems are discussed briefly next, touching on both experimental challenges and research opportunities.

In the magnetic data storage industry many of the technological advances made have depended directly on innovations in magnetic materials. The read/write heads in magnetic hard disks utilize advanced magnetic thin films and multilayers. For example, soft magnetic materials with very high saturation magnetization are required for the write heads in order to generate the gap fields needed to reverse domains on high-performance media. While nanocrystalline materials such as FeTaN have recently been developed and appear to be adequate, alternative materials might be identified by a combinatorial-type approach. One specific goal might be the identification of magnetic materials with a resistivity higher than that of FeTaN (about 100 $\mu\Omega$-cm) combined with a high saturation magnetization (over 1.5 T). In addition, materials with higher and/or more controllable anisotropies might also be useful, although anisotropies generally depend strongly on microstructure, which is not easy to control or vary in a combinatorial-type study.

New colossal magnetoresistance (CMR) materials may potentially be identified and studied using combinatorial-style searches. Elucidating the structural and electronic phase diagrams of these fascinating materials would be a particularly valuable contribution that might utilize rapid-throughput approaches. Such studies could help sort out the apparently complex physics that underlies CMR behavior. An indication of the value of these approaches was recently provided by Yoo et al. (17).

Superconducting materials have been sought in extensive investigations since the initial discovery of the phenomenon in 1911. Most recently, a massive search involving investigators worldwide was sparked by the discovery of "high-temperature" superconductivity in 1986. Thus, finding new superconductors is a daunting challenge. While it is true that a number of new superconductors, with higher and higher transition temperatures, have been found, it also is clear that most of these require very specific and narrow processing windows for phase formation, rendering them extremely difficult to capture, even using combinatorial-style approaches. Still, the payoff in terms of basic scientific understanding as well as possible commercially important materials is potentially very large. Methods to evaluate superconducting properties, including rapid and convenient scanning approaches, are well understood. By elucidating phase diagrams and well as providing new insights into the underlying physics of superconductivity, combinatorial-type approaches may prove to be valuable tools in superconductivity research.

The potential to revolutionize a variety of heating and cooling technologies has driven the search for a breakthrough thermoelectric material. Again, the likelihood of discovering a breakthrough seems low because so many studies of candidate materials have already been concluded. Still, many of the ingredients needed for a successful combinatorial-type search are in place. A well-accepted figure of merit for thermoelectric materials exists already, although this FOM may be difficult to measure in a thin film geometry because of the difficulty of measuring thermal conductivity in a thin film. It is possible that a more crude (and easier to measure) yet indicative FOM could be used to identify promising candidates. Also, some interesting ideas concerning factors that lead to high-FOM materials could provide guidance concerning materials systems worthy of investigation.

Piezoelectric materials have a variety of commercial applications, and a large number of such materials are known. To examine such materials in thin-film form, highly textured crystalline thin films are likely to be needed, so microstructure control is therefore an issue. Screening approaches should be straightforward. Because of the commercial importance of these materials, new phases, perhaps with wider processing windows or larger coefficients, remain of interest.

The variety of applications possible for ferroelectric materials have resulted in intensive investigations in the past. Again, textured crystalline thin films are likely to be needed, which implies that microstructure control would be an important issue. Screening and measurement approaches are straightforward. A major opportunity for ferroelectrics may be in nonvolatile electronic memory schemes, such as FeRAM. For this application, compatibility with Si processing is a paramount issue that may be hard to quantify. Other important applications of ferroelectric materials include use in multilayered ceramic capacitors and as electromechanical actuator materials.

The recent explosion of interest in optical communications has yielded a variety of problems that fall under the heading of optical materials, including new glass compositions, transparent conductors, phosphors, and amplifier materials. Screening approaches might be straightforward, but the match between performance and application requirements must be carefully understood in order to conduct a meaningful investigation. Also, it is advantageous if materials are screened in a form related to that in which they will be used. Thus, where bulk single crystals are required for implementation, evaluation of thin-film samples may miss significant and relevant information.

Catalysts are so important in chemical industry that combinatorial-type approaches are already being utilized as tools in the identification and optimization of both heterogeneous and homogeneous catalysts. Also, the CCS approach may allow exploration of materials in forms that have not been extensively investigated in the past, including metastable and amorphous materials. It is important to understand that many engineering considerations are critically important in producing commercial catalytic processes. The likelihood of finding a commercially important material among these is not high, but it is likely that significant scientific knowledge could be accrued by thorough study of these materials enabled by combinatorial-type approaches.

VI. CONCLUDING COMMENTS

A powerful feature of the combinatorial approach as applied to the problem of high-dielectric-constant thin-film materials relative to a point-by-point investigation is its propensity to discover interesting properties in unexpected regions of phase space. Conventional approaches would necessarily be guided by prior studies, often of crystalline phases, because of the time and expense involved in preparing multiple targets and examining the resulting films. Our CCS approach has allowed us to investigate broad areas of phase space with fewer preconceived constraints and has lead to some specific cases of materials with excellent properties found in unexpected regions of composition space.

ACKNOWLEDGMENTS

The authors would like to thank their many colleagues and collaborators who have contributed to the work reported here. We would particularly like to acknowledge L. Stirling and G. B. Alers for allowing us to use unpublished data on the crystallization of TaO_x, R. M. Fleming for providing RBS data on the oxide films, and R. J. Felder for technical support.

REFERENCES

1. Kennedy, K. et al. J. Appl. Phys. 1965, 36, 3808.
2. Sawatzky, E.; Kay, E. Cation deficiencies in RF sputtered gadolinium iron garnet films. IBM J. Res. Dev. 1969, 13, 696–702.
3. Hanak, J. J. The 'multi-sample concept' in materials research: synthesis, compositional analysis and testing of entire multicomponent systems. J. Mat. Science. 1970, 5, 964–971.
4. Hammond, R. H.; Ralls, K. M.; Meyer, C. H.; Snowden, D. P.; Kelly, G. M.; Pereue, J. H., Jr. Superconducting properties of $Nb_3(Al,Ge)$ prepared by vacuum vapor deposition of the elements. J. Appl. Phys. 1971, 43, 2407–2413.
5. van Dover, R. B.; Hong, M.; Gyorgy, E. M.; Dillon, J. F., Jr; Albiston, S. D. Intrinsic anisotropy of Tb-Fe films prepared by magnetron cosputtering. J. Appl. Phys. 1985, 57, 3897–3899.
6. van Dover, R. B.; Gyorgy, E. M.; Frankenthal, R. P.; Hong, M.; Siconolfi, D. J. Effect of oxidation on the magnetic properties of unprotected TbFe thin films/citetitle> . J. Appl. Phys . 1986, 59, 1291–1296.
7. Middelhoek, S. Ferromagnetic domains in thin Ni-Fe films; Drukkerij Wed. G. van Soest N. V: Amsterdam, 1961, 20.
8. Kwo, J.; Geballe, T. H. Phys. Rev. B. 1981, 23, 3230–3239.
9. Dayem, A. H.; Geballe, T. H.; Zubeck, R. B.; Hallak, A. B.; Hull., G. W., Jr J. Phys. Chem. Solids. 1978, 39, 529.
10. Unguris, J.; Celotta, R. J.; Pierce, D. T. Observation of two different oscillation periods in the exchange coupling of Fe/Cr/Fe(100). Phys. Rev. Lett. 1991, 67, 140–144.
11. Celotta, R. J.; Pierce, D. T.; Unguris, J. SEMPA studies of exchange coupling in magnetic multilayers. MRS Bulletin. 1995, 20(10), 30–33.
12. Hartsough, L. D.; Hammond, R. H. The synthesis of low-temperature phases by the co-condensation of the elements: a new superconducting A15 compound, V3Al. Sol. St. Comm. 1971, 9, 885–889.
13. van Dover, R. B.; Hessen, B.; Werder, D.; Chen, C-H.; Felder, R. J. Investigation of ternary transition-metal nitride systems by reactive cosputtering. Chem. of Materials. 1993, 5, 32–35.
14. Hanak, J. J.; Gittleman, J. I. Iron-nickel-silica feromagnetic cermets. AIP Conf. Proc. 1973, 10, 961–965.

15. van Dover, R. B.; Schneemeyer, L. F.; Fleming, R. M. Discovery of useful thin-film dielectric using a composition-spread approach. Nature. 1998, 392, 162.

16. Xiang, X.-D. Combinatorial synthesis and high throuput evaluation. Bull. Am. Phys. Soc. 1999, 44, 103.

17. Yoo, Y. K.; Chang, H.; Dong, Y.; Duewer, F.; Li, J.; Xiang, X-D.; Martin, M.; Isaacs, E. D. Combinatorial synthesis and characterization of $Ln_{1-x}B_xMnO_3$. Bull. Am. Phys. Soc. 1999, 44, 103.

18. Xiang, X-D.; Sun, X.; Briceno, G.; Lou, Y.; Wang, K-A.; Chang, H.; Wallace-Freedman, W. G.; Chen, S-W.; Schultz, P. G. A combinatorial approach to materials discovery. Science. 1995, 268, 1738–1740.

19. Gallop, M. A.; Barrett, R. W.; Dower, W. J.; Fodor, S. P. A.; Gordon, E. M. Applications of combinatorial technologies to drug discovery. 1. Background and peptide combinatorial libraries. J. Med. Chem. 1994, 37, 1233–1251.

20. Gordon, E. M.; Barrett, R. W.; Dower, W. J.; Fodor, S. P. A.; Gallop, M. A. Applications of combinatorial technologies to drug discovery. 2. Combinatorial organic synthesis, library screening strategies, and future directions. J. Med. Chem. 1994, 37, 1385–1401.

21. Doyle, P. M. Combinatorial chemistry in the discovery and development of drugs. J. Chem. Tech. 1995, 64, 317–324.

22. Reddington, E.; Sapienza, A.; Gurau, B.; Viswanathan, R.; Sarangapani, S.; Smotkin, E. S.; Mallouk, T. E. Combinatorial electrochemistry: a highly parallel, optical screening method for discovery of better electrocatalysts. Science. 1998, 280, 1735.

23. Theurer, H C.; Hauser, J. J. Getter sputtering for the preparation of thin film interfaces. Trans. AIME. 1965, 233, 588–591.

24. Miyajima, H.; Sato, K.; Mizoguchi, T. Simple analysis of torque measurement of magnetic thin films. J. Appl. Phys. 1976, 47, 4669–4671.

25. Sato, K. Measurement of magneto-optical Kerr effect using piezo-birefringent modulator. Jpn. J. Appl. Phys. 1981, 20, 2403–2409.

26. Basceri, C.; Streiffer, S. K.; Kingon, A. I.; Waser, R. The dielectric response as a function of temperature and film thickness of fiber-textured $(Ba,Sr)TiO_3$ thin films grown by chemical vapor deposition. J. Appl. Phys. 1997, 82, 2497–2504.

27. Hellman, F.; van Dover, R. B.; Nakahara, S.; Gyorgy, E. M. Magnetic and structural investigation of the composition dependence of the local order in amorphous Tb-Fe. Phys. Rev. B. 1989, 39, 10591–10605.

28. Dillon, J. F., Jr; van Dover, R. B.; Hong, M.; Gyorgy, E. M.; Albiston, S. D. Magneto-optical study of uranium additions to amorphous Tb_xFe_{1-x}. J. Appl. Phys. 1987, 61, 1103–1107.

29. The National Technology Roadmap for Semiconductors. Sematech. 1994, 123.

30. El-Kareh, B.; Bronner, G. B.; Schuster, S. E. The evolution of DRAM cell technology. Solid State Technology, May 1997, 89–101.

31. Park, S. O.; Hwang, C. S.; Cho, H-J.; Kang, C. S.; Kang, H-K-.; Lee, S. I.; Lee, M. Y. Fabrication and electrical characterization of $Pt/(Ba,Sr)TiO_3/Pt$ capacitors for ultralarge-scale integrated dynamic random access memory applications. Jpn. J. Appl. Phys. 1996, 35, 1548–1552.

32. Chaneliere, C.; Autran, J. L.; Devine, R. A. B.; Ballard, B. Tantalum pentoxide (Ta_2O_5) thin films for advanced dielectric applications. Mat. Sci. and Eng., R. 1998, 22, 269–322.

33. Wolfram, G.; Göbel, H. Existence range, structural and dielectric properties of $Zr_x Ti_y Sn_z O_4$ ceramics $(x + y + z = 2)$. Mat. Res. Bull. 1981, 16, 1455.

34. Iddles, D. M.; Bell, A. J.; Moulson, A. J. Relationships between dopants, microstructure and the microwave dielectric properties of $ZrO_2-TiO_2-SnO_2$ ceramics. J. Mater. Sci. 1992, 27, 6303.

35. Nakagawara, O.; Toyoda, Y.; Kobayashi, M.; Yoshino, Y.; Katayama, Y.; Tabata, H.; Kawai, T. Electrical properties of $(Zr,Sn)TiO_4$ dielectric thin film prepared by pulsed laser deposition. J. Appl. Phys. 1996, 80, 388.

36. Ramakrishnan, E. S.; Cornett, K. D.; Shapiro, G. H.; Howng, W-Y. Dielectric properties of radio frequency magnetron sputter deposited zirconium titanate-based thin films. J. Electrochem. Soc. 1998, 145, 358.

37. van Dover, R. B.; Schneemeyer, L. F. Deposition of Uniform Zr–Sn–Ti–O films by on-axis reactive sputtering. IEEE Electron Device Letters. 1998, 19, 329.

38. Schneemeyer, L. F.; van. Dover, R. B.; Fleming, R. M. In: Ferroelectric Thin Films, VII; Jones, R. E., Schwartz, R. W., Summerfelt, S. R., Yoo, I. K., eds.; Mat. Res. Soc. Symp. Proc; Vol. 541, 567–572.

39. Schneemeyer, L. F.; van Dover, R. B.; Fleming, R. M. High dielectric constant Hf–Sn–Ti–O thin films. Appl. Phys. Lett. 1999, 75, 1967–1969.

40. Sreemoolanadham, H.; Ratheesh, R.; Sebastian, M. T.; Rodrigues, N.; Philip, J. Synthesis, characterization and properties of Hf-substituted $Zr_{.8}Sn_{.2}TiO_4$ dielectric ceramics. J. Phys. D. 1997, 30, 1809–1814.

41. Mackay, R.; Sleight, A. W.; Subramanian, M. A. Structure, dielectric properties and thermal expansion of a new phase, $Hf_{0.75}Sn_{0.25}O_2$. J. Solid State Chem. 1996, 121, 437–442.

42. van Dover, R. B. Amorphous lanthanide-doped TiO_x dielectric films. Appl. Phys. Lett. 1999, 74, 3041–3043.

43. Roy, P. K.; Kizilyalli, I. C. Stacked high-ε gate dielectric for gigascale integration of metal-oxide-semiconductor technologies. Appl. Phys. Lett. 1998, 72, 2835–2837.

44. van Dover, R. B.; Fleming, R. M.; Schneemeyer, L. F.; Alers, G. B.; Werder, D. J. Advanced dielectrics for gate oxide, DRAM and rf capacitors. 1998 IEDM Technical Digest. 1999, 823.

4

Combinatorial Approach to Ferroelectric/Dielectric Materials

Hauyee Chang
University of California, Berkeley, Berkeley, California, U.S.A.

Ichiro Takeuchi
University of Maryland, College Park, Maryland, U.S.A.

I. INTRODUCTION

Dielectric and ferroelectric materials are an important class of insulating materials used in a wide variety of technological applications. Examples of their properties and relevant applications include the piezoelectric effect for transducer/actuator and microelectromechanical systems (MEMS) device applications, the electro-optical (e-o) effect for fiber-optic communication applications, high dielectric constants and remanent polarization for microelectronic memory device applications, electric-field tunability at microwave frequency for radar and microwave device applications, and the pyroelectric effect for infrared imaging.

Many of these properties are often closely correlated. For instance, a material with a high dielectric constant will often have a high piezoelectric coefficient and dielectric tunability. High tunability is often associated with a large electro-optic coefficient. Because it is nontrivial to implement rapid characterization measurements for some of these properties, correlated functional properties often provide a helpful indirect means to perform high-throughput screening of ferroelectric properties. In some instances, it is possible to perform screening of multiple materials properties on the same library. Combinatorial libraries and phase diagram chips described in this chapter are characterized mainly by the scanning evanescent microwave probe (SEMP) technique, and an enhanced dielectric constant is used as an indicator of pronounced ferroelectric properties.

Owing to over 50 years of worldwide research in the field, a large body of knowledge exists collectively for ferroelectric and dielectric materials. Physical mechanisms of ferroelectricity and dielectricity are also relatively well understood. It is well known that slight materials modifications can lead to drastic changes in their properties. Historically, modification studies have been carried out in old-fashioned trial-and-error experiments.

It is known that solid solutions of perovskite compounds with slightly different crystal structures can give rise to compositions with morphotropic phase boundaries (MPBs). MPB compositions usually have dramatically enhanced piezoelectric coefficients, dielectric constants, and other properties related to ferroelectricity. In the $Pb(Zr,Ti)O_3$ (PZT) system, the MPB is where the crystal structure changes from the Zr-rich ferroelectric rhombohedral structure to the Ti-rich ferroelectric tetragonal structure at the Ti:Zr ratio of 48:52 (1). The MPB compounds with a compositional transition between a relaxor (a diffused ferroelectric with a broadened temperature-dependent paraelectric-to-ferroelectric transition) and a ferroelectric phase usually have even more enhanced ferroelectric properties. In addition to the large dielectric constant and piezoelectric coefficient, these compositions often display very large quadratic e-o coefficients, large microwave tunabilities, and low losses (1). It is known that substitution of La into PZT changes the ceramic from opaque to transparent at visible wavelengths, making it functional for electro-optical applications. In addition, it also leads to the formation of a new MPB associated with a relaxor–ferroelectric transition different from the original ferrolectric–ferroelectric MPB in PZT. The 8.5/65/35 composition of $(Pb,La)(Zr,Ti)O_3$ (PLZT) (formed by substituting 8.5% of La_2O_3 for PbO and having the ratio of $PbZrO_3$ to $PbTiO_3$ of 65:35) has a quadratic e-o coefficient an order of magnitude higher than that of a slightly different compound, 9/65/35, and at least two orders of magnitude higher than other known ferroelectric materials (1). Identifying these narrow phase regions in a binary and ternary system requires performing up to 100 and 10,000 experiments, respectively. There are at least a few hundred candidate systems that are of interest. Painstakingly going through them using the traditional one-by-one synthesis method would require about 10,000 scientist workyears. The continuous phase diagram mapping technique discussed next is suitable for such a project. We will discuss several examples of such experiments.

Another type of modification of materials involves the addition of a small amount of elemental or compound dopants. It is known that properties such as dielectric and mechanical losses can be dramatically improved by this type of modification. For instance, titanates such as $BaTiO_3$ and PZT display very complex defect chemistry, and the introduction of dopants can strongly influence their insulating behavior. To perform this type of study in a traditional way, experiments are expected to be even more tedious and time-consuming than phase diagram mapping. Combinatorial libraries with discrete samples are ideal for this type of exploration.

II. SYNTHESIS OF COMBINATORIAL LIBRARIES AND CONTINUOUS PHASE DIAGRAM CHIPS USING AMORPHOUS PRECURSORS

One of the major distinctions between the combinatorial chip fabrication technique discussed here and other techniques, such as diffusion multiples (2) and composition-spread through codeposition techniques (3)– (7) is that in the present technique, multicomponent compounds are formed from amorphous precursor multilayers. The goal here is to survey as diverse a compositional variation as possible in individual experiments using mathematically designed masking strategies. This technique uses a series of high-precision shadow masks to define the layout of multilayers of amorphous precursors. Elemental metals, simple metal oxides, fluorides, and carbonates are often used as precursors. For instance, in order to make $(Ba,Sr)TiO_3$ (BST), a multilayer consisting of amorphous layers of BaF_2, SrF_2, and TiO_2 are used (Figure 1). The substrate is usually held at room temperature or very low temperature ($100-200°C$) during the layer-by-layer precursor deposition. Postannealing and heat treatment are used for the diffusion of amorphous precursors and phase formation.

There are different combinatorial mask configurations to effectively create and search through large compositional phase spaces. Figure 2 illustrates the quaternary combinatorial masking scheme (8). This scheme involves n different masks, which successively subdivide the substrate into a series of self-similar patterns of quadrants. The rth ($1 \leq r \leq n$) mask contains 4^{r-1} openings, where

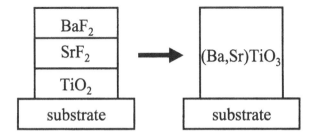

Figure 1 A schematic of the phase formation process from an amorphous precursor multilayer. A controlled thermal treatment turns the amorphous multilayer consisting of BaF_2, SrF_2, TiO_2 into a single-phase $(Ba,Sr)TiO_3$.

each opening exposes one-quarter of the area exposed in the preceding mask. Within each opening, there are an array of 4^{n-r} gridded sample sites. Each mask is used in up to four sequential depositions, where for each deposition the mask is rotated by 90°. This process results in up to 4^n different combinations of precursors created after $4 \times n$ precursor depositions. This can be effectively applied to survey a large number of different compositions, each consisting of up to n elemental components, and each component is selected from a group of up to four precursors.

Implementation of masking schemes is accomplished using either photo-lithographic lift-off steps or physical shadow masking. Because of its high spatial resolution and alignment accuracy, the lift-off method is particularly suitable for generating chips containing a high density of sites. Figure 3 is a photograph (taken under daylight) of a 1024-member-library chip (on the left in the figure) designed to search for new high-ε_r dielectric oxides. This library was laid out by lift-offs using five quaternary masks on a 1-cm $\times$ 1-cm Si substrate (9). Different sites all appear different, due to the varied thicknesses and optical indexes of precursor multilayer films. The variation reflects the diversity one can achieve using this technique.

Continuous phase diagram chips are fabricated by using "shutter" masks that move linearly at a controlled rate during each precursor deposition. For example, a ternary phase diagram is synthesized from gradient thickness depositions of three precursors using a high-precision linear shutter mask (Figure 4)

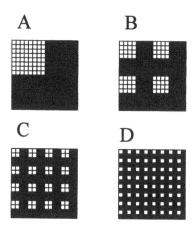

Figure 2   Shadow mask patterns used for quaternary design of combinatorial libraries. Each mask is used in up to four depositions on a square substrate with four different orientations rotated 90° from each other.

Figure 3 A library of Ta_2O_5 and TiO_2-based oxide library deposited on a 1-cm $\times$ 1-cm Si substrate. One thousand twenty-four different compositions of metal oxides were created in search of high-ε_r materials. A penny is shown for size comparison.

and subsequent ex situ postannealing. This technique can be used to generate precisely controlled stoichiometric profiles/layouts within a small area.

Phase formation by annealing amorphous mutilayers of precursors is far from conventional for making stoichiometric compounds. In the beginning, it was not clear that this synthesis approach would generate meaningful crystalline compounds. For instance, in annealing a multilayer, nucleation could occur at each interface between precursors, resulting in the formation of multiple binary phases instead of the desired multielement single phase. In fact, we had made many failed attempts at converting the precursors into crystalline compounds before an effective multistep annealing process was found (10). This approach was inspired by the work of Johnson and colleagues at the University of Oregon on metal precursor interdiffusion (11). It takes advantage of competition between interdiffusion and nucleation at the interfaces of precursor layers. For every bi-layer combination, there exists a critical thickness, determined by the kinetics, below which interdiffusion at the interface is dominant over nucleation. Therefore, it is possible to interdiffuse the precursor layers to form an intermediate amorphous state close in stoichiometry to the desired crystalline phase before the nucleation of thermodynamically stable phases starts. We found that most metal alloys or oxide compounds can be formed from multilayers of metal and/or metal oxide precursors through controlled thermal treatments. An extended period of low-temperature annealing under different atmospheres is often necessary for proper interdiffusion of metal or oxide precursors and stress release and to prevent

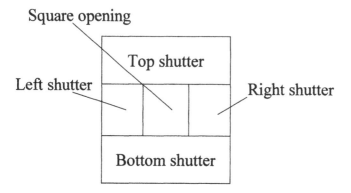

Figure 4 A schematic of a two-dimensional shutter system. Four shutters are attached to four feedthroughs that control their motion. Substrates are placed behind the shutters. A rectangular opening of any shape and size can be placed anywhere within a given substrate, typically less than 1 in. $\times$ 1 in.

evaporation of nonreacted layers. High-pressure/high-vacuum ovens with controlled atmospheres have been found to be useful for proper thermal treatments. When metal precursors are used, in situ postannealing under ultrahigh vacuum is desired. Various analysis tools, such as Rutherford backscattering spectroscopy (RBS), x-ray diffraction (XRD), Auger spectroscopy, ion scattering spectroscopy, and microprobe, can be used to monitor the diffusion and crystalline growth of the films for optimizing the layering sequences as well as the annealing procedures. Typically, prior to synthesizing a full library, a few sampling experiments are performed for a representative composition in order to arrive at appropriate processing conditions for the library.

When volatile elements such as Pb are involved, one needs to exercise extra caution because they can evaporate before the interdiffusion of other nonvolatile elements is even close to completion. Thus, for example, to make $Pb(Zr,Ti)O_3$ and related compounds such as $[Pb(Ni_{1/3}Nb_{2/3})O_3]_x[PbZrO_3]_y$ $[PbTiO_3]_{1-y-x}$ (PNNZT), where $0 \leq x+y \leq 1$, an extra step has been added. Nonvolatile precursors such as Ti and Zr are first annealed at 800°C for interdiffusion in vacuum. A layer of PbO is subsequently deposited onto the alloyed metal, and the multilayer is heated at 500°C together with extra PbO powder to diffuse PbO into the underlying metallic alloy (Figure 5) (12). The extra PbO powder is used primarily to prevent Pb deficiency in the end compound. After the layers are uniformly diffused, the film is heated in flowing oxygen at 800°C to form the crystalline perovskite phase. Figure 6 shows an XRD pattern of resulting crystalline films, indicating high purity of the perovskite phase.

Precursor metal
layers

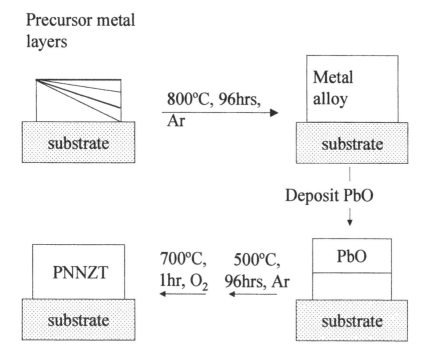

Figure 5   Schematic showing fabrication steps of lead-containing compounds. Gradient thicknesses are deposited for creating composition spreads.

We have found that the multistep annealing of amorphous precursor multilayers not only yields predominantly single-phase materials, but can also produce *epitaxial* thin films on lattice matched substrates. Figure 7a is a $\theta/2\theta$ XRD pattern of a $BaTiO_3$ (BTO) film on a $LaAlO_3$ (LAO) substrate fabricated from a BaF_2/TiO_2 bilayer. The film shows a clear (100) orientation. One can hardly see any impurity phases in the logarithmic intensity scale. Figure 7b shows a ϕ-scan of the (101) planes, displaying a clear fourfold symmetry and indicating that the film is in-plane aligned with the substrate. The ability to epitaxially grow entire libraries is crucial in many fields and applications where material properties are closely tied to the crystalline quality of the films. This technique can be extended to synthesize epitaxial films of other classes of materials, such as metal alloys and nitrides.

While XRD can be used to assess the overall macroscopic crystalline properties of materials, in order to understand the microstructural details of materials, one needs techniques that allow investigation of materials at a microscopic level. Ferroelectric/dielectric properties are notorious for being dominated by micro-

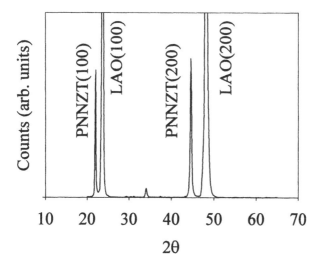

Figure 6   θ/2θ X-ray diffraction pattern of a PNNZT composition fabricated from precursors.

structure. In order to examine the microstructural details of films fabricated from amorphous precursors, we have performed high-resolution transmission electron microscopy (TEM) on $(Ba,Sr)TiO_3$ films made from layers of TiO_2 (deposited first), BaF_2, and SrF_2 on LAO substrates (13,14). The annealing condition was 200°C for a week, followed by 900°C for 1.5 hours in flowing O_2. We have observed that the films do indeed consist of large epitaxial grains that had nucleated at the film/substrate interface. Figure 8a is a high-resolution TEM image of an interface region of one such $Ba_{0.8}Sr_{0.2}TiO_3$ film. The image is taken along the [010] direction, and it clearly shows an atomically sharp interface between the film and the substrate. Figure 8b shows a selected-area electron diffraction of the same region, which further attests to the cube-on-cube orientation of the film and the substrate. Periodically present misfit dislocations seen at the interface are consistent with the lattice match between $Ba_{0.8}Sr_{0.2}TiO_3$ and LAO. In some regions (outside the range of Figure 8a), however, we have also observed small pockets of polycrystalline and amorphous grains. Such regions were found to be typically tens of nanometers in size and take up a very small fraction of the entire film. These minority phases have likely formed as a result of incomplete interdiffusion and/or local stoichiometric deviation (13,14). Other films we have looked at also revealed large epitaxial grains with similar microstructure. Dielectric properties of these films were measured using interdigitated electrodes at 1 MHz and by a SEMP at 1 GHz, and they were found to be very similar to those

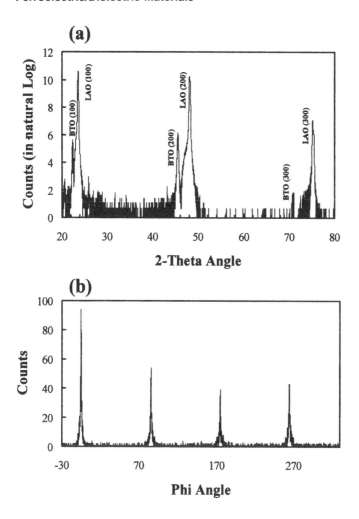

Figure 7   X-ray diffraction of a BaTiO$_3$ film made from precursors. (a) θ/2θ scan showing strong peaks from (100) BaTiO$_3$; (b) φ scan of the (101) plane of BaTiO$_3$. The fourfold symmetry indicates in-plane alignment.

of in situ grown epitaxial films of the same compositions deposited on heated substrates (13). This indicates that, at least in these BST films, the presence of pockets of minority phases and related defects do not affect the macroscopic dielectric properties too much.

For fabrication of libraries and phase diagram chips, proper selection of substrate materials is critical. Refractory single-crystal substrates are usually used to minimize substrate–sample reactions. Lattice matching and thermal expansion coefficients are often taken into consideration so as to allow growth of epitaxial thin-film libraries in some cases. Common substrates we use include MgO, LAO, sapphire, $SrTiO_3$, and Si.

Previously we used mainly pulsed laser deposition (PLD) and RF sputtering for deposition of precursors. An ultrahigh-vacuum (UHV) combinatorial ion-beam sputtering system has also been developed for depositing different types of materials, e.g., metal alloys, nitrides, hydrides, and oxides (including even air-sensitive or volatile precursors). Originally, this system was designed only for pursuing metal alloy systems. However, it soon became apparent that it could play an important role in fabricating many oxide systems and, in particular, many ferroelectric materials. For example, during an initial attempt to fabricate PZT, we found that layers of TiO_2 and ZrO_2 remain undiffused even at temperatures of 800°C for 2 days. In order to interdiffuse the nonvolatile components for PZT, precursors of elemental Ti and Zr were used, as mentioned earlier. Due to the high reflectance and high thermal conductivity of metal targets, laser ablation is not suitable for metal deposition. Metal precursor targets, including air-sensitive precursors such as Ca, Sr, and Ba, are easily available, compared to the oxide targets used in PLD. Some metals interdiffuse more readily than their oxide counterparts, and thus they serve as ideal precursors. In the ion-beam sputtering system, oxidation, nitration, and hydration can be performed in situ as the postdeposition annealing steps.

The ion-beam sputtering chamber is equipped with an automated eight-target carousel and an x–y precision physical masking/shutter system. It is maintained in ultrahigh vacuum ($\sim 10^{-10}$ torr) by a cryopump and a Ti sublimation (or an ion) pump suitable for metal alloy deposition. During sputtering, deposition rates may vary as the ion-beam current fluctuates. Using a feedback control of the ion-beam current from a beam power supply with the thickness monitor signal, one can keep a constant deposition rate that varies less than 0.001 Å/s. In order to calibrate and monitor the deposition rate before and during deposition, a thickness monitor with a flexible feedthrough is positioned either in front of the substrate or at the side of substrate. The $x - y$ masking/shutter system is highly instrumental in the fabrication of continuous phase diagram chips with quick turnaround. The system can also easily be equipped with a series of discrete combinatorial masks, such as the quaternary masks (Figure 2), for the fabrication of discrete site combinatorial libraries.

III. PHASE DIAGRAM MAPPING

A. $(Ba_{1-x-y}Sr_xCa_y)TiO_3$ System

We have made a continuous phase diagram chip to explore the dielectric properties of $Ba_xSr_yCa_{1-x-y}TiO_3$ (BSCT), where $0 \leq x+y \leq 1$ (15). An equilateral

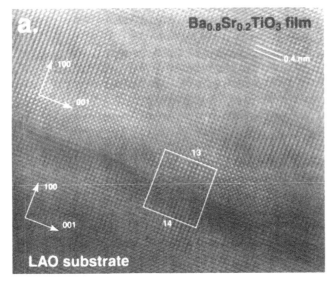

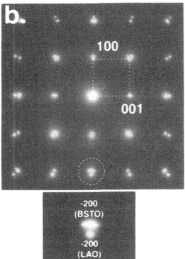

Figure 8 High-resolution TEM of a $Ba_{0.8}Sr_{0.2}TiO_3$ film made from precursors on $LaAlO_3$. (a) High-resolution TEM image of an interface region showing an epitaxial relationship between the film and the substrate. A Burgers circuit shows the presence of a misfit dislocation. (b) Corresponding selected-area diffraction of the substrate and the film. The relationship is cube-on-cube. An enlargement of the (200) reflection is shown below.

triangle–shaped LAO substrate with the triangle height of 1 in. is used as the substrate. The bottommost precursor layer is a uniform 750-Å-thick TiO_2. A shutter in the deposition system was moved with constant velocity during deposition to create linear gradient thicknesses of $SrCO_3$ (0–1475 Å), $BaCO_3$ (0–1647 Å), and $CaCO_3$ (0–1255 Å) on top of the TiO_2 layer in three directions, which are rotated 120° from each other. This provides a design of a composition spread where a ternary phase diagram of BSCT is created on the triangular chip. This chip was heated at 400°C for 24 hours for premixing of the precursors, followed by 900°C for 1.5 hours for crystallization under flowing oxygen. The CO_2 from the precursors is removed during the annealing process.

The XRD patterns of large individual samples of various selected compositions from the BSCT phase diagram showed that the films were (100) oriented and *epitaxial* with respect to the (100) LAO substrates. And RBS results on similar samples deposited on MgO indicated that the precursor layers had been interdiffused. Atomic force microscopy measurements of the triangle phase diagram chips show that the average grain size is 150×150 nm^2 across the chip, with no signs of off-stoichiometric outgrowths at the surface.

Dielectric properties of the phase diagram chip were mapped using a SEMP, which allows nondestructive quantitative characterization of dielectric properties with high spatial resolution. (The SEMPs are described in detail in Chapter 11.) Figure 9 presents ε_r and tan δ maps of the BSCT triangle at 1 GHz. The composition of $SrTiO_3$ displays the lowest ε_r (≈ 150), and $BaTiO_3$ has $\varepsilon_r \approx 400$ at 1 GHz. Shown in the loss tan δ map, tan δ for $BaTiO_3$ is 0.4, which is the same value as that of a single crystal, whereas tan δ for $SrTiO_3$ is the lowest, as expected. The variation of the phasespread in both ε_r and tan δ is predominantly due to changes in the composition of the thin films rather than to extrinsic effects such as differences in microstructure and strain in the film.

On the left (vertical) and lower edges of the ε_r map, where the composition is $(Sr_{1-x}Ca_x)TiO_3$ and $(Ca_{1-x}Ba_x)TiO_3$ ($0 < x < 1$), respectively, we found that ε_r decreases with increasing amounts of Ca. Similar trends have been observed previously in ceramic studies of such compositions (16,17). Along the upper edge of the ε_r map, the composition is $(Ba_{1-x}Sr_x)TiO_3$ ($0 < x < 1$). A line scan of ε_r versus composition along the upper edge gives a maximum ε_r (≈ 500) at $Ba_{0.7}Sr_{0.3}TiO_3$. This is consistent with values reported for bulk samples, where a transition occurs at $x = 0.3$ at room temperature from a pseudo-cubic paraelectric phase to a tetragonal ferroelectric phase. This observation is significant in that we have demonstrated that a crystalline-phase boundary could be reliably identified with a continuous phase diagram chip. We have also measured the electro-optical coefficients of a $(Ba_{1-x}Sr_x)TiO_3$ composition spread (18). Figure 10 shows the linear (γ) and quadratic (s) coefficients as a function of x. The nonzero value of the linear e-o coefficient for compositions with x larger than 0.3 is perhaps indicative of the fact that the transition from ferroelectric–paraelec-

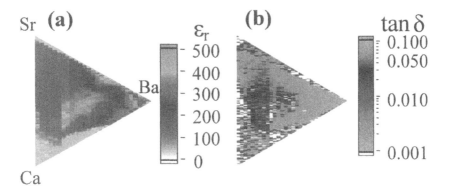

Figure 9 Dielectric constant mapping of a $(Ba,Sr,Ca)TiO_3$ phase diagram chip. Data are converted from SEMP images. (a) ε_r mapping; (b) tan δ mapping

tric phase boundary is not complete at $x = 0.3$. This diffusive nature of the boundary may also be an indication of the coexistence of ferroelectric and cubic phases in this film.

The grayishwhite regions in Figure 9b are the areas with tan δ less than 0.02. Selecting regions of high ε_r with low tan δ from Figure 9, we find that compositions within the regions of $Ba_{0.12-0.25}Sr_{0.35-0.47}Ca_{0.32-0.53}TiO_3$ have lowest tan δ and are the best candidates for dielectric applications with ε_r between 130 and 160. The values of tan δ for this region are even less than those for $SrTiO_3$, whereas the values of ε_r are about the same or larger.

B. $Pb(Zr_{1-x}Ti_x)O_3$

A family of lead-based ferroelectrics exhibits many interesting properties. They include $Pb(Mg_{1/3}Nb_{2/3})O_3$, $Pb(Zn_{1/3}Nb_{2/3})O_3$, $Pb(Fe_{1/2}Ta_{1/2})O_3$, and $Pb(Co_{1/2}W_{1/2})O_3$, which are classified as relaxor ferroelectrics. These compounds have the perovskite structure (ABO_3), in which the B sites are occupied by two types of ions having chemical valences different from 4^+. Table 1 lists many possible B-site element combinations for forming complex relaxor compounds. They display diffuse paraelectric–ferroelectric phase transitions because of the fluctuation in the ratio of B1 to B2 ion concentration (1). The diffused and metastable phases can be electrically driven from the paraelectric to the ferroelectric phase. This property is extremely useful in many applications, including e-o devices in telecommunication and tunable microwave and radar devices.

By adding one of these complex compounds to a PZT binary system, a ternary solid solution with a perovskite structure can be formed. Figure 11 shows

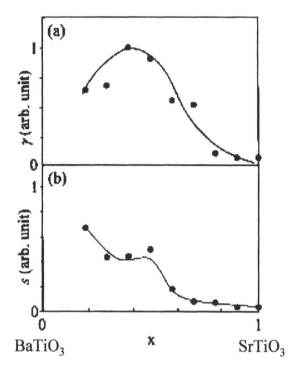

Figure 10 Electro-optical coefficients measured on a $Ba_{1-x}Sr_xTiO_3$ composition spread. (a) linear electro-optical coefficient; (b) quadratic electro-optical coefficient. Solid lines are guides.

Table 1 Complex Metal–Ion Combination in Some Ternary Piezoelectric Ceramics

$Mg_{1/3} Nb_{2/3}$	$Co_{1/3} Nb_{2/3}$	$Li_{1/4} Ta_{3/4}$	$Mn_{1/2} Nb_{1/2}$	$Ni_{1/3} Sb_{2/3}$
$Zn_{1/3} Nb_{2/3}$	$Fe_{1/3} Nb_{2/3}$	$Cu_{1/4} Nb_{3/4}$	$Mn_{2/3} W_{1/3}$	$Ni_{1/3} Bi_{2/3}$
$Mg_{1/3} Ta_{2/3}$	$Fe_{1/3} Sb_{2/3}$	$Sb_{1/2} Nb_{1/2}$	$Mn_{1/3} W_{3/2}$	$Ni_{1/2} W_{1/2}$
$Zn_{1/3} Ta_{2/3}$	$Ni_{1/3} Nb_{2/3}$	$In_{1/2} Nb_{1/2}$	$Mn_{1/2} W_{1/2}$	$Fe_{1/2} Sb_{1/2}$
$Sb_{1/3} Nb_{2/3}$	$Sn_{1/3} Nb_{2/3}$	$Co_{1/2} W_{1/2}$	$Mn_{1/2} Ta_{1/2}$	$Sb_{1/2} Ta_{1/2}$
$Mn_{1/3} Nb_{2/3}$	$Li_{1/4} Sb_{3/4}$	$Cd_{1/2} W_{1/2}$	$Mn_{1/2} Sb_{1/2}$	$Al_{1/2} Te_{1/2}$
$Mn_{1/3} Ta_{2/3}$	$Li_{1/4} Nb_{3/4}$	$Co_{1/3} Nb_{2/3}$	$Mn_{1/3} Bi_{2/3}$	$In_{1/2} Te_{1/2}$
$Cd_{1/3} Nb_{2/3}$	$Fe_{1/2} Ta_{1/2}$	$Mg_{1/2} W_{1/2}$	$Mn_{1/3} Sb_{2/3}$	$Te_{1/3} Fe_{2/3}$
$Y_{1/2} Nb_{1/2}$	$Zn_{1/3} Ta_{2/3}$	$Ni_{1/5} Fe_{1/5} Nb_{3/5}$		

an example of a ternary phase diagram at room temperature (1). As in PZT, compositions near MPBs in the ternary system exhibit large values of the dielectric constant and the piezoelectric coefficient, especially for compositions near the pseudo-cubic (PC)–tetragonal–rhombohedral transition. If the composition is on the PC side, the compounds are expected to be relaxors with a large quadratic e-o effect and microwave tunability. However, most of these systems are yet to be investigated. A tremendous amount of work will be required if conventional materials synthesis methods are used to explore them. Thus, great opportunities exist in systematically pursuing these materials systems using the continuous phase diagram mapping technique.

Synthesis of these materials is in itself a very challenging task. The thermodynamically stable perovskite phases are usually "contaminated" with the kinetically favorable but undesirable pyrochlore phases. Several methods of fabrication for both bulk and thin films have been developed to overcome this hurdle (19). We developed a technique that allows formation of pyrochlore-free lead perovskite compounds using metal precursors. For creating $Pb(Zr,Ti)O_3$, first, 200 Å of Zr and 150 Å of Ti (to obtain the Zr:Ti ratio of 1:1) were deposited and annealed at 800°C under 1 atm of Ar for 4 days to interdiffuse the metals. Then 670 Å of PbO was deposited on top of the alloyed thin film and heated under 10 atm Ar at 400°C for 4 days, to enable diffusion of Pb into the metal alloy (Figure 5). To confirm that significant mixing of the different components had taken place, RBS was used. The film was then further annealed under flowing oxygen at 700°C for half an hour to form the crystalline perovskite compound. X-ray diffraction data showed only Pb-based perovskite compound peaks, with no observable pyrochlore peaks (Figure 6).

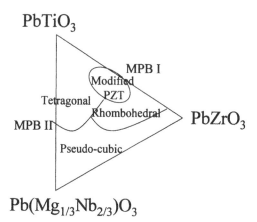

Figure 11 $Pb(Mg_{1/3}Nb_{2/3})TiO_3$–$PbTiO_3$–$PbZrO_3$ ternary structural phase diagram.

Figure 12 is a SEMP dielectric constant line scan of a $Pb(Zr_{1-x}Ti_x)O_3$ composition spread. ε_r peaks near the Ti:Zr ratio of 48:52, the well-known MPB in PZT [1]. This again demonstrates that MPBs in ferroelectric materials systems can be reliably mapped out using composition spreads. Using this technique, we can now begin to rapidly explore a variety of relaxor-PT-PZ ternary systems.

IV. DOPANTS STUDY

In order to study the effects of different dopants on the dielectric properties of $(Ba_{1-x}Sr_x)TiO_3$ (BST), a library consisting of four different stoichiometries of BST thin films ($x = 1.0, 0.8, 0.7,$ and 0.5) as hosts was generated. The hosts were doped with different combinations of up to three out of nine different metallic elements, with each dopant added in excess of 1 mol% with respect to the BST host (20). The quaternary combinatorial masking scheme was used to generate $4^4 = 256$ different dopant combinations in 16 steps. To fabricate the library, TiO_2 (870 Å) was deposited first to generate an array of 256 samples, each 650×650 μm^2, on a (100) LAO substrate. This was then followed by deposition of

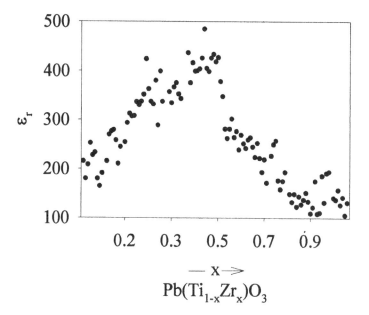

Figure 12   Dielectric constant mapping of a $Pb(Ti_{1-x}Zr_x)O_3$ composition spread. The peak is observed at $x \cong 0.5$ near the morphotropic-phase boundary.

BaF_2, SrF_2, and different dopants in the following sequence: B_1:Fe_2O_3 (7 Å); B_2: W (5 Å); B_3:CaF_2 (12 Å); C_1:Cr (4 Å); C_2:Mn_3O_4 (7 Å); C_3:CeO_2 (12 Å); D_1: MgO (7 Å); D_2:Y_2O_3 (10 Å); D_3:La_2O_3 (12 Å); A_1:BaF_2 (1640 Å); A_2:SrF_2 (270 Å) + BaF_2 (1320 Å); A_3:SrF_2 (410 Å) + BaF_2 (940 Å); A_4:SrF_2 (680 Å) + BaF_2 (830 Å), where A, B, C, and D denote the quaternary masks used in the deposition steps, each subscript (i) indicates an ($i - 1$) × 90° clockwise rotation of the mask relative to the orientation of the mask shown in Figure 2, and the numbers in parentheses indicate the film thickness. The resulting dopant map of the library (for each host compound) is shown in Figure 13. The dopant layers were sandwiched between TiO_2 and the fluoride materials to prevent evaporation during subsequent annealing steps. The library was then heated at 400°C in flowing oxygen for 24 h to initiate mixing of the precursors (TiO_2, BaF_2, and SrF_2) as well as diffusion of the dopants. This was followed by a further annealing in flowing oxygen at 900°C for 1.5 h. X-ray diffraction of separately prepared individual composition samples has confirmed that this procedure would result in the formation of epitaxial films in the library.

We used a SEMP for nondestructive dielectric characterization of the thin-film library. Dielectric constant images (at 1 GHz) of the BTO-containing region of the library, consisting of 64 samples, are shown in Figure 14. Each square corresponds to a different thin-film sample site. The data have been converted to averaged ε_r and tan δ values for each site from images of resonant frequency and Q shifts of the SEMP.

The ε_r image of the library Figure 14a, in which a lower dielectric constant is represented by a darker shade, shows that samples in the upper right-hand quadrant have a lower value of ε_r relative to pure BTO (marked sample 1 in Figure 14b. All these sites contain 1 mol% W, which indicates that doping with W lowers the dielectric constant of BTO. This trend is also observed in other BST host composition regions of the library. In general, most dopants seem to decrease ε_r, except for a few dopants, including La (sample 2) and Ce (sample 3). A detailed look at samples that differ only in being doped with either La or Y but that are otherwise identical reveals that the samples doped with the larger La ion have higher ε_r, than with those doped with Y. This is evident in the alternating light and dark sites in the even-numbered columns (counting from the leftmost column) of the ε_r image for samples with the BTO host Figure 14a. We have found that the differences between La-doped and Y-doped samples become less pronounced in hosts containing more Sr. These results are summarized in Table 2.

In the tan δ image Figure 14b, a lighter shade implies a lower loss tangent (tan δ). Thus all samples in the W-doped quadrant, which have slightly reduced ε_r, also have lower loss tangent in comparison with pure BTO. For example, tan δ of W-doped BTO (sample 6) is 0.1, compared to 0.42 for undoped BTO (sample 1), whereas its ε_r is reduced to 406 from 593 (for BTO) at 1 GHz. In many device applications, having low tan δ is of most importance. In particular, for tunable

Fe Mg Cr	Fe Y Cr	Fe Mg Mn	Fe Y Mn	W Mg Cr	W Y Cr	W Mg Mn	W Y Mn
Fe Cr	Fe La Cr	Fe Mn	Fe La Mn	W Cr	W La Cr	W Mn	W La Mn
Fe Mg	Fe Y	Fe Mg Ce	Fe Y Ce	W Mg	W Y	W Mg Ce	W Y Ce
Fe	Fe La	Fe Ce	Fe La Ce	W	W La	W Ce	W La Ce
Mg Cr	Y Cr	Mg Mn	Y Mn	Ca Mg Cr	Ca Y Cr	Ca Mg Mn	Ca Y Mn
Cr	La Cr	Mn	La Mn	Ca Cr	Ca La Cr	Ca Mn	Ca La Mn
Mg	Y	Mg Ce	Y Ce	Ca Mg	Ca Y	Ca Mg Ce	Ca Y Ce
	La	Ce	La Ce	Ca	Ca La	Ca Ce	Ca La Ce

Figure 13 Dopant map of the discrete site $(Ba,Sr)TiO_3$ library. The same map was applied to different host materials. Arrangement of the dopants was achieved using the quaternary masks.

microwave devices, the figure of merit is given by $2(\Delta\varepsilon_r/\varepsilon_r)/\tan\delta$. The reduction in $\tan\delta$ by doping with W was also observed in the other host materials.

The effect of the addition of tungsten on the dielectric constant and $\tan\delta$ are summarized in Table 3. A separate tunability $(\Delta\varepsilon_r/\varepsilon_r)$ study has shown that tungsten-doped samples do not display significantly reduced tunability compared to undoped BST (Table 3). Thus, the figure of merit (FOM) in BST has been increased (by a factor of ≈ 2) by doping with tungsten. Following this work, a careful study of high-quality in situ grown films of W-doped BSTO by J. Horwitz at the Naval Research Laboratory has yielded a record low microwave loss tan

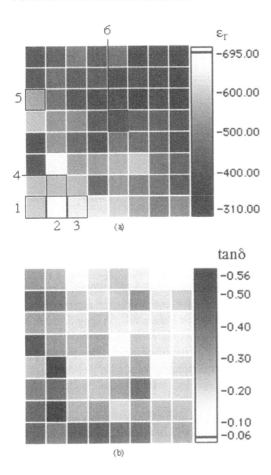

Figure 14 Dielectric constant mapping of doped $BaTiO_3$ films. (a) ε_r mapping; (b) tan δ mapping. Refer to Figure 4.13 for dopant map.

δ of 0.005 at room temperature (21). This represents a significant improvement in the material for tunable microwave device applications.

V. CONCLUDING REMARKS

The success and integrity of combinatorial thin-film synthesis using the precursor technique depends on how well one can map out the trend with which the compositional variation changes the physical properties across the library. This does not

Table 2 Dielectric Constant (ε_r) and Loss (tan δ) at 1 GHz of the Highlighted Films (Fig. 14).

Films	Sample	ε_r	tan δ
$BaTiO_3$	1	593	0.42
$BaTiO_3/La_2O_3$	2	695	0.35
$BaTiO_3/CeO_2$	3	634	0.45
$BaTiO_3/Y_2O_3$	4	576	0.54
$BaTiO_3/Fe_3O_4,MgO$	5	559	0.28
$BaTiO_3/W$	6	406	0.10
$SrTiO_3$		215	0.02
$SrTiO_3/W$		177	0.01

Pure and doped samples of $SrTiO_3$ are included for comparison.

necessarily translate to having to fabricate materials with absolutely no minority phases and microstructural defects within each compositional region. For instance, in libraries of potentially superconducting materials, as long as there are sizable superconducting regions within a fixed compositional area, most measurement techniques can detect their presence. On the other hand, there are materials properties that are critically affected by small amounts of impurity phases. Applicability of the precursor technique to a specific materials system needs to be determined on a case-by-case basis.

In this chapter, we have demonstrated its applicability to ferroelectric materials. Development of new materials in this field is always pushed to the limit by technological demands. At the time of this writing, there are exciting developments, for instance, in the search for lead-free ferroelectrics and novel ferroelectrics with layered structures for nonvolatile-memory applications. Given the immense potential for discovering new materials and the size of the compositional phase space yet to be explored in this field, we believe techniques such as the ones

Table 3 Summary of the Effect of W doping into (Ba, Sr) TiO_3

Film	Tunability (%)	ε_r at 0 kV/cm	tan δ at 0 kV/cm	FOM
$Ba_{0.5}Sr_{0.5}TiO_3$	56	1187	0.030	37
W-doped $Ba_{0.5}Sr_{0.5}TiO_3$	40	706	0.012	67
$Ba_{0.7}Sr_{0.3}TiO_3$	45	1106	0.034	25
W-doped $Ba_{0.7}Sr_{0.3}TiO_3$	39	1001	0.017	46

described here will be found more and more among the mainstream experimental techniques in the future.

REFERENCES

1. Xu, Y. Ferroelectric Materials and Their Applications: North-Holland; 1991.
2. Hasebe, Mitsuhiro; Nishizawa, Taiji Application of Phase Diagram in Metallurgy and Ceramics. NBS special pub. 496 : Washington, DC, 1977, 2, 911.
3. Kennedy, K.; Stefansky, T.; Davy, G.; Zacky, V. F.; Parker, E. R. J. Appl. Physics. 1965, 36, 3808.
4. Carl Miller, N.; Shirn, George A. Appl. Physics Lett. 1967, 10, 86.
5. Hanak, George A.; Gittleman, J. I.; Pellicane, J. P.; Bozowski, S. Physics Lett. 1969, 30A, 201.
6. Sawatzky, E.; Kay, E. IBM J. Res. Develop. Nov. 1969, 696.
7. Hanak, J. J. J. Mater. Sci. 1970, 5, 964.
8. Xiang, X.-D.; Shultz, P.G. Physica C. 1997, 428, 282–287.
9. Takeuchi, I. unpublished.
10. Xiang, X.-D.; Sun, X.; Briceño, G.; Lou, Y.; Wang, K.-A.; Chang, H.; Wallace-Freedman, W.G.; Chen, S.-W.; Schultz, P.G. Science. 1995, 268, 1738–1740.
11. Fister, L.; Novet, T.; Grant, C. A.; Johnson, D. C. Adv. Synthesis Reactivity Solids. 1994, 2, 155.
12. Chang, H.; Yu, K.-M.; Dong, Y.; Xiang, X.-D. Appl. Physics Lett. 2002, 81, 2062–2064.
13. Takeuchi, I.; Chang, K.; Sharma, R. P.; Bendersky, L. A.; Chang, H.; Xiang, X.-D.; Stach, E. A.; Song, C.-Y. J. Appl. Physics. 2001, 90, 2474–2478.
14. Bendersky, L. A.; Lu, C. J.; Scott, J. H.; Chang, K.; Takeuchi, I. J. Mater. Res. 2002, 17, 2499–2506.
15. Chang, H.; Tacheuchi, I.; Xiang, X.-D. Appl. Physics Lett. 1999, 74, 1165.
16. Durst, G.; Grotenuis, M.; Barkow, A.G. J. Am. Ceram. Soc. 1950, 33, 133.
17. Mitsui, T.; Westphal, W. B. Phys. Rev. 1961, 124, 1354.
18. Li, Jinwei; Duewer, Fred; Gao, Chen; Chang, Hauyee; Xiang, X.-D. Appl. Physics Lett. 2000, 76, 769–771.
19a. Chen, Y.-F.; Yu, T.; Chen, J.-X.; Shun, L.; Li, P.; Ming, N.-B. Appl. Physics Lett. 1996, 66, 148.
19b. Song, Y.J.; Zhu, Y.; Desu, S.B. Appl. Physics Lett. 1998, 72, 2686.
19c. Lin, Y.; Zhao, B.R.; Peng, H.B.; Xu, B.; Chen, H.; Wu, F.; Tao, H.J.; Zhao, Z.X.; Chen, J.S. Appl. Physics Lett. 1998, 73, 2781.
20. Chang, H.; Gao, C.; Takeuchi, I.; Yoo, Y.; Wang, J.; Schultz, P. G.; Xiang, X.-D.; Sharma, R. P.; Downes, M.; Venkatesan, T. Appl. Physics Lett. 1998, 72, 2185.
21. Chang, W.; Horwitz, J. S.; Kim, W.-J.; Pond, J. M.; Kirchoefer, S. W.; Chrisey, D. B. In: Ferroelectric Thin Films, VII; Jones, R. E., Schwartz, R. W., Summerfelt, S. R., Yoo, I. K., eds.; Materials Research Society Symposium Proceedings: Warrendale: PA, 1999; Vol. 541, 699.

5
Parallel Synthesis of Artificially Designed Lattices and Devices

Mikk Lippmaa
University of Tokyo, Kashiwa, Japan

Masashi Kawasaki
Tohoku University, Sendai, Japan

Hideomi Koinuma
Tokyo Institute of Technology, Yokohama, Japan

I. INTRODUCTION

The Merrifield method for high-throughput synthesis of organic compounds employs sequential bond formation at the interface between a homogeneous solution and an insoluble solid resin (1,2). The method has been extended to the synthesis and discovery of new drugs and other organic materials by locating a large number of resin beads at spatially resolved sites and subjecting them to a series of different reactions (3). Combinatorial chemistry, developed in the drug industry, has started to attract the interest of other fields, with the aim of discovering new or improved materials, including catalysts, polymers, and ceramics. The coupling of thin-film deposition techniques and combinatorial chemistry has emerged as one of the most promising methods for high-throughput synthesis of new functional materials (4–7). The conventional combinatorial material process is composed of three basic steps: (a) several thin films with different compositions are deposited sequentially at room temperature on a substrate segmented with masks, (b) annealing of the films at a low temperature to obtain a homogenized amorphous phase, and (c) further heating of the films to promote crystallization (8). This process is essentially a one-pot synthesis at each reaction site using the classical sintering process. Hence, the products obtained are generally in thermally equilibrated polycrystalline phases.

Recent progress in thin-film technology has made it possible to fabricate atomically controlled lattices and heterojunctions even if they are not in thermo-dynamically stable phases. Products prepared by such a process attract much interest, since they could exhibit new and exotic properties and quantum-size effects. The formation of atomically defined structures, however, requires a lot of time due to the necessity of optimizing the composition and various other reaction parameters that influence thin-film growth, if we employ the conventional one-by-one deposition procedure on a single substrate. In order to obtain a breakthrough in this situation, we have developed a new technology for the parallel synthesis of molecular layers under in situ monitoring of the reflection high energy electron diffraction (RHEED) pattern and intensity while rapidly scanning the electron beam over the growing film surface (9).

This chapter is devoted to the description of the necessity of advanced combinatorial technology, the basic concept and design of combinatorial laser MBE systems for the parallel fabrication of lattice-engineered materials, and the typical result of oxide film libraries made with these systems. Another example of combinatorial sequential fabrication of thin films and junctions in a nonequilibrium phase is also included: combinatorial plasma chemical vapor deposition and its application to the optimization of the structure and deposition conditions of amorphous silicon thin-film transistors (10).

Due to the nonequilibrium nature of reactions occuring in the system and the availability of additional combinatorial parameters, such as temperature, pressure, and evaporation energy source and power, the variation of material properties within a library can vary widely, from amorphous to quasi-equilibrated crystals, films doped with impurities exceeding solubility limits, artificially designed lattices and superlattices, heterojunctions, and devices.

II. COMBINATORIAL CHEMISTRY AND REACTION COORDINATES

Combinatorial chemistry addresses the need to synthesize large numbers of samples while still allowing individual characterization of each sample. The word *sample* has a very wide meaning here and can refer to anything ranging from organic molecules or bulk solid-state materials to thin films or even complete electronic devices. The development of new materials or structures having specific desired properties often requires a large amount of experimental work, in order to find out how the various compositional, structural, or fabrication-related parameters affect the physical or chemical properties (11).

Combinatorial techniques have been developed since the early 1990s mainly in the field of organic synthesis with the aim of reducing the amount of time required to find new bioactive molecules. Two different methods are commonly used. A large number of samples can be rapidly synthesized in both cases, but the methods for separating individual reaction products are different. The first method is known as *mix and split*, illustrated in Figure 1a. The figure illustrates a case where we have three families of compounds (A, B, and C). Each family can have several members (1, 2, and 3). The aim is to synthesize all possible combinations of these compounds. The order in which the reactions are performed is generally also important. The process starts with some form of carriers, usually polymer beads. In the first reaction step the A compounds are attached to the carrier beads in separate reactions. This ensures that each carrier holds only a single member of the A family, such as $\bigcirc$ -A_1 or $\bigcirc$ -A_2, but not mixtures, such as $\bigcirc$ -A_1-A_2. The products are mixed together and the mixture is divided into three samples. In step 2, each sample is allowed to react with a compound from the B family. The products of these reactions are again mixed together, resulting in a mixture that has all possible combinations of the form $\bigcirc$ -A_i-B_j, where i, j = 1, 2, 3. The mixture is again divided into three components, and in step 3 each one is allowed to react with the compounds belonging to the C family. The number of families can be arbitrarily large, and the number of compounds in each familiy can, of course, be much larger than three.

At the end of the synthesis process the molecules are detached from the carrier beads and the mixture can be analyzed for the desired chemical or biological activity. This method makes it easy to synthesize a large number of related compounds. The testing will tell if there are any compounds in the final mixture with the desired properties. If not, a new set of starting materials can be selected, or the order of reactions can be changed. However, even if the testing shows that a compound with the desired properties is present, it can be difficult to determine from this experiment alone which particular combination of the form A_i-B_j-C_k is responsible for the positive test result. Various tagging techniques have been developed that can help in identifying which precursor combination is attached to any particular carrier bead.

This question is easier to answer when the parallel synthesis method is used. In this case all reaction products are kept in separate reactors. This can be done by using a set of test tubes or, more conveniently, by using special plastic multiwell plates that can be handled by automated laboratory robots. In the first step, components of the A family are placed in individual rows of wells. In the second step, members of the B family are added in different columns of the same well array. This produces all possible combinations of the memebers of the A and B families of precursors. More components can be added by using additional multiwell plates or by duplicating rows or columns in the first steps. This method has the advantage of making it easy to identify a compound if desired activity

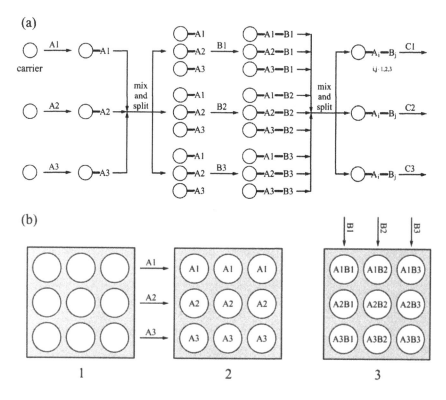

Figure 1   (a) The principle of the mix-and-split method. (b) The parallel synthesis method.

is detected in a particular well during screening. The screening process itself, however, is more time-consuming and is typically highly automated in order to achieve acceptable throughput.

The advantages of combinatorial chemistry can be appreciated if one compares combinatorial synthesis with traditional synthesis. All chemical reactions, including both organic materials and inorganic materials, in liquid or solid form, are influenced by a number of factors. These include time, choice of reaction precursors, temperature, pressure, presence of catalysts, and external energy in the form of heat or light. In a conventional "test tube" case, the only reaction coordinate is time. Factors such as composition, temperature, pressure, external supply of energy, and catalysis, are fixed for the duration of the reaction. This is illustrated in Figure 2a.

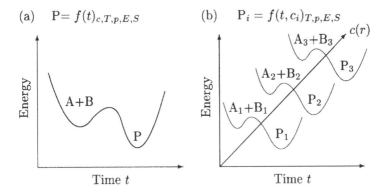

Figure 2 (a) The reaction coordinate of a single chemical reaction. (b) The reaction coordinate space of a combinatorial experiment has at least one additional dimension, such as the set of reaction precursors c. Many reactions take place in parallel.

Combinatorial solid-state synthesis adds one more dimension to the reaction parameter space, as shown in Figure 2b, by adding the capability of producing many samples with different compositions in parallel. The reaction parameters are now time t and the composition c. This type of combinatorial synthesis has been developed by Xiang and coworkers (4,8).

It is possible to extend this idea further and to reduce the number of parameters that remain constant in a single synthesis step. In an ideal case the synthesis process would be described by

$$P = f(t,c,T,p,E,S). \tag{1}$$

The last four parameters (temperature, pressure, external energy, and catalysts) are usually considered to be process parameters that are not of primary importance in the task of finding new materials (12). These parameters are, however, of great importance when combinatorial synthesis methods are applied to more complex structures, such as thin films, superlattices, nanodots, nanowires, or complete electronic devices. The reaction temperature and ambient pressure have a dramatic influence on the growth process of thin films. The following sections describe a number of approaches we have explored. We work mostly with transition metal oxides, although some other examples, related to amorphous silicon, are presented as well. All work is done on thin films, and the particular choice of deposition techniques is dictated mostly by the needs of oxide deposition, i.e., the need to work at temperatures exceeding 1000°C in a pure oxygen atmosphere.

III. COMBINATORIAL SOLID-STATE CHEMISTRY

A. The Concept

Combinatorial thin-film synthesis is compared with the original Merrifield synthesis path in Figure 3. Organic synthesis (Figure 3a) starts with a carrier bead (step 1), which serves as a starting point for the growth of a linear molecule. A linker is first attached to the bead (step 2), followed by repeated reactions where a larger molecule is grown one unit at a time (step 3). Between individual reactions the intermediate reaction products can be mixed or divided, as described in the previous section. When the desired number of reaction steps have been completed (step 4), the linker can be severed, releasing the molecule from the carrier bead (step 5).

B. Combinatorial Thin-Film Deposition and Sintering

There are two different routes that can be used to fabricate a thin-film library. The route shown in Figure 3b was developed by Xiang et al. (4). In this case steps 2 and 3 are repeated with different masks, generating a stack of component materials as shown in step 4. The depositions are performed at room temperature,

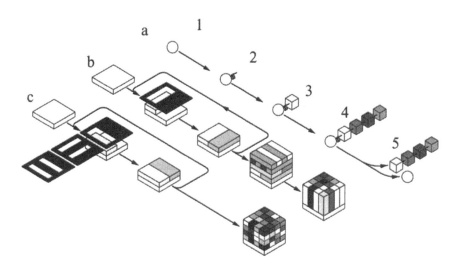

Figure 3   Comparison of the combinatorial Merrifield organic synthesis (a) and combinatorial thin-film growth techniques. Route (b) can be used to synthesize thermodynamically equilibrated compounds. Route (c) can be used to grow nonequilibrated phases or artificial crystal structures.

using contact masks if necessary. The sample can then be annealed, allowing conventional solid-state reactions to occur in each cell of the library. A major benefit of this technique is the ability to use contact masks due to the room-temperature deposition. This means that even on fairly small oxide substrates it is possible to achieve very high cell densities. The minimum cell size is typically limited more by the needs of the library characterization process than by the deposition procedure itself.

C. Combinatorial Lattice Engineering

Combinatorial lattice engineering uses the parallel synthesis path, where a number of different lattices are deposited on a single substrate (Figure 3c) under conditions tuned for epitaxial growth of each layer individually. In this case the substrate surface takes the role of the multiwell plastic reactor. Instead of individual plastic beads, the reaction centers are defined on the substrate surface using movable masks (step 2).

Epitaxial thin films typically grow directly from the gas phase, and hence it is possible to omit the separating walls required for liquid-phase synthesis. Physical separation of the various cells on the substrate surface, each containing a film with a different composition or structure, is required if contact-mode, such as four-point resistivity, or magnetoresistance measurements are needed to characterize the material properties. This can most easily be achieved by using various stencil or contact masks to define the individual cell positions on the substrate surface. Movable shutters can then be used to select which parts of the substrate are open for deposition. If contact-mode characterization is not needed, the cell definition stencil mask can be omitted.

In the process shown in Figure 3c, a stencil mask is positioned close to the substrate surface. Thin films are selectively grown in only certain cells of the library at a time by using changeable masks. The main advantage here is that the growth temperature, background oxygen pressure, or any of the other process parameters affecting film growth can be freely adjusted for each component layer. The final annealing step can also be skipped. The main advantage of this route is the ability to control better the structure of the film and thus to grow structures that cannot be formed in thermodynamic equilibrium conditions, such as superlattices and metastable phases. Several examples of this technique are described in this chapter. The main disadvantage is the limited library density. The cell boundary definition mask cannot be placed in contact with a hot substrate, limiting the masking accuracy. A typical deposition configuration is shown in Figure 4. In practice, 1 mm × 1 mm is a useful cell size.

There are several possible approaches to parallel synthesis of thin films. The type of lattice engineering described here relates mostly to pulsed laser deposition (PLD) and laser molecular beam epitaxy (LMBE) in which PLD is performed

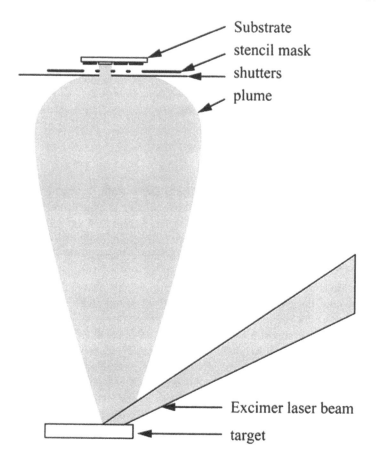

Figure 4 Principle of combinatorial laser deposition. A stencil mask can be used to define isolated cells on a substrate surface. Movable shutters can be used to open some of the cells for deposition.

under MBE conditions, at pressures below 10^{-5} torr (13). These deposition techniques use solid targets with a predetermined composition. During ablation, the stoichiometry of the material reaching the substrate surface is for all practical purposes the same as that of the target. An additional advantage of PLD is the ability to deposit films in the presence of a background gas at pressures of up to about 1 torr. Various radical sources can also be used to improve reactivity of the precursors on the film surface. Typical PLD chambers use a multiple-target carousel that can be used to switch quickly between the deposition targets.

Probably the most obvious choice for a combinatorial PLD experiment is to deposit a film with a different composition in each cell on the substrate. This is typically done to study doping effects. In this type of experiment the main lattice remains unchanged. Only small amounts of additives are periodically deposited on the growth front, by switching deposition targets and moving the masking shutters to expose only some parts of the substrate. An experiment can be designed to study the effects of different dopants or different dopant concentrations.

The main advantage of this type of experiment is the great speedup of the deposition process. Several tens or even hundreds of different films can be grown during a single deposition run. In addition to improved fabrication speed, the parallel synthesis eliminates small variations in process parameters that are inevitably present if a number of films are grown in a series of experiments (14). This, in turn, improves the reliability of data when conductivity or some other parameter is measured as a function of the doping level. The number of different films that can be fabricated is limited by several factors, including substrate size, masking accuracy, and the requirements of the characterization techniques. Typical single-crystal oxide substrates are available in sizes of up to 15 mm $\times$ 15 mm. The use of larger substrates is problematic due to the difficulty of heating them uniformly to temperatures of 1000°C and above. The shape of the deposition plume in PLD also limits the maximum sample size, because only the center part of the plume can be used to deposit films with uniform thickness. The substrate rotation techniques that can otherwise be used to improve PLD film thickness uniformity cannot be used in combinatorial synthesis, due to the use of masks. Various film characterization techniques also have different minimum sample size requirements. Typically contact-mode resistivity measurements require the largest cell sizes to accommodate the deposition of metal contacts on the film surface. The masking accuracy that can be achieved with noncontact stencil masks is about 0.3 mm, meaning that a realistic minimum cell size is 1 mm $\times$ 1 mm.

A special case of combinatorial film synthesis is the *composition spread technique* (15). In this case the film composition can be varied over the surface of the substrate while still maintaining epitaxial growth at all points. This is obviously possible only in certain cases where continuous substitutions are possible, such as the $Ba_{1-x}Sr_xTiO_3$ system in the range $0 \leq x \leq 1$. Composition spread films can be grown either by periodically switching deposition targets or by using a special deposition setup where several targets can be ablated simultaneously with several excimer lasers.

The basic two-dimensional parallel thin-film synthesis techniques described earlier can be developed further to take advantage of the epitaxial growth of thin films. If the reaction conditions, notably the substrate temperature and the background gas pressure, are chosen suitably, layer-by-layer or step-flow growth can be achieved (16). In this case the film surface remains flat on the atomic scale, and it is possible to deposit multilayer structures and even superlattices.

The physical properties of superlattices in particular are very sensitive to the actual compositions of the individual layers and the overall periodicity of the superlattice. Combinatorial synthesis techniques can be used to deposit large numbers of superlattice samples on a single substrate. Each cell on the substrate can have a different composition or periodicity. Combinatorial deposition of superlattices requires an epitaxial film growth technique. The PLD technique is nearly ideally suited to this type of work.

The superlattice design techniques can also be used to develop parallel device synthesis methods. In this case large numbers of electronic devices, such as thin-film transistors, can be fabricated in parallel by varying either compositional or structural parameters of each device. Combinatorial device fabrication is the most complex method described here, in terms of deposition chamber design. Large numbers of special masks need to be fabricated for each device type, and the masks must be manipulated during deposition with high accuracy. The increased complexity of the deposition system is, however, offset by the great increase in the efficiency of optimizing device characteristics. The effects of the individual layer compositions, doping levels, thicknesses, and deposition conditions on device performance can be rapidly mapped using this technique.

IV. COMBINATORIAL SUPERLATTICE CHAMBER

We describe here an oxide thin-film growth chamber that, in its essential components, is similar to the combinatorial LMBE system described in the following sections. This particular design was adapted for the growth of a variety of composition spread or superlattice libraries. A general view of the system is shown in Figure 5.

The deposition chamber itself* is relatively small; it has a height of 350 mm and a diameter of 203 mm. Instead of a traditional halogen lamp heater (13), the sample is heated with a Nd:YAG laser operating at a wavelength of 1.064 μm (17). The YAG laser's focusing lens is visible at the top of Figure 5. Laser heating makes it possible to reduce the chamber size and increase the achievable temperature range. The laser light enters the chamber through a normal Pyrex viewport and is focused on the back side of the sample holder. There is no need for a bulky heater unit or for water-cooled shrouds in vacuum. Only the sample holder is heated, and it is therefore possible to achieve temperatures of up to 1400°C in an oxygen atmosphere. The maximum temperature is limited mainly by the melting point of the sample holder. Typical heating power required to heat a 15-mm × 15-mm substrate to 1000°C is 150 W. A commercial 300-W welding

---

* Available from Pascal Ltd., http://www.pascal-co-ltd.co.jp/

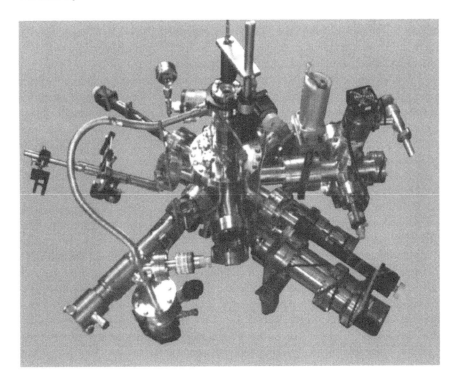

Figure 5 External view of the combinatorial superlattice chamber. The loading chamber is visible on the right side of the main chamber. The RHEED gun is on the front left side. The mask manipulators are located on either side of the chamber, 90° from the RHEED axis. Heating laser focusing optics are visible at the top of the chamber. The deposition laser light enters the chamber from the left.

laser has enough power to reach the sample holder's melting point. The heat load on the chamber is greatly reduced compared to a halogen lamp heater, which would require more than 1000 W to heat the sample to similar temperatures.

A KrF excimer laser is used for deposition. The laser operates at a wavelength of 248 nm and can be pulsed at a maximum rate of 20 Hz. A typical laser pulse energy is 100 mJ, which translates to a fluence of approximately 5 J/cm^2 on the sample surface. The laser light is focused with a single lens. The position of the lens can be changed to achieve the desired fluence on the target surface.

The target manipulator is visible at the bottom of the chamber. The motorized target carousel can be used to switch targets quickly during deposition.

Laser heating makes it easier to maintain a low back pressure in the chamber. Only the sample holder is heated, which reduces degassing from other surfaces. A typical back pressure in this chamber is 10^{-9} torr, if pumped with a single 250 L/s turbo pump. Operation in the 10^{-10}-torr range is possible if a titanium sublimation pump is added to the system.

The distance between the ablation target and the sample holder is variable and can be as large as 80 mm. This is somewhat larger than what would normally be used in a PLD setup, but it helps to achieve a more homogeneous film thickness on the sample surface. The chamber has two linear feedthroughs that can manipulate a stencil mask and a shutter. A typical mask-and-shutter design is shown in Figure 6. A cross section of the chamber in Figure 6a shows the positioning of the mask and shutter relative to the substrate and targets. A top view of the cell-definition mask is shown in Figure 6b. The figure also shows the position of the movable shutter, which can be used to allow deposition only in certain cells. A schematic diagram of the deposition process can be seen in Figure 6c. The RHEED beam is scanned over the sample surface during film growth in order to monitor the growth rate and growth mode at several points simultaneously. The need for RHEED monitoring puts a limit on the minimum distance between the substrate surface and the masks. In this chamber, the stencil mask is at a distance of 2 mm from the sample surface and the shutter is 1 mm below the stencil mask. The stencil mask pattern shown in Figure 6b has seven 1-mm-wide slits and four 1.75-mm-wide slits. Positioning either set in front of the sample makes it possible to deposit either four or seven different samples on a single substrate. More elaborate masking patterns are also available, but the minimum cell size is usually limited by the requirements of the film characterization techniques. Cell definition on the substrate is not ideal. The 2-mm distance between the stencil mask and the substrate means that there is a transition region at the edge of the cell where the film thickness decreases gradually from the desired value in the center of a cell to zero between cells. The width of the transition region can reach approximately 300 μm when the ambient oxygen pressure is in the mtorr range. This limits the minimum cell width that can be fabricated with this setup to about 1 mm. A potential problem with masking is the sputtering of material from the mask edges by the high-energy particles in the plume. This would result in contamination of the film by material that has been deposited previously. This effect can be minimized by using thin steel plates for shuttering. The masks in this chamber are made from 0.2-mm-thick stainless steel, and we have not observed cross-contamination in the sample due to the presence of the masks.

The chamber operation is fully automated. The sample temperature is monitored either with an optical pyrometer or a thermocouple at the back of the sample holder. An optical pyrometer is more convenient to use than a thermocouple since it does not require physical contact with a sample holder. A computer monitors the sample temperature and adjusts the Nd:YAG laser power to achieve the desired temperature. The pyrometer cannot be used while masks are in use; during that

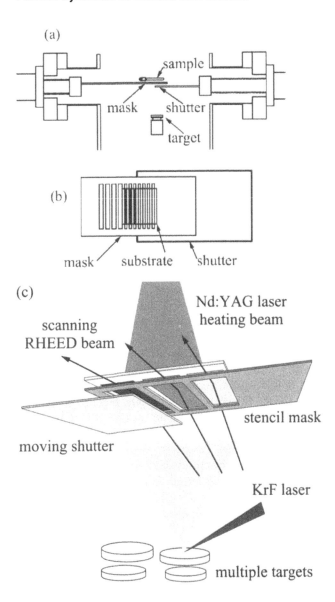

Figure 6 (a) Masking system consisting of movable pattern definition and shadow masks. The pattern mask is used for sample area definition, and the shadow mask is used to choose which substrate regions are open for deposition (shaded areas) (b). A schematic diagram of the deposition process and the position of the scanning RHEED beam is shown in (c).

time the Nd:YAG laser is operated at a constant power level. Masks can be withdrawn periodically to obtain a new temperature reference and to adjust the laser power, if necessary.

A deposition program is created before the start of a deposition, and during film fabrication the computer is responsible for controlling all chamber functions. This includes sample temperature, sample rotation, oxygen pressure, mask positions, target position, selecting targets, and controlling the power, frequency, and pulse count of the excimer laser.

An important part of this chamber is the scanning RHEED system. The RHEED technique is useful for monitoring the film growth mode and growth rate. This information is usually obtained by monitoring the specular spot intensity as a function of time. The specular spot intensity depends on the atomic-scale roughness of the surface. An atomically smooth surface would normally give the highest specular spot intensity. During layer-by-layer growth the surface roughness changes periodically as new layers are nucleated and gradually filled. Each oscillation of the RHEED specular intensity corresponds to the growth of a single molecular layer. Counting the number of oscillations after the start of a deposition makes it possible to control accurately the film thickness.

The substrate surface is divided into a number of cells during combinatorial deposition, with each cell having a different composition. Changes in composition can also translate into differences in growth rates. It is therefore important to monitor film growth at each point on the sample surface independently. This can be done by scanning the RHEED beam periodically over the sample surface. Normal RHEED guns have a set of four beam-tilting coils that can be used to direct the electron beam at any point on the sample surface. During intensity monitoring the electron beam is usually oriented along a high-symmetry crystal direction. The RHEED pattern is extremely sensitive to the electron beam orientation with respect to the crystal axes. Using only a single set of beam-tilting coils therefore results not only in beam movement, but also in a change of the azimuthal orientation of the beam, as shown in Figure 7a. The change of the azimuthal angle changes the RHEED pattern and also the specular intensity. A second set of beam- deflection coils have to be added to the RHEED gun to obtain parallel beam scanning, as shown in Figure 7b. In this case it is possible to measure the specular spot intensity accurately at any point on the substrate surface. An industrial video camera is used for measuring the intensity of RHEED patterns. The video camera provides a video frame 30 times a second. If the substrate is divided into 10 cells, each point can be measured at most three times per second. This is an adequate rate for monitoring even a fairly high-growth-rate process. In usual cases, measuring the specular intensity once a second is sufficient. Plotting the intensity oscillations for each cell in real time can help the operator to decide exactly how many times the excimer laser needs to be fired to obtain the desired layer thickness.

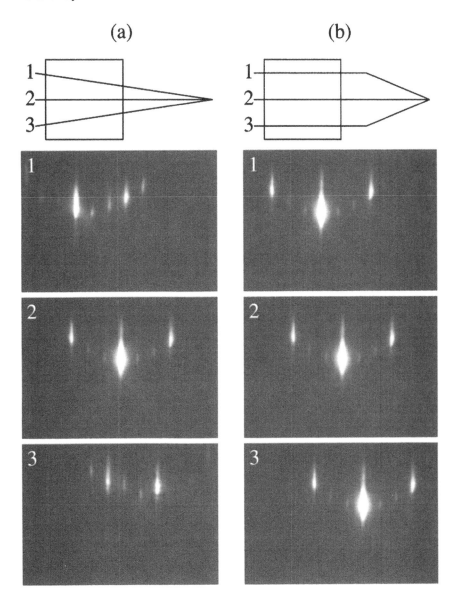

Figure 7 Single-coil scanning (a) causes the electron beam to tilt away from the [100] orientation, distorting the RHEED image. No such distortion occurs when parallel beam motion is used (b).

In addition to scanning the beam over the sample surface (Figure 8a), the double-coil deflection system can also be used to scan over incident angles (Figure 8b). The RHEED intensity behavior during film growth is sensitive to the incident angle, which is usually selected as either in phase or out of phase. A 30-kV electron beam has a de Broglie wavelength of 0.07 Å. For a perovskite, such as SrTiO₃ with a lattice constant of 3.905 Å, the in-phase angles where the Bragg condition is satisfied are 0.53°, 1.1°, etc. A rocking curve of a SrTiO₃ substrate is shown in Figure 9. The out-of-phase and in-phase angles are marked with dashed lines and continuous lines, respectively. The scattering of electrons cannot be analyzed by assuming a purely kinetic process, as is the case with weakly interacting x-rays. Electron interactions with the surface atoms are stronger, and dynamic scattering effects are significant. This is why the rocking curve does not show clear peaks exactly at the points where the Bragg condition would be fulfilled. In most experiments the incident angle was set to approximately 1.1°, i.e., the second in-phase point. This angle is small enough for the electron beam to pass between the substrate and the stencil mask while providing sufficient intensity for growth monitoring.

The time dependence of the specular spot intensity at various incident angles is shown in Figure 10. A homoepitaxial SrTiO₃ film was deposited at a temperature of 600°C, and the incident angle of the electron beam was continuously

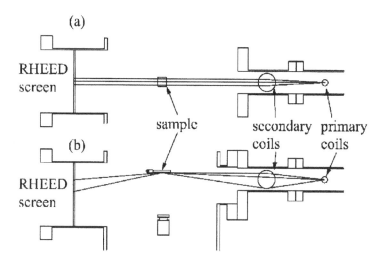

Figure 8 Double-coil beam deflection of the scanning RHEED system can be used to monitor several points on a sample surface, top view (a), or to control the incident angle of the electron beam, side view (b).

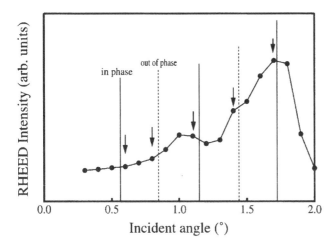

Figure 9   Rocking curve of an etched SrTiO$_3$ substrate.

scanned between 0.4° and 2°. Intensity oscillations can be seen at all incident angles, but clearly the 1.1° and 1.7° points are the most useful for monitoring the growth of very thin layers, for example, during deposition of superlattices. Interestingly there is a phase shift between the oscillations at 1.1° and 1.7°. Accurate selection of the incident angle is thus important in order to be able to determine the thickness of the growing film accurately.

V. PARALLEL SYNTHESIS OF PEROVSKITE SUPERLATTICES

Combinatorial deposition techniques can be used to speed up many types of experiments. Typical applications include material compositions, but other types of studies can also benefit. As an example of a different type of combinatorial synthesis we describe the growth of several superlattices in parallel on a single substrate. The physical properties of superlattices depend on a number of factors in addition to the particular choice of materials. The aim of the experiments described here was to explore the effect of variable layer thicknesses and superlattice periodicities on transport properties. A schematic view of the types of structures that have been grown are shown in Figure 11. The figure shows four different superlattices deposited on a single substrate. The actual number of lattices can at the moment be up to 10 on a 15-mm-wide substrate. These structures were grown by periodically switching between two deposition targets while using the

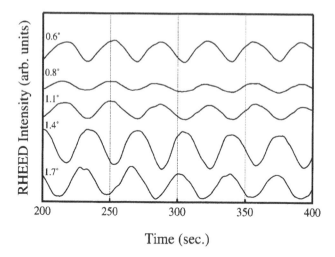

Time (sec.)

Figure 10   RHEED specular intensity oscillation during the initial stage of homoepitaxial SrTiO₃ growth, measured at various incident angles of the electron beam.

combinatorial masks to control each layer thickness. In type (a) only the periodicity of the lattice is changed. In type (b) the periodicity is changed while keeping the layer thickness of one of the materials in the superlattice constant. Type (c) is useful when physical properties depend on the ratio of layer thicknesses while keeping the total film thickness constant. The combinatorial approach offers several advantages. Most importantly, several different lattices can be fabricated simultaneously on the same substrate. It is therefore possible to obtain a whole set of resistivity graphs for various layer thicknesses by making only a single deposition. Individual layers in a superlattice typically have thicknesses of less than 10 monolayers. The superlattice properties are sensitive to the film quality, growth temperature, oxygen pressure, surface morphology of films, etc. Fabricating many superlattices on a single substrate in a single deposition ensures that the deposition conditions of all superlattices on a substrate are exactly the same. This improves the reliability of data. Preparing a set of samples, one in each deposition, always introduces an element of uncertainty, because it is difficult to control all the process parameters with sufficient accuracy between depositions.

A typical superlattice deposition sequence is shown in Figure 12. The figure shows a SrTiO₃/BaTiO₃ sequence, but the description also applies to various other superlattice structures, including titanates, vanadates, manganites, ruthenates, etc. A stencil mask is used for defining individual cells on the substrate surface. This is necessary because the individual cells must be isolated from each other for

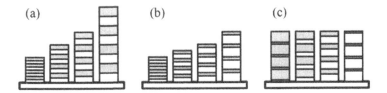

Figure 11 Three types of superlattice structures used in the experiments. Types (a) and (b) are used to study the influence of superlattice periodicity on the physical properties. Type (c) is useful when the ratio of layer thicknesses is important.

later measurement of transport properties of each superlattice. Using a stencil mask complicates the deposition process, but a later etching procedure can be avoided this way. Only a single shutter is sufficient for controlling which cells are open for deposition.

The substrate size in the experiment can be either 5 mm × 10 mm or 5 mm × 15 mm. The sample is mounted on a stainless steel holder and heated to 900°C while the masks are withdrawn to provide a clear view of the substrate for the pyrometer. The optical pyrometer used in these experiments operated at a wavelength of 2–2.5 µm. Undoped $SrTiO_3$ and various other substrate materials

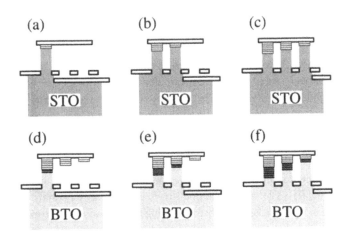

Figure 12 Deposition sequence showing the stencil mask and shutter positions during $SrTiO_3$ and $BaTiO_3$ deposition. Two molecular layers were grown during each of the six phases.

are nearly transparent in this range, and the pyrometer is therefore also measuring infrared emission from the material behind the substrate crystal. A 5-μm-thick gold film was pressed between the sample holder and the substrate to obtain good thermal contact. Platinum paste was also used but required longer degassing. The emissivity factor of the pyrometer was therefore adjusted accordingly. Temperature calibration can be done by depositing a layer of gold on a substrate surface and heating the sample until the gold starts to melt. The pyrometer is at the same time used to measure the temperature at a point that is not covered by the gold layer. In this way it is possible to find the correct emissivity setting for each substrate/mounting combination. The sample is clamped from two edges, leaving the [001] crystal direction free for RHEED observation.

The deposition sequence shown in Figure 12 starts with the growth of two molecular layers of $SrTiO_3$ in the leftmost cell. The shutter is then moved and two more layers are deposited. After step (c) the film thicknesses in the three cells are six layers, four layers, and two layers. The target is then switched to $BaTiO_3$ and the same process repeated. After step (f) the first period of a superlattice is finished and the whole sequence, starting from step (a), is repeated again. A total of 20 or 30 periods are typically needed for x-ray and transport measurements. This sequence results in a superlattice such as shown in Figure 11a. Other types of superlattices can be deposited by selecting appropriate layer thicknesses. The sequence of mask movements and target changes remains the same in all cases.

Successful fabrication of a set of superlattices requires accurate knowledge and control of the individual layer thicknesses. There are two main problems that need to be addressed: film thickness uniformity and measurement of film thickness. Laser deposition can be problematic in terms of film thickness uniformity. Substrate rotation or laser beam scanning are not realistic options due to the presence of masks. The only way to ensure uniform deposition is to increase the deposition distance and to limit the sample size. In practice it is possible to achieve adequate uniformity at deposition distances in the range of 60–80 mm. The sample size is limited to 15 mm × 15 mm not only by the uniformity requirement, but also by the availablility of single-crystal substrates, such as $SrTiO_3$.

Film thickness measurement can be done during deposition by using RHEED. The scanning RHEED setup described in Section IV is essential for measuring the growth in each cell on the substrate during deposition. If the growth temperature is selected so that the film grows in the layer-by-layer mode, each specular spot intensity oscillation corresponds to the growth of a single molecular layer. Care must be taken if the growth temperature is high enough for step flow growth to contribute. In this case the true growth rate can be slightly higher than what would be estimated from the RHEED data alone. The final superlattice periodicity can be verified by the analysis of x-ray diffraction patterns, as shown later.

The time dependence of the RHEED specular spot intensity observed during a deposition of a $SrTiO_3/BaTiO_3$ superlattice is shown in Figure 13. The three traces shown in the figure correspond to the three superlattices shown in Figure 12. The bottom trace corresponds to the thickest superlattice, located on the left edge of the substrate. The topmost trace corresponds to the thinnest superlattice, which was at the right-hand edge of the substrate. The labels in Figure 13 correspond to the sequence steps shown in Figure 12. Two monolayers were deposited during each step in the sequence. Accordingly, the RHEED data show two oscillations at each step. The deposition target was switched from $SrTiO_3$ to $BaTiO_3$ between steps (c) and (d). The chamber pressure was also switched at the same time if necessary. Titanates can be deposited at 10^{-6} torr of oxygen, but various other oxides, such as $La_{1-x}Sr_xMnO_3$ and $SrRuO_3$, require higher oxygen pressures. The scanning RHEED gun has only a single differential pumping stage, and the maximum pressure was therefore limited to 10^{-3} torr. During each step the operator can observe the RHEED oscillation and, if necessary, change the number of laser pulses for the next step, thus making small corrections to the amount of material deposited in each layer.

The superlattice structure was verified by x-ray diffraction. X-ray measurements were performed with a four-circle diffractometer (Philips MRD) using Cu

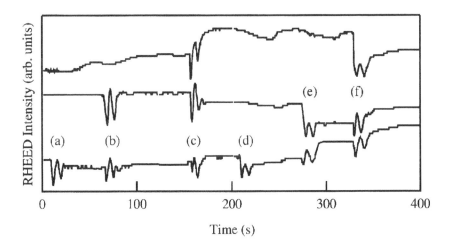

Figure 13 RHEED specular spot intensity oscillation during one cycle of the $SrTiO_3/BaTiO_3$ superlattice deposition process. The letters (a–f) correspond to the deposition sequence labels in Figure 12. Similar oscillations were observed when oxides other than $BaTiO_3$ were used in combination with $SrTiO_3$.

K_α radiation. The sample was mounted on an XYZ stage in the diffractometer, and θ–2θ scans were measured at each cell of the combinatorial superlattice chip. The use of a concurrent x-ray machine is also possible, as decribed later in this chapter. The θ–2θ scans of three $[(SrTiO_3)_n/(BaTiO_3)_n]_{20}$ superlattices with $n = 2$, 4, and 6 are shown in Figure 14.

The substrate peaks are at 23° and 46°. Fundamental superlattice peaks are visible immediately to the left of the substrate peaks. Other satellite peaks can be seen on either side of the fundamental peaks. Up to third order peaks could be identified in the diffraction patterns, confirming that the superlattice deposition was successful. The gradual decrease of superlattice peak intensities shows that the interfaces between $SrTiO_3$ and $BaTiO_3$ layers had a certain degree of roughness. This is an inevitable consequence of the layer-by-layer growth mode during LMBE growth, because although there is periodic surface smoothness variation,

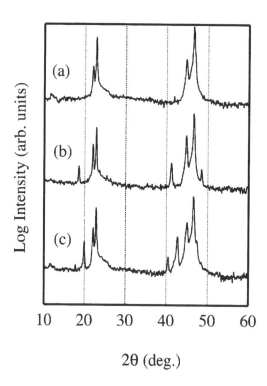

Figure 14   X-ray diffraction patterns of $[(SrTiO_3)_n/(BaTiO_3)_n]_{20}$ superlattices with (a) $n = 2$, (b) $n = 4$, and (c) $n = 6$.

as shown by RHEED oscillations, more than one unfinished layer is always present on the film surface. Nucleation of a new molecular layer starts before the previous layer is completely filled. The only way to avoid this effect is to use step-flow growth at higher temperature, if lattice mismatch between materials is small enough to avoid island growth on the heterointerfaces (16). The drawback of step-flow growth, however, is the loss of accurate film thickness control, because RHEED oscillations can no longer be used for this purpose.

The superlattice periodicity measured from the x-ray diffraction patterns was $\Lambda_{exp} = 1.615$ nm in Figure 14(a). The design value was $\Lambda_{cal} = 1.589$ nm. A similar difference was seen in all measurements, showing that the layer thickness of the film was $\approx 1.7\%$ larger than the calculated value, indicating that step-flow growth was contributing slightly at the 900°C growth temperature.

The results of transport measurements of a superlattice library are shown in Figure 15. In this case a set of $SrTiO_3/SrRuO_3$ superlattices were grown on a $SrTiO_3$ substrate using a slightly modified deposition sequence. The mask movements were the same as those shown in Figure 12, but the layer thicknesses were modified to obtain superlattices of the type shown in Figure 11c. In this case the oxygen pressure in the chamber was also changed for each target, 10^{-6} torr for $SrTiO_3$ and 10^{-4} torr for $SrRuO_3$. The superlattices can be described with the formula $[(SrTiO_3)_6/(SrRuO_3)_m]_{30}$, where $m = 2, 4, 6, 8, 10$. The thickness of the $SrTiO_3$ layer was fixed at six monolayers. The films were terminated with the $SrRuO_3$ layer. Gold contacts were evaporated on the sample surface after deposition at room temperature, and four-wire resistivity measurements were performed by inserting the sample slowly into a He storage dewar. As the plot shows, the resistivity of the superlattices [Figures 15(a–e)] was always higher than that of a 50-nm-thick $SrRuO_3$ film [Figure 15(f)]. In addition to a general increase in resistivity with the gradual decrease of the $SrRuO_3$ layer thickness, there is a marked temperature dependence change, as has also been observed by Izumi et al. (18). The resistivity was calculated using the total thickness of the superlattice film. Assuming that conductivity is confined to the $SrRuO_3$ layers, part of the resistivity increase is due to the smaller fraction of $SrRuO_3$ in the thinner films. This, however, does not explain the factor-of-10 changes observed between resistivity curves (d–e) and (a–b). Both pairs were deposited on the same substrate, so variation due to a change in some other process parameters can be ruled out. It is therefore our understanding that this increase is dominated by effects related to two-dimensional confinement of the current flow in the $SrRuO_3$ layers. Another factor that can affect the film conductivity is the oxygen defect density in the $SrTiO_3$ layers. It is known that $SrTiO_3$ films grown at low oxygen pressures are highly oxygen deficient and without additional oxygenation can have significant conductivity.

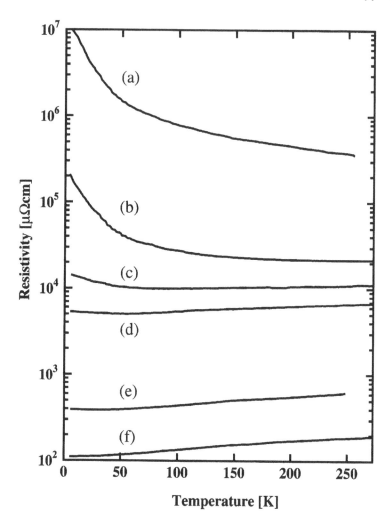

Figure 15 Temperature dependence of $[(SrTiO_3)_6/(SrRuO_3)_m]_{30}$ superlattices. The $SrRuO_3$ layer thickness was (a) 2 ML, (b) 4 ML, (c) 6 ML, (d) 8 ML, and (e) 10 ML. The resistivity of a $SrRuO_3$ film is shown in (f).

VI. COMBINATORIAL LASER MBE

Combinatorial laser molecular beam epitaxy (CLMBE) is an extension of the well-established molecular beam epitaxy and pulsed laser deposition techniques. The emphasis in CLMBE is on parallel deposition of a large number of films on a single substrate in order to fabricate combinatorial libraries and thus speed up the mapping of the physical properties of thin films. The basic ability to deposit combinatorial libraries can be achieved with a fairly simple masking system and can thus be retrofitted, even in existing PLD chambers. Specially designed CLMBE chambers can, however, offer other advantages, including simultaneous deposition from several targets, accurate mask-positioning carousels, and simultaneous processing of several substrates. The chamber design described here includes all of these features (14).

An outline of the CLMBE chamber is shown in Figure 16. The system actually includes three independent subchambers: a preheating chamber, a deposition chamber, and a postannealing chamber, shown in Figure 17. Each component chamber has independent pressure controls and gas supplies. A sample exchange chamber is located directly above the three subchambers. The exchange chamber houses three lamp heaters on a continuously rotatable carousel. Sample holders are attached to the heaters, not the small subchambers, which means that samples can be moved from one subchamber to another without changing sample temperature by lifting the heater carousel, rotating it by 120°, and lowering it in a new position. Chamber pressures, of course, have to equalized for a brief moment during this rotation operation.

During normal operation the carousel is lowered so that the heater units seal each of the three subchambers with water-cooled O-ring seals. The exchange chamber is continuously pumped with a 1000-L/s turbo pump, making it possible to operate some of the three subchambers in the mtorr range while others are at UHV levels. Gas pressures are equalized only during sample movement from one subchamber to another, but the desired pressures can be restored in all subchambers once the heater units have been lowered.

A typical deposition sequence is shown in Figure 18. Substrates are first entered into the loadlock chamber (step a), also shown in the side view of the chamber in Figure 17. The substrates are attached to a disk-shaped sample holder, which can be transferred into the heater unit in the preheating chamber (step b). During transfer, all heater units are lifted out of the three subchambers, and the heater carousel can be freely rotated to select a free heater, move it above the preheating chamber, and insert the sample holder into the heater unit. The heaters are then again lowered into the normal operating position. Substrate annealing is done in the preheating chamber in order to achieve a clean and atomically smooth substrate surface before deposition (19,20). The precise annealing temperature and oxygen pressure depend on the substrate material, but typically

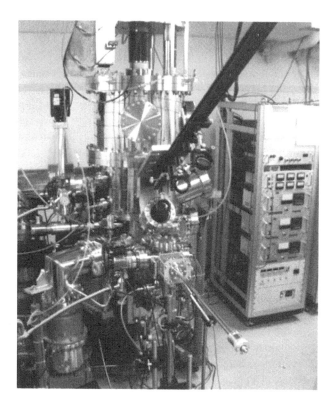

Figure 16 Front view of the CLMBE chamber (Pascal Co.). Electrical feedthroughs and the heater manipulator are visible at the top of the chamber. The bulk of the chamber consists of the heater carousel space. Excimer laser optics and the target exchange loadlock are visible in the front.

a 2-hour anneal is needed at 900°C to obtain a high-quality surface. This is a fairly time-consuming process, and having a separate preheating chamber that can operate in parallel with the deposition chamber can significantly speed up sample fabrication. The substrate can be monitored through a viewport during annealing. Temperature reference is normally obtained from a thermocouple in the heater unit, but an optical pyrometer can also be used for confirming that the sample surface temperature is at a desired level. The smaller preheating and postannealing chambers are pumped with 300-L/s turbo pumps, and the oxygen pressure can be independently controlled with variable leak valves. The chamber has an additional port for the installation of a RHEED gun for substrate surface monitoring during annealing. Electron diffraction is a useful tool for determining

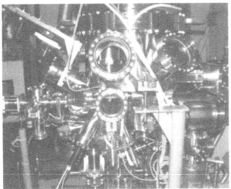

a

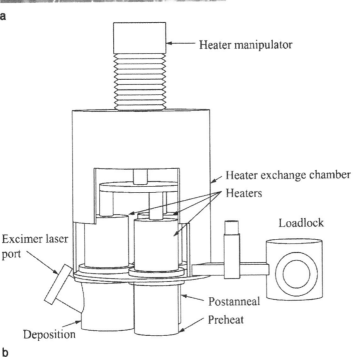

b

Figure 17 (a) Side view of the CLMBE chamber. The deposition subchamber is visible at the bottom. The loadlock can be seen on the right side of the main chamber. (b) A schematic view of the CLMBE chamber, showing the loadlock, the three subchambers, and the three heaters in the normal operating position. The heaters can be lifted and rotated to move samples from one subchamber into another. Heaters must also be lifted when samples are loaded or removed from the heater, which is located above the preheating subchamber.

the substrate surface quality and can be used to reject a defective substrate before further processing. Typical processing conditions for $SrTiO_3$ substrates are T_{sub} = 900°C, P_{O_2} = 10^{-6} torr, and annealing time t_{ann} = 2 hours.

After annealing, the substrate can be transferred to the deposition chamber without intermediate cooling. To do this, all three heaters are again lifted and the heater carousel is rotated by 120°. A new substrate can be entered from the loadlock into an empty heater unit, which is now above the preheating chamber. After lowering the heaters, the first substrate is now in the deposition chamber and a new substrate can be annealed in the preheating chamber (step c).

After deposition and preheating have finished, the heaters are again raised and rotated, shifting the freshly deposited film into the postannealing chamber and bringing the second substrate into the deposition chamber. A third substrate

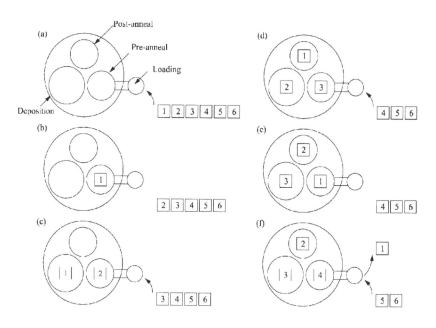

Figure 18 Sample movement sequence in the CLMBE chamber. A substrate is (a) first placed in the loading chamber and (b) then moved into the preheating chamber. After annealing, the heaters are rotated, moving the first substrate to the deposition chamber (c). A second substrate is loaded into the preheating chamber. (d) The heaters rotate again, and the first substrate moves into the postannealing chamber. (e) After one more heater rotation, the first substrate is moved back into the preheating chamber, from which it can be unloaded and a new substrate can be loaded into the preheating chamber (f).

can be loaded into the preheating chamber at the same time (step d). After finishing the second deposition the samples are again moved (step e). This time the original substrate can be unloaded into the loadlock and a fresh substrate moved from the loadlock to the heater unit, which is now above the preheating chamber (step f). Three substrates can thus be processed in parallel. Considering that each substrate can have 9, 16, or even 25 cells, it is clear that the deposition process is no longer the limiting step in the fabrication and analysis chain. Assuming a typical deposition time of 2 hours, the system can produce a new combinatorial library every few hours. Feeding the deposition chamber and measuring the properties of fabricated films also requires parallel techniques with similarly high throughput. These techniques are discussed in Section IX.

The deposition chamber is the largest subchamber and includes a high-precision mask-alignment carousel and a multitarget stage capable of holding up to eight tragets. The deposition chamber also has a scanning RHEED gun that can be used to scan the electron beam across the sample surface under computer control. Film growth can in this way be monitored simultaneously at several parts of the substrate. The scanning RHEED system is described in greater detail in Section IV. The deposition chamber also has an independent loadlock for changing targets in the target stage. Gas pressure can be controlled during deposition with variable leak valves. A 800-L/s turbo pump ensures that adequate gas flow can be maintained even at high pressures while making it also possible to work in UHV conditions.

Two KrF excimer lasers operating at a wavelength of 248 nm can be used for deposition, each abalating a different target on the target stage. The plumes from the two targets must overlap at the sample surface. Using targets with a 10-mm diameter allows them to be placed close enough to each other for this type of codeposition to work. The advantage of this setup is the increased flexibility in controlling film compositions. The pulse frequency, pulse energy, and timing can all be controlled by computer and synchronized with the movement of the target stage and the combinatorial mask carousel. A deposition program must therefore be prepared before deposition can start.

The mask carousel can hold eight mask patterns. A typical set of masks used in doping experiments is shown in Figure 19a. This mask set can be used to define nine cells on a substrate surface, as shown at the top of Figure 19b. In addition to changing masks, it is possible to rotate the sample during deposition. This adds another degree of freedom in selecting the masking pattern. A cell layout that makes use of the sample rotation is shown at the bottom of Figure 19b. In this case 16 cells can be defined on the sample surface. This configuration has the added benefit of providing more space around each cell. This is useful if the films need to be patterned and etched for characterizing transport properties or for defining device structures, such as Josephson junctions.

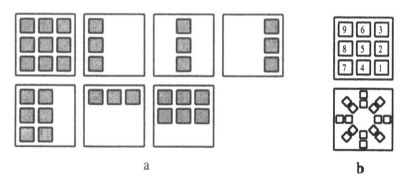

Figure 19 (a) A set of eight masks used in doping studies. This mask set defines a total of nine different cells, shown in the upper part of (b). Lower pattern in (b) shows a different masking technique, where sample rotation is also used.

VII. APPLICATION OF COMBINATORIAL LASER MOLECULAR BEAM EPITAXY

This CLMBE system is ideally suited to doping studies. The aim of the experiment described here was to determine how the bandgap of ZnO can be modified by adding various dopants to ZnO films. There is a large choice of potential dopants, including Mg, Sc, Ti, V, Cr, Mn, Fe, Co, Ni, Cu, W, Eu, Er, etc. Finding the best element and determining the bandgap shift as a function of dopant concentration requires a very large number of samples.

The CLMBE chamber was used to prepare combinatorial libraries with nine cells on each substrate. The Mg-doping experiment is described here in greater detail (14), although the whole list of dopants mentioned earlier was studied. The deposition procedure was the same in each case. The observed bandgap changes were, of course, different for each dopant.

High-quality c-axis-oriented films were grown on polished 16-mm $\times$ 16-mm αAl_2O_3 (0001) substrates at 600°C. The background oxygen pressure was 10^{-6} torr. Two targets were used in each experiment, a pure ZnO ceramic target with a purity of 99.999% and a 10 mol% Mg-doped ZnO target. The KrF excimer laser pulse repetition rate was 5 Hz, and the fluence on the target surface was 3 J/cm². The deposition sequence is shown in Figure 20. Nine cells were defined on the substrate surface by using eight masks. The open regions of each mask are shown in Figure 19. The whole sequence was repeated 80 times.

The doping level in each cell can, in principle, be determined by counting the total number of laser pulses used for the deposition of clean ZnO and doped ZnO. Unfortunately, the stoichiometry of the target was not perfectly transferred to the film. A certain degree of Mg enrichment was observed in the film. To

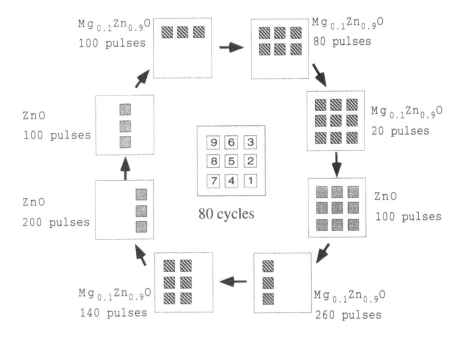

Figure 20 Combinatorial masks and deposition sequence of fabricating Mg-doped ZnO films.

calibrate the enrichment factor, films were deposited by using only the doped target, and the dopant concentration was determined using Rutherford backscattering (RBS). It was found that although the Mg content in the target was 10 mol%, the final film contained nearly 20 mol% of Mg, indicating that an enrichment factor of 2 should be used when selecting the pulse counts for the deposition of Mg-doped films. Counting deposition laser pulses was used only to achieve doping levels close to the desired values. The cells in these combinatorial libraries were large enough for RBS measurement and the actual Mg content was determined in each cell. The ratio of $Mg_{0.1}Zn_{0.9}O$ ablation pulses for each cell is shown in Figure 21a. The cell numbers are shown in the center of Figure 20. The nine cells covered a range of pulse ratios from 0.06 to 0.86. This translated into an actual Mg content variation from 2 mol% to 19 mol% in the films, as measured by RBS (Figure 21b).

The lattice constants of the doped films were determined by x-ray diffraction analysis. All Mg-doped ZnO films showed a single wurtzite phase. The added Mg did not appear to affect the crystal structure, even at 19 mol% doping levels.

Lippmaa et al.

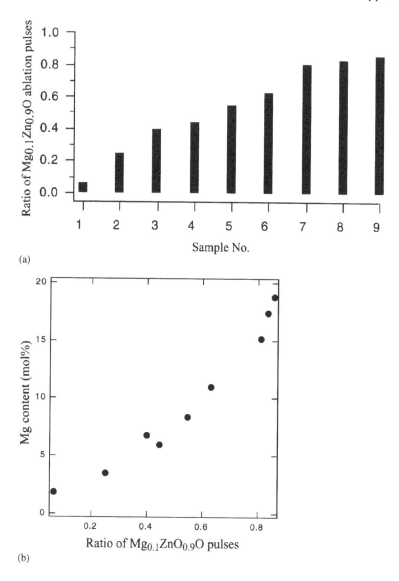

(a)

(b)

Figure 21 (a) Ratio of laser pulses used to ablate the $Mg_{0.1}Zn_{0.9}O$ target for each cell. The cell numbers are shown in Figure 20. (b) The actual Mg content in each cell as determined by Rutherford backscattering.

The c-axis length of the films as a function of Mg content is shown in Figure 22a. It can be seen that the dependence is linear. A linear dependence on the doping level was also observed for the bandgap, which is shown in Figure 22b. The figures also show previously obtained data (plotted with open squares). The earlier data were obtained from traditional experiments in which only a single sample was prepared in a single deposition. The earlier data suggested that there is little or no lattice constant change at low doping levels. The combinatorial sample allowed nine films to be prepared in a single experiment and showed a linear c-axis change even at the lowest doping levels.

This experiment demonstrates two of the main advantages of combinatorial synthesis: improved throughput and improved data reliability. The whole Mg-doping study consisted of a single deposition. Indeed, the technique was used to rapidly map a dozen elements and determine the doping effects for each one. The data obtained from this experiment were also more reliable, because there are no variations in growth temperature or cooling procedures, all of which can affect the strain in the film and thus change the observed c-axis values or the bandgap. The CLMBE chamber has also been used for parallel fabrication and characterization of ZnO/MgZnO superlattices (21,22) and transition metal–doped ZnO (23).

VIII. OTHER FORMS OF COMBINATORIAL SYNTHESIS

The principles of combinatorial laser deposition were described in detail in the previous two sections. This technique is suitable for the growth of various materi-

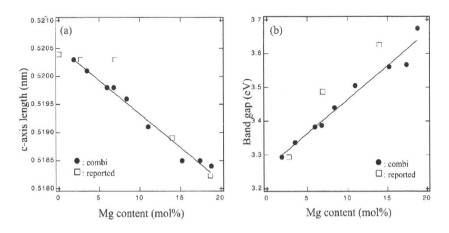

Figure 22 (a) c-Axis length of Mg-doped ZnO films as a function of the doping level. (b) Bandgap dependence on the Mg content. The plots include data from a combinatorial library (•) and conventional film growth experiments (□).

als in thin-film form. The main thrust in using combinatorial synthesis is to develop new materials faster. This usually involves the optimization of deposition conditions and mapping the physical properties of various compositions. There are a number of other applications where solid-state combinatorial synthesis can also be valuable. Some of these fields use adaptions of the basic pulsed laser deposition technique, while others use totally different deposition methodology. The final goal is usually to develop a material or structure with particular properties, and no single deposition technique can be guaranteed to be optimal. We therefore describe in this section another deposition method that has been adapted for combinatorial deposition work: chemical vapor deposition (CVD).

A. Design of Catalysts

The development of catalysts involves the optimization of a wide range of parameters. Catalysts often involve a support material, such as SiO_2 or Al_2O_3, and various metals, such as Pt, Rh, Cu, and Pd. The catalytic activity and the density of active sites on the surface of the catalyst depend on the particular mixing ratios of the various compounds, particle size, and the treatment procedures. The application of combinatorial techniques can help in mapping the wide parameter space and quickly focusing on the most promising combinations (24–27).

We describe here combinatorial synthesis of TiO_2 films doped with various transition metals (28). Titanium dioxide is a well-known photocatalyst, which can be used for the breakup of organic pollutants in water or for dissociating water itself. The surface area or grain size aspects of catalyst design were practically eliminated in this work by using thin-film samples. The aim was to study the efficiency of electron hole pair formation under optical illumination, which is the first step in a catalytic reaction.

Laser MBE can be used to grow two types of TiO_2 films. The c-axis-oriented anatase phase, $TiO_2(A)$, grows epitaxially on the $SrTiO_3(001)$ surface. It is also possible to grow the a-axis-oriented rutile phase $TiO_2(R)$ on $Al_2O_3(0001)$ substrates. The rutile form of TiO_2 is thermodynamically more stable, and a spontaneous phase transition occurs at approximately 1000 K in bulk TiO_2. It is therefore difficult to control the precise crystal structure of TiO_2 at high temperatures. Thin-film samples do not suffer from this limitation, due to the template effect of the substrate crystal. It is possible to grow $TiO_2(A)$ films on $SrTiO_3$ that maintain structural stability even at high temperatures.

The CLMBE chamber was used to fabricate a series of combinatorial libraries, covering all of the $3d$ transition metal ions. The purpose was to change the electronic structure of TiO_2 and thus the photocatalytic efficiency. Both anatase and rutile forms of TiO_2 were mapped. Each library included nine cells with varying dopant concentrations. The $3d$ ions included Sc, V, Cr, Mn, Fe, Co, Ni, and Cu, resulting in a total of 144 films in 16 library chips.

The valence states of the dopant ions were estimated from the Ellingham diagrams using the known oxygen pressure in the chamber during deposition (29). The expected valence states were $+4$ for V, $+3$ for Sc and Fe, and $+2$ for Co and Ni. A $+3/+2$ mixture was expected for Mn and a $+2/+1$ mixture for Cu. The actual dopant concentration in the films was calibrated by electron microprobe analysis (EPMA). The measured dopant concentration was found to follow closely the design value. The masking sequence and cell layout on the substrate crystal were described in Section VII.

The maximum doping levels used in this experiment reached 50% (Figure 23). This is well above the typical solubility limit of the $3d$ transition metals in either anatase or rutile forms of TiO_2. X-ray analysis was therefore used to first map the solubility limits for each dopant. It was found that V and Fe have the highest solubility limits in $TiO_2(R)$. This follows a general trend, which is also known from bulk samples, that the rutile phase is more tolerant of dopants. The only exception is Co, which suprisingly has a much higher solubility in anatase. The solubility data for bulk samples cannot be directly translated to thin-film samples because the lattice strain can have a large effect on the solubility limits.

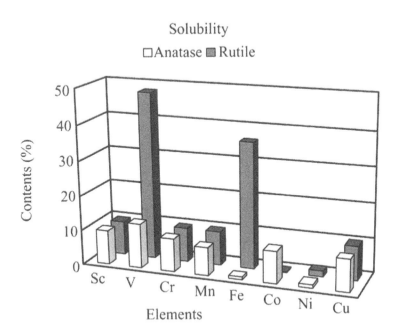

Figure 23   Solubility limit of each transition metal element in TiO_2 thin films of anatase and rutile structures.

Once the combinatorial catalyst libraries had been fabricated, it became necessary to develop an equally efficient analysis method. Measuring the efficiency of each one of the 144 films individually was clearly impractical. A two-dimensional pH imaging method was therefore developed that relies on the appearance of a distribution of pH values around the anode and the cathode in a solution. A diagram of the measurement setup is shown in Figure 24a. The sensor is made of semiconducting silicon covered with silicon nitride. The proton density in the solution is sensed locally by the same mechanism as is used with conventional potentiometric electrodes, such as in a pH field-effect transistor (30,31). The photocatalytic activity of each cell in the library was determined by adding a small amount of oxidizing or reducing agent to water and scanning a 780-nm 5-mW laser beam across the sample. In the presence of Fe^{3+} ions, acting as an oxidizing agent, a series of chemical reactions occur on the TiO_2 surface, which can be summarized as:

$$2H_2O + 4Fe^{3+} \rightarrow 4Fe^{2+} + 4H^+ + O_2 \uparrow \qquad (2)$$

The generation of hydrogen is expected to be accompanied by decomposition of water. The pH distribution generated by the decomposition of water above a Co-doped TiO_2 rutile film library is shown in Figure 24b. The Co content was lowest in cell 1 and highest in cell 9. The figure shows that there is a distinct photoactivity change as a function of Co concentration in the film.

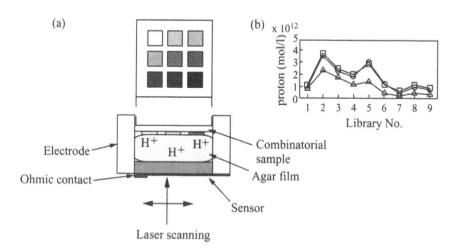

Figure 24   (a) Electrochemical cell used for mapping photocatalytic activity. (b) pH distribution on the Co-doped TiO_2 rutile thin films. The library numbers refer to the nine cells in the library. Co concentration is increasing from 1 to 9.

B.   Combinatorial Plasma Chemical Vapor Deposition

Hydrogenated amorphous silicon (a-Si:H)-based devices can be manufactured on large substrates at a low cost. Probably the two most important application areas are thin-film transistors (TFTs) for flat-panel display devices and large solar cell arrays. Plasma-enhanced chemical vapor deposition (PECVD) is an efficient technique for growing a-Si:H devices. The actual device performance depends on the precise deposition conditions and device characteristics, such as individual layer thicknesses. Optimizing the operation of a device, such as a solar cell or a thin-film transistor, is a time-consuming task that can benefit greatly from the combinatorial synthesis techniques.

The basic idea of a combinatorial CVD experiment is similar to what was described earlier for pulsed laser deposition. The main difference here is that the sample size can be much larger, up to about 100 mm $\times$ 100 mm in our experiments. The aim of the experiment descibed here was to optimize the a-Si:H and a-SiN:H layer thicknesses in a bottom-gate-inverted staggered-type thin-film transistor (TFT). The masking setup is shown in Figure 25a. Individual cells on the substrate were defined with a contact mask. A sliding slit mask was placed above the contact mask. The sliding mask opened a single row of cells at a time for deposition, as shown in Figure 25b. The deposition started with the growth of a-SiN:H at a temperature of 380°C. The rf power density was 140 mW/cm^2. A reaction pressure of 75 mtorr was obtained by mixing SiH_4 and N_2-diluted NH_3 (10% of NH_3) at flow rates of 1 sccm and 25 sccm, respectively. The contact mask divided the substrate surface into 49 cells, organized in seven rows and seven columns. Each column had a constant thickness of the a-SiN:H layer, and each row had a constant thickness of the a-Si:H layer. The deposition times of a-SiN:H in each column were 0, 20, 40, 60, 80, 100, and 120 minutes. The growth rate was 3000 Å/hour, resulting in a film thickness range of 0–6000 Å. The plasma was switched off for approximately 1 minute while the slit mask was shifted to the next column of cells.

After a-SiN:H deposition the masks were withdrawn and the anode electrode, which also served as a sample holder, was rotated by 90° (Figure 25c). The masks were then positioned again above the substrate and the deposition sequence was repeated (Figure 25b). The slit mask now exposed a row of cells for deposition of a-Si:H. The deposition conditions were also changed while the sample was being rotated. The power density was reduced to 50 mW/cm^2 and the reaction pressure was set at 30 m torr. The SiH_4 flow rate was 10 sccm. The sample temperature was also slightly lower at 250°C. The deposition times of a-Si:H in the seven rows were 0, 10, 20, 30, 40, 50, and 60 minutes, giving up to 3000-Å-thick films. The growth rate was nearly the same as in the case of a-SiN:H.

Device fabrication was finished by depositing aluminum electrodes on the film surface by thermal evaporation. The channel length of the TFTs was

(a)   (b)   (c)

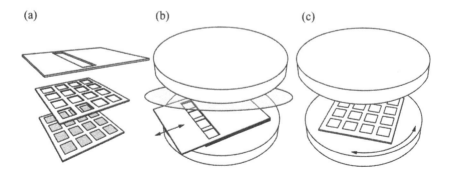

Figure 25 (a) Schematic diagram of a substrate, a contact mask for cell boundary definition, and a sliding slit mask. (b) Deposition geometry used for growing a-Si:H and a-SiN:H films. One row of cells is open for deposition at a time. The position of the plasma is shown between the mask and the top electrode; (c) the substrate is rotated 90° between a-Si:H and a-SiN:H depositions while the masks are withdrawn.

200 μm; the width was 3000 μm. The resultant TFT structure is shown in Figure 26a. The ITO layer on the substrate served as a gate electrode. A photograph of the combinatorial TFT library is shown in Figure 26b. Each TFT in the library had a different combination of a-Si:H and a-SiN:H layer thicknesses. The cell in the lower left corner has no deposited film at all; the color is due to the ITO layer on the substrate. The lowest row contained only an a-SiN:H film, and the leftmost column had only an a-Si:H layer. These regions were used to verify that the quality of the component films is adequate for TFT operation. The optical bandgap of the a-Si:H film was 1.7 eV, the hydrogen content was 10%, and the photosensitivity was 10^4. The dielectric breakdown field strength of the a-SiN:H layer was in the MV/cm range. Typical leakage current density was 10^{-10} A/cm^2 at 1 MV/cm.

The main performance parameters that are of interest in these TFT structures are the on/off current ratio and the threshold voltage. The on/off ratio is a direct measure of the field effect and the switching capability of the transistor. All TFTs in the library had an on/off ratio exceeding 10^4, which is adequate for basic display applications.

The threshold voltage V_T of a TFT is determined mainly by the combination of deep gap states, interface states, and fixed charges localized at or near the gate insulator/a-Si:H interface. The variation of V_T in the TFT library is due mostly to the trapped charges in the bulk of the gate dielectric layer. Other factors, such as interfacial charge at the a-SiN:H/a-Si:H interface, are not expected to vary appreciably among the TFTs in a single library. The value of V_T was determined

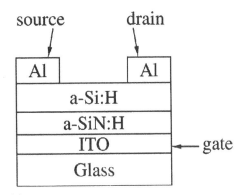

(a)

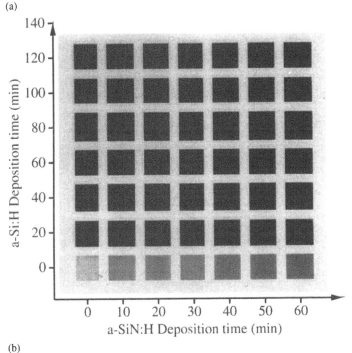

(b)

Figure 26 (a) Schematic diagram of the bottom-gate-inverted staggered-type TFTs. The thickness of the a-Si:H and a-SiN:H layers are different in each TFT in the combinatorial library. (b) Photo of the TFT library. The a-SiN:H layer thickness is constant in each column. The a-Si:H layer thickness is constant in each row.

for each TFT by measuring the $I_D(V_G)$ curves and finding the V_G value at the $I_D = 0$ intercept. The variation of the V_T values is shown in Figure 27. The fact that the threshold voltage appears to depend on both a-Si:H and a-SiN:H layer thicknesses is intriguing and indicates that the microstructure of the a-SiH layer, and thus the bulk density of states, depends also on the thickness of the bottom a-SiN:H layer.

This experiment demonstrated that it is possible to use combinatorial synthesis also to speed up device structure optimization. The results shown here were obtained from a simple TFT structure. We are currently exploring the extension of these techniques to the fabrication of more complex structures, such as field-effect solar cells.

IX. HIGH-THROUGHPUT CRYSTALLOGRAPHY OF EPITAXIAL FILM LIBRARIES

Combinatorial synthesis addresses the need to create large numbers of slightly different materials or structures in parallel. The aim is to save time and enable one to scan through a much larger number of possibilities than one could with conventional techniques. To make the combinatorial approach work, it is just as important to have efficient techniques for characterizing the combinatorial librar-

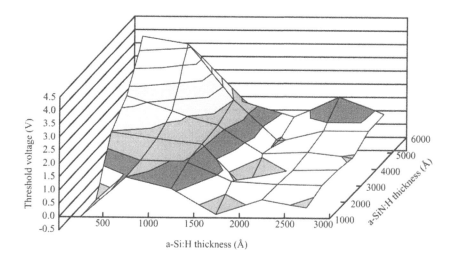

Figure 27 Variation of the TFT threshold voltage with a-Si:H and a-SiN:H thickness in a single combinatorial device library.

ies. Characterization of thin-film samples usually requires structural characterization by x-ray diffraction and various temperature-dependent measurements, such as resistivity, magnetization, dielectric properties, and optical properties. Parallel measurement setups are needed to achieve high throughput in the whole synthesis-and-characterization cycle. Some measurements can be performed in parallel. These include optical measurements that can make use of various imaging techniques. The luminescence intensity and spectral properties of phosphors, for example, can be viewed with a color camera. Other properties, such as detailed optical spectroscopy, x-ray diffraction, and transport measurements, are by their very nature sequential. In this case automation can help, and various scanners are used to connect different individual samples in a combinatorial library to the measurement device. This can be done either by mechanically stepping over a sample surface and measuring at one point at a time or by making connections to each cell in a library. In either case the need for electrical contact puts a limit on the number of cells that can be placed in a single library. Practical limit appears to be approximately 100 cells in a single library when four-wire resistivity measurements are required. This is a fairly small number when compared with other examples of combinatorial synthesis. If only optical observation is required, the number of cells in a library can be in the range of thousands. In contrast, a split-and-mix liquid-phase library might include 10^5 or more different compounds.

It is clear that new analysis techniques must also be developed before the combinatorial synthesis technique can be expected to have a significant impact on the discovery of new materials. We describe here one such method, concurrent x-ray diffraction, in greater detail.

A. Concurrent X-Ray Diffraction

It is desireable to reduce the cell size in a thin-film combinatorial library as much as possible in order to achieve the maximum benefit from combinatorial synthesis. There are a number of factors limiting the minimum cell size. As mentioned earlier, the masks used in the deposition chamber do not provide ideally sharp cell edges. At a sample-to-mask distance of 1–2 mm, we have observed a transition region with a width of ≈ 300 μm at the edge of a cell. This puts a limit of about 1 mm $\times$ 1 mm on the cell size. This limit is not yet the practical limit because most of the analysis techniques require much larger regions. X-ray diffraction is an important tool in thin-film characterization and is used for determining lattice constants and the presence of undesired impurity phases and for measuring the actual superlattice periodicities. The conventional x-ray machines used for thin-film characterization have a spot diameter of several millimeters on the sample surface. It is possible to measure combinatorial libraries if the cells are large enough, and the diffractometer is equipped with a programmable sample-positioning stage. With such a setup it is possible to measure diffraction patterns from each cell sequentially in an automated fashion. The number of cells in a single library is limited to about 10.

The x-ray beam diameter can be reduced dramatically if a synchrotron source is available. It has been shown by Isaacs et al. (32) that a focused microbeam can be scanned over a combinatorial library and one can perform x-ray fluorescence, diffraction, and near-edge x-ray absorption spectroscopy. The cells in the superlattice library had a size of 1 mm × 2 mm, and the x-ray beam spot on the sample was less than 20 μm in diameter. This technique is extremely versatile and can be used to scan over the library rapidly due to the high intensity of the x-ray beam. The drawback is the limited access to synchrotron sources.

We describe here a modified x-ray diffractometer that addresses the most important question in thin-film x-ray analysis, namely, the measurement of the lattice constant (33). The machine is a modification of a standard single-crystal diffractometer, using a curved monochromator to produce a wedge-shaped x-ray beam and a two-dimensional detector for measuring a large number of diffraction patterns in parallel. A schematic diagram of the measurement setup is shown in Figure 28. The instrument uses a 1.2-kW rotating-anode fine-focus copper x-ray generator with an apparent source size of 0.1 mm × 0.1 mm. A curved Johan-type monochromator is used to eliminate Cu K_{α_2}, Cu K_β, and white radiation. The Cu K_{α_1} radiation is focused on the sample surface irradiating an area of 0.1 mm × 10 mm. The convergence angle is ≈2°. The sample and a two-dimensional detector are placed on an ω 2θ stage of a two-circle goniometer. The detector can be placed at a Bragg-peak position to measure lattice constants. It is also possible to measure reflectivity of a sample by using the detector in the low-angle configuration. This is useful for the analysis of the periodicity of superlattice samples.

Either an imaging plate or a special CCD camera can be used as a detector. The images shown here were obtained with an imaging plate. Until now, imaging plates have offered a wider dynamic range, but recent improvement in the quantum efficiency and reduction of the background noise levels of CCD detectors have practically eliminated this advantage. The use of a CCD camera is planned for future work with this instrument.

The detector can see a diffraction pattern that has the width equal to the convergence angle of the monochromator, which is presently 2°. Thin films typically have lattice constants close to the lattice constant of the substrate. The detector is therefore placed so that one of the substrate peaks is visible on the detector. Any peaks originating from the film would be observed close to the substrate peak. The observation of the rocking curve of a Bragg reflection from a nearly perfect crystal usually requires an accurate setting of the ω and 2θ angles. The use of an area detector eliminates this need, making it much easier and faster to set up and perform the diffraction measurement.

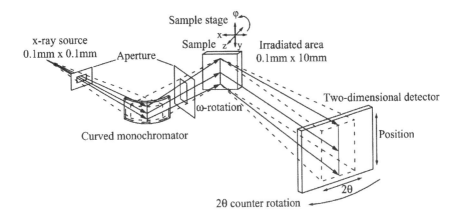

Figure 28   Schematic view of the concurrent x-ray diffractometer. A curved monochromator produces a line-shaped convergent Cu $K\alpha_1$ beam on the sample surface. Diffracted x-rays are detected with a two-dimensional detector.

An example of how the concurrent x-ray diffraction system can be used to characterize superlattice samples is shown in Figure 29. A set of superlattices was grown on a $SrTiO_3$ substrate. Each superlattice had 30 periods consisting of $SrTiO_3$ and $BaTiO_3$ layers. The thickness of the $SrTiO_3$ layer was kept constant at 10 monolayers. The thickness of the $BaTiO_3$ layer was different in each superlattice cell. The superlattices can be described with a general formula $[(SrTiO_3)_{10}/(BaTiO_3)_n]_{30}$, where n is the thickness of the $BaTiO_3$ layer in monolayers. The x-ray beam covered three superlattices on the substrate. Each superlattice cell was 1.75 mm wide, and there was a 1-mm space between individual cells. Two different measurement geometries were used. Figure 29a shows a reflectivity image obtained from the imaging plate. The graphs below the image show three cross sections taken along the arrows shown in the photo. The $BaTiO_3$-layer thicknesses in the three cells measured here were $n = 5$, 10, and 15. As expected, the low-angle superlattice peaks can be seen to shift to higher angles as the superlattice period gets smaller.

The diffraction images measured close to the (001) and (002) Bragg peaks are shown in Figures 29b and c. The images show clear satellite peaks due to the superlattice structure. The actual superlattice period can be calculated from the satellite peak positions. The intensities of the satellite peaks can be used to determine the quality of the superlattice. The strong continuous line in the images is the substrate peak. The line is not straight due to bending of the substrate while it was clamped to the sample holder during superlattice deposition.

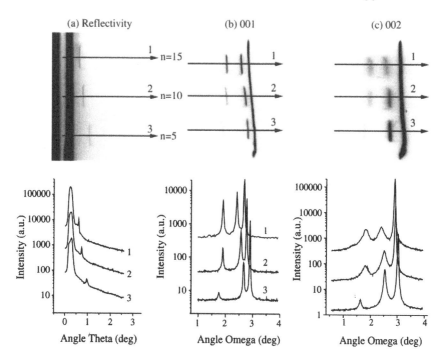

Figure 29 Concurrent x-ray measurements of a $[(SrTiO_3)_{10}/(BaTiO_3)_{\,n}]_{30}$ superlattice library. Diffraction from three cells with $n = 5, 10, 15$ are visible in the images. Part (a) shows a reflection image, (b) was measured close to the 001 reflection, and (c) was measured close to the 002 substrate peak.

The vertical resolution of the x-ray image is approximately 0.1 mm. This value is much smaller than what is realistically achievable with the present deposition systems. The concurrent x-ray machine has therefore made it possible to use much smaller cells and to increase the number of cells on a single substrate. Each image requires an aquisition time of about 1 minute. A single image gives all the necessary information to analyze a single row of cells in a library. Using a CCD image detector, it is possible to step through all rows in a library under computer control and thus to obtain all relevant diffraction data in a matter of minutes. The images and cross sections in Figure 29 also show that the angular resolution of 0.02° is sufficient.

Another example of how the concurrent x-ray machine can be used is shown in Figure 30. In this case a composition spread sample was deposited on a $SrTiO_3$ (001) substrate. The film was grown by depositing periodically $SrTiO_3$ and $BaTiO_3$ and using a movable mask during deposition to control the amount of either $SrTiO_3$ or $BaTiO_3$ deposited at each point on the substrate. The stencil mask was not used in this experiment, resulting in a continuous transition without cell boundaries.

The horizontal axis in the image is the diffraction angle. The bright vertical line at 46.4° is the $SrTiO_3$ (002) substrate peak. The vertical axis shows the position on the 9-mm-wide substrate. Only $SrTiO_3$ was deposited at the bottom 2 mm of the substrate. The film peak can be seen as a vertical line at approximately 46.2°. The homoepitaxial film had slight compressive in-plane strain, resulting in slight elongation of the c-axis. This is why the film peak can be seen separately from the substrate peak. The strain in the homoepitaxial $SrTiO_3$ was found to decrease and disappear completely as the deposition temperature was increased from 600 to 900°C. The top part of the substrate was covered by a $BaTiO_3$ film, which shows a fairly wide peak at 44.8°. The center part of the substrate was covered by a $Sr_{1-x}Ba_xTiO_3$ film with $x = 0$ at the 2-mm vertical position and reaching $x = 1$ at the 8-mm position. The film peak can be seen to shift to lower

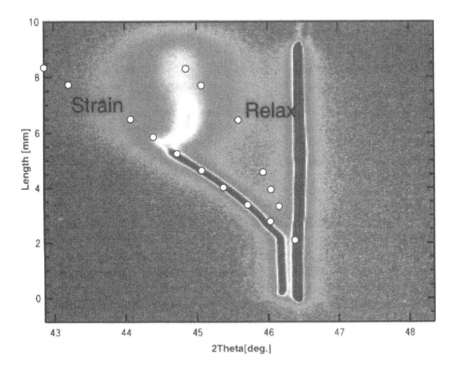

Figure 30 Strain in an SBTO composition spread sample.

diffraction angles as x starts to increase. The peak width does not change, showing that the film grows coherently and the c-axis gets gradually larger as the Ba content in the film increases. Misfit dislocations appear in the film at approximately $x = 0.54$, resulting in relaxation of the film. At the same point the width of the diffraction peak increases, showing that the presence of the misfit dislocations also decreases the crystallinity of the film. The two rows of dots in the image show the expected film peak positions for a strained film and a relaxed film. The peak positions for the relaxed film were calculated by assuming that the lattice constant of the film would change linearly from 3.905 Å for $x = 0$ to 4.04 Å for $x = 1$. The points corresponding to a strained film were calculated assuming that the in-plane lattice constant of the film matches that of the substrate, resulting in a more rapid increase of the c-axis length.

This type of experiment can be performed very efficiently when combinatorial sample synthesis is coupled with concurrent x-ray diffraction. It is possible to deposit several composition spreads on a single substrate at either different oxygen pressures or different substrate temperatures. The concurrent x-ray measurement provides the lattice constant data for each composition spread stripe at a time in a measurement lasting approximately 1 minute. A single experiment can thus give all relevant data on how strain in $Sr_{1-x}Ba_xTiO_3$ films depends on x and possibly other parameters, such as PO_2, or T_{sub}.

B. Other Characterization Methods

The combinatorial thin-film synthesis techniques described in this chapter can offer an efficient tool for the discovery of new materials and for the optimization of crystal growth conditions and resulting physical properties. Many of the technological difficulties related to parallel film growth have already been solved, and the main effort on this front is currently directed at increasing the number of cells in a single library. As mentioned earlier, however, in many cases the library density is limited more by the characterization needs than by the deposition process.

While an essentially one-to-one correspondence exists between the original Merrifield process of combinatorial organic synthesis and the thin-film growth procedures described here, few similarities can be found between the characterization needs and methods. In the case of organic compounds it is possible to screen a vast number of samples rapidly for specific desired biological activity. The main assumption that has to be made is that the sequence of reaction steps used during the synthesis of a library really reflects the actual molecular structures that are present in the individual library cells. In case of solid-state synthesis, this correspondence cannot be guaranteed. This problem is particularly acute when films are grown far from the thermodynamic equilibrium conditions, as is the case in the PLD libraries described in this chapter. The main consequence is that screening a library for a particular characteristic, such as electrical conductivity, cannot be expected to give useful information if other characteristics of the materials are not determined at the same time.

One particularly important parameter for oxides is the precise oxygen stoichiometry. Let us assume, as an example, that we are preparing a library containing copper oxides, with several other elements added in various proportions. Even if a particular cell happens to contain $YBa_2Cu_3O_x$, there is no guarantee that the cell would actually be superconducting when the library is screened for superconductivity, due to the unknown value of x. Indeed, it is impossible to guarantee a priori that proper oxygen stoichiometry has been achieved for new and as yet unknown materials. The optimum value of x would likely be different for different cells in the library anyway. This shows that combinatorial thin-film synthesis as a complete process leading to useful discoveries is a far more complex process than what is apparent if one looks only at the fabrication of a library.

It is for this reason that we are directing much of our work at developing other characterization techniques that can be applied to combinatorial libraries. Results of the concurrent x-ray system were already described. Other techniques under development are microwave microscopy and various scanning optical and magnetic measurement techniques.

Microwave microscopy can be used to measure the dielectric constant and microwave loss or surface conductivity of thin films (34). The machine typically operates at a frequency of a few GHz, which is a technologically important region of the spectrum. Also, by using a scanning microwave cavity with an evanescent wave probe, it is possible to reach submicron spatial resolution, which is more than adequate for thin-film library characterization. In addition to the dielectric constant, the instrument can yield information on thin-film conductivity. This is particularly important, because it makes it possible to eliminate contact-mode resistivity measurements in many cases.

Another technique under development is a scanning SQUID magnetometer. This instrument can measure the magnetic field strength on the sample surface at various temperatures with a spatial resolution of ≈ 2 μm (Seiko). This instrument opens up the possibility of using combinatorial techniques for the development of magnetic materials. The SQUID measurements have been successfully used to map the magnetic phase diagrams of various doped manganite compositions by using composition spread libraries.

We have applied combinatorial synthesis to superlattices, heterojunctions, and simple devices such as transistors. In such cases an additional parameter affecting the physical properties of the lattice or device is the quality of the interface between different layers. Periodic thin-film structures can be analyzed by using diffraction techniques. Aperiodic structures, such as transistors, are also sensitive to interface properties. While some of the electrical properties of interface layers can be deduced from voltage–capacitance measurements, it is often necessary to look at the interface on an atomic scale with a transmission electron microscope (TEM) to understand how the growth process parameters affect the interface structure. In order to make TEM an effective tool for combinatorial libraries, we have set up a rapid sample preparation system using a focused ion beam (FIB) machining tool. The FIB can be used to cut a miniature TEM sample out of several places in a library (35). The TEM sample preparation time can be reduced to approximately 1 hour, making TEM a valuable tool also for combinatorial interface development.

ACKNOWLEDGMENTS

We wish to thank several of our colleagues for performing many of the experiments described here: T. Koida, D. Komiyama, M. Murakami, Z. Jin, Dr. H. N. Ayer, and Dr. N. Matsuki. We thank A. Nara for help with translations. This work is being financed by CREST-JST and STA-COMET.

REFERENCES

1. Merrifield, R. B. J. Am. Chem. Soc. 1963, 85, 2149.
2. Merrifield, R. B. J. Org. Chem. 1964, 29, 3100.
3. Plunkett, M. J.; Ellman, J. A. Sci. Am. April (1997), 54.
4. Xiang, X. D.; Sung, X.; Briceño, G.; Lou, Y.; Wang, K. A.; Chang, H.; Wallace-Freedman, W. G.; Chen, S. W.; Schultz, G. Science. 1995, 268, 1738.
5. Sun, X.-D.; Gao, C.; Wang, J.; Xiang, X.-D. Appl. Phys. Lett. 1997, 70, 3353.
6. Briceno, G.; Chang, H.; Sun, X.; Schultz, P. G.; Xiang, X.-D. Science. 1995, 270, 273.
7. Danielson, E.; Golden, J. H.; McFarland, E. W.; Reaves, C. M.; Weinberg, W. H.; Wu, X. D. Nature. 1997, 389, 944.
8. Xiang, X.-D. Annu. Rev. Mater. Sci. 1999, 29, 149.
9. Koinuma, H.; Koida, T.; Ohnishi, T.; Komiyama, D.; Lippmaa, M.; Kawasaki, M. Appl. Phys. A. 1999, 69[Suppl], S29.
10. Koinuma, H.; Aiyer, H. N.; Matsumoto, Y. Sci. Tech. Adv. Mat. 2000, 1, 1.
11. Koinuma, H.; Tsuchiya, R.; Kawasaki, M. SPIE Proc. 1996, 2696, 525.
12. Koinuma, H. Solid State Ionics. 1998, 108, 1.
13. Koinuma, H.; Kawasaki, M.; Yoshimoto, M. Proc. Mat. Res. Soc. Symp. 1996, 397, 145.
14. Matsumoto, Y.; Murakami, M.; Jin, Z.; Ohtomo, A.; Lippmaa, M.; Kawasaki, M.; Koinuma, H. Jpn. J. Appl. Phys. 1999, 38, L603.
15. van Dover, R. B.; Schneemeyer, L. F.; Fleming, R. M. Nature. 1998, 392, 162.
16. Lippmaa, M.; Nakagawa, N.; Kawasaki, M.; Ohashi, S.; Koinuma, H. Appl. Phys. Lett. 2000, 76, 2439.

17. Ohashi, S.; Lippmaa, M.; Nakagawa, N.; Nagasawa, H.; Koinuma, H.; Kawasaki, M. Rev. Sci. Instrum. 1999, 70, 178.
18. Izumi, M.; Nakazawa, K.; Bando, Y. J. Phys. Soc. Jpn. 1998, 67, 651.
19. Kawasaki, M.; Takahashi, K.; Maeda, T.; Tsuchiya, R.; Shinohara, M.; Ishiyama, O.; Yonezawa, T.; Yoshimoto, M.; Koinuma, H. Science. 1994, 266, 1540.
20. Yoshimoto, M.; Maeda, T.; Ohnishi, T.; Koinuma, H.; Ishiyama, O.; Shinohara, M.; Kubo, M.; Miura, R.; Miyamoto, A. Appl. Phys. Lett. 1995, 67, 2615.
21. Ohtomo, A.; Tamura, K.; Kawasaki, M.; Makino, T.; Segawa, Y.; Tang, Z. K.; Wong, G. K. L.; Matsumoto, Y.; Koinuma, H. Appl. Phys. Lett. 2000, 77, 2204.
22. Ohtomo, A.; Makino, T.; Tamura, K.; Matsumoto, Y.; Segawa, Y.; Tang, Z.; Wong, G. K. L.; Koinuma, H.; Kawasaki, M. Proc. SPIE. 2000, 3941, 70.
23. Jin, Z. W.; Murakami, M.; Fukumura, T.; Matsumoto, Y.; Ohtomo, A.; Kawasaki, M.; Koinuma, H. J. Cryst. Growth. 2000, 214/215, 55.
24. Moates, F. C.; Somani, M.; Annamalai, J.; Luss, J. T. R. D.; Willson, R. C. Ind. Eng. Chem. Res. 1996, 35, 4801.
25. Senkan, S. M. Nature. 1998, 394, 350.
26. Reddington, E.; Sapienza, A.; Gurau, B.; Viswanathan, R.; Sarangapani, S.; Smotkin, E. S.; Mallouk, T. E. Science. 1998, 280, 1735.
27. Cong, P.; Doolen, R. D.; Fan, Q.; Giaquinta, D. M.; Guan, S.; McFarland, E. W.; Poojary, D. M.; Self, K.; Turner, H. W.; Weinberg, W. H. Angew. Chem. Int. Ed. 1999, 38, 484.
28. Matsumoto, Y.; Murakami, M.; Jin, Z.; Nakagawa, A.; Yamaguchi, T.; Ohmori, T.; Suzuki, E.; Nomura, S.; Kawasaki, M.; Koinuma, H. Proc. SPIE. 2000, 3941, 19.
29. Gaskell, D. R. Introduction to Metallurgical Thermodynamics, 2nd ed.: Hemisphere: New York; 1981, 287.
30. Yoshinobu, T.; Iwasaki, H.; Nakao, M.; Nomura, S.; Nakanishi, T.; Takamatsu, S.; Tomita, K. Jpn. Appl. Phys. 1998, 37, L353.
31. Nomura, S.; Nakao, M.; Nakanishi, T.; Takamatsu, S.; Tomita, K. Anal. Chem. 1997, 69, 977.
32. Isaacs, E. D.; Marcus, M.; Aeppli, G.; Xiang, X.-D.; Sun, X.-D.; Schultz, P.; Kao, H.-K.; III, G. S. C.; Haushalter, R. Appl. Phys. Lett. 1998, 73, 1820.
33. Omote, K.; Kikuchi, T.; Harada, J.; Kawasaki, M.; Ohtomo, A.; Ohtani, M.; Ohnishi, T.; Komiyama, D.; Koinuma, H. Proc. SPIE. 2000, 3941, 84.
34. Gao, C.; Xiang, X.-D. Rev. Sci. Instrum. 1998, 69, 3846.
35. Minowa, K.; Takeda, K.; Tomimatsu, S.; Umemura, S. J. Cryst. Growth. 2000, 210, 15.

6
Combinatorial Synthesis of Display Phosphors

Ted X. Sun
General Electric Company, Schenectady, New York, U.S.A

I. INTRODUCTION

A. Overview

Phosphors are key materials used to convert electronic information into visible signals. The rapid development of various full-color, electronic, flat-panel displays (FPDs), including field-emission displays (FEDs), electroluminescent displays (ELDs), and plasma display panels (PDPs), demands timely discovery of advanced phosphors. The specific excitation media and operating environment of various FPDs require different sets of tricolor phosphors for desirable contrast, operation lifetime, and color properties. Although the photophysical process of phosphors under regular ultraviolet (UV) excitation is relatively well known, the fluorescent process of phosphors under excitation by an electron beam (FED, projection TV), charge injection (EL), or vacuum UV (plasma diaplay) are not well understood. The development of various FPD technologies is, to a large extent, limited by the lack of efficient tricolor phosphors.

To speed up the discovery of phosphors for display applications, a combinatorial approach has been developed and applied to the synthesis and screening of phosphors. This chapter will review the combinatorial methodologies and their applications to phosphor materials. Phosphor libraries can be made in either thin-film or powder forms, using masking strategies and a liquid dispensing system, respectively. High areal density libraries with 100–1000 discrete phosphor compositions on a 1-inch-square substrate can be routinely made. Both compositions and synthesis temperatures can be optimized in a high-throughput mode. Detailed examples of high-throughput synthesis (HTS) and screening of phosphors are given. These methods provide general combinatorial tools for HTS of other solid-state materials as well. A few highly efficient tricolor phosphors discovered with combinatorial methods have been reproduced in bulk forms, and their photo- and cathodoluminescent properties have been measured.

B. Display Phosphors

There is considerable interest in the development of phosphors for applications in advanced display technologies, including PDPs, FEDs, ELDs, and cathode ray tube (CRT) projection displays. A list of state-of-the-art commercial phosphors for various display technologies is given in Table 1. All of these phosphors were discovered by the traditional trial-and-error methods, and most of them were discovered decades ago (1). There are continuing needs for phosphors with improving luminescence properties (1–3), and finding new phosphors with better properties is essential for the advancement of display technologies.

Oxides provide attractive host materials for the development of advanced phosphors, due largely to their relative ease of synthesis and their stability. Most state-of-the-art phosphors used in various displays and lamp applications are oxides (Table 1). For example, efficient refractive oxide phosphors, Y_2O_3:Eu^{3+} (red) (4), $Y_3Al_5O_{12}$:Tb^{3+} (green) (5) and $BaMgAl_{10}O_{17}$:Eu^{2+} (blue) (6), have found applications in projection TVs and plasma displays, due primarily to their high luminescent quenching temperature and their stability under harsh operational conditions, such as in intense plasma, in vacuum UV, or under electron beam excitations. Extensive research has been carried out on rare earth activated oxide phosphors because of their superior color purity and good chemical and thermal stability compared to d- or s-series element activated nonoxides (1–7).

Table 1 Tricolor Phosphors Used in Various Full-Color Display Technologies

Display type	Red	Green	Blue
CRT (cathode ray tube)	Y_2O_2S:Eu^{3+}	ZnS:Au,Al,Cu	ZnS:Ag^+, Cl^-
PDP (plasma display panel)	$(Y,Gd)BO_3$:Eu^{3+}	Zn_2SiO_4:Mn^{2+}	$BmMgAl_{14}O_{23}$:Eu^{2+}
FED (field-emission display)	Y_2O_3:Eu^{3+}	$SrGa_2S_4$:Mn^{2+}	$SrGa_2S_4$:Ce^{3+}
TFEL (thin-film electroluminescence)	ZnS:Mn^{2+}	ZnS:Tb^{3+},Cl^-	SrS:Ce, or $SrGa_2S_4$:Ce
PTV (projection TV)	Y_2O_3:Eu^{3+}	$Y_3(Al,Ga)_5O_{12}$:TB^{3+}	ZnS:Ag^+

Source: S. Shionoya and W. Yen, eds. *Phosphor Handbook*: Boca Raton, FL: CRC Press, 2000.

However, despite these efforts the number of oxide phosphors with ideal properties is limited. $Y_2O_3:Eu^{3+}$, discovered decades ago (4) as a major breakthrough (1, 2), is still the best red phosphor in many display applications due to its high efficiency (97%), its color purity (with nearly a single emission peak at ≈ 610 nm), and its high thermal and chemical stability. It is currently used in tricolor lamps (1), field-emission displays (8), projection TV displays (1), etc. The only drawback of $Y_2O_3:Eu^{3+}$ is that its emission peak is at 610 nm (corresponding to an orange-red color), which limits the color space of a full color display (9).

The search for advanced phosphors with multiple superior qualities for display applications is a nontrivial task. Although photophysical processes leading to luminescence are relatively well understood, the specific spectral properties, luminescence efficiencies, and operational lifetimes of a phosphor system depend on complex interactions among the excitation source, the host lattice, the sensitizer, and the luminescent center. Luminescence properties of a phosphor are usually sensitive to the changes in composition, stoichiometry, dopants, and processing conditions (1). Consequently, identification of phosphors that are optimally suited to the requirements of a given display technology is highly empirical (7). In order to meet the ever-increasing demands from display technologies for new high-performance phosphors, a combinatorial approach was developed to speed up the discovery of new phosphors.

The quick discovery and optimization of phosphors requires both a comprehensive combinatorial method and the rational design of experiments. Some empirical knowledge of phosphors was taken into account in the design of phosphor libraries. Among the criteria were:

a. Start with refractory metal oxide host materials. As discussed previously, refractory oxide phosphors with high quenching temperatures, high chemical stability, and resistance to plasma, vacuum UV, or electron-beam damage are desired in displays.
b. Use rare earth (RE) activators for better color purity. (RE activators are well known for their narrow emission line in f–f transitions (1).
c. Choose hosts with crystal structures into which the RE dopants can fit. A high RE activator concentration is desirable for applications such as projection TV, which requires a high saturation current (1). Host structures also determine the point symmetry of the activators, and they change the intensity of emission by affecting selection rules.
d. Choose host ions with appropriate size, valance, and electronegativity in order to tune the crystal field strength on activators. These factors affect the fluorescent color and intensity of RE activators (1).

Perovskites (with the standard chemical formula ABO_3) are known for their structural flexibility. Various metal cations can be accommodated into A or B sites to form perovskites or related compounds (10). Aluminates are well known for their refractory nature. Thus, aluminate-based perovskites were chosen as the focus of the initial combinatorial research. To increase the solubility of RE activators in the host, an A^{3+} cation would have to be close in ionic radii to the activator cations, and La^{3+} was selected. Gd^{3+} was also used, because it was recently reported that Gd^{3+} can transfer UV energy effectively to some RE activators (7). To accommodate Eu^{2+} (a wide-band activator) in aluminates, its charge needs to be compensated. The lattice has to be modified with cations similar to Eu^{2+} in the host lattice, and Sr^{2+} was selected. Taking these into account, we chose to study a series of compounds with compositions $La(Gd)_m(Sr)_nAlO_x:Eu^{2+,3+}$ using the combinatorial method. Other RE activators were also investigated, including Tb^{3+} and Ce^{3+}. There had been some studies on the spectral properties of some nominal compositions from this series of compounds (11)– (14). However, there is little detailed information on the compositional and processing effects on the luminescence properties of this series. Even after the empirical filtering with the reasoning described earlier, we still found a tremendous amount of possible compositions to study in this series. A combinatorial approach was developed and applied to quickly screen this series of compositional spaces for efficient phosphors.

II. COMBINATORIAL METHODOLOGY

Phosphors can be synthesized in either thin-film or powder forms. We have developed the combinatorial methods to allow a parallel or fast sequential synthesis of phosphors in both forms. A combination of thin-film deposition with a set of shadow masks was used to generate a thin-film "spatially addressable library," where each as-deposited sample on the library is a precursor multiple layer (15, 16). Following the deposition, a thermal annealing step was necessary to diffuse layers of precursors before the high-temperature synthesis of the final phases. In the synthesis of powder-form phosphor libraries, solutions of elemental precursors were dispensed using an automatic liquid injector with an accurate control of delivery volume on an array of cells (17). In each cell, the soluble form of the elements mix on a molecular scale before the coprecipitation, drying, and high-temperature synthesis processes.

Various high-vacuum thin-film deposition apparatuses (sputtering, laser ablation, thermal and e-beam evaporation, etc.) were employed in making thin-film libraries. An integral part of making a thin-film library is the selective-area deposition, which is achieved by using shadow masks. There are different types of masking strategies, which can be used to create diverse compositional variations in phosphor libraries. These include using a series of stationary masks with varying opening patterns (15, 16, 18) and movable shutter masks (19). The movable shutter masks are often used for composition optimization once a set of promising phosphor compositions is identified.

Selective-area depositions using a series of shadow masks result in the generation of different combinations of precursor layers deposited at different sites on a substrate. The pattern of the masks applied plus the corresponding amount of deposition for a precursor determine the stoichiometry of samples in the phosphor library.

The choice of masking patterns depends on the design of a thin-film library. We have extensively used binary masking (Figure 1a), quaternary masking (Figure 1b), and their combinations. In the binary masking, one-half of the total substrate area is always covered for each deposition step, and all of the binary masks have different spatial patterns of coverage Figure 1a. In this scheme, after n steps of depositions, a total of 2^n different combinations of precursors can be generated.

To increase the flexibility in the design of the library, we also devised the quaternary masking scheme Figure 1b. In this design, the patterns in the masks successively subdivide the substrate into a series of self-similar patterns of quadrants. In using n masks, the rth ($1 \leq r \leq n$) mask contains 4^{r-1} windows, where each window exposes one-quarter of the area deposited with the preceding mask. Within each window there are an array of 4^{n-r} sample sites separated by grids Figure 1b. Each mask is used in up to four sequential depositions, where each time the mask is rotated by 90°. This process produces 4^n different compositions with $4 \times n$ deposition steps. A series of five masks is necessary to make $4^5 = 1024$ different compositions.

For some applications, these masks can be implemented in the form of photoresist patterns using the standard photolithography. While this usually requires a liftoff process after each deposition, the mask alignment is much more accurate than that of physical shadow masks, and arbitrarily high-density phosphor libraries can be fabricated.

In the following section, specific examples of phosphors libraries will be discussed. Except for some phosphors used in thin-film electroluminescent displays, most phosphors used in the display industry are in powder forms, which can be prepared from solution precursors. To make powder phosphor libraries, we have developed a scanning multiple-channel solution-dispensing system capable of accurately and rapidly delivering down to nanoliter volumes of reagent (17).

III. THIN-FILM PHOSPHOR LIBRARIES

A. Thin-Film Libraries: La(or $GdF_3)_m(Sr_n)AlO_x:Tb_y^{3+}(Ce_z^{3+})$: $Eu_n^{2+/3+}$ Made with Binary Masking Method

1. Combinatorial Synthesis

A combination of radio frequency (RF) sputtering and physical masking steps was used to generate libraries of thin-film phosphors of different composition

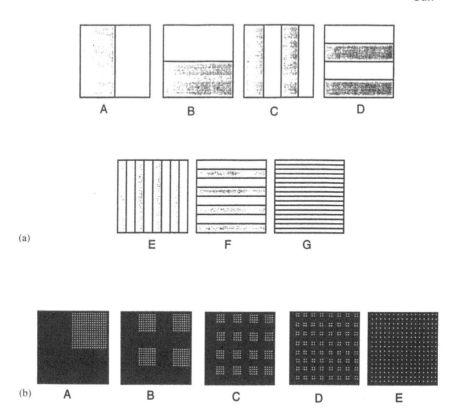

Figure 1 (a) Scheme of a series of binary mask patterns. After seven sequential masking and deposition steps, up to 2^7 (128) samples with different elemental compositions can be generated. (b) Scheme of a series of quaternary masks. Using each mask, up to four different masking patterns can be generated by rotating the mask by 90°.

with aluminate hosts (15, 16). The sputtering targets used (>99.9% purity) included La_2O_3, GdF_3, $SrCO_3$, Al_2O_3, EuF_3, Tb_4O_7, and CeO_2. Libraries were deposited on polished (100) LaAlO3, (100) MgO, and Si single crystal substrates. The size of the library is typically 1 inch square. Using the binary masking scheme, a total of 128 different compositions were created, where each site had the area of 1 mm by 2 mm. The amount of precursor deposited was monitored in situ with a quartz crystal thickness monitor. A profilometer study revealed that film thickness varied less than 5% over a 2-inch-diameter deposition area for each target. Indeed, identical composition sites within a library resulted in the same relative luminescent intensities and chromaticities after synthesis, attesting to the thickness uniformity of thin-film deposition. Seven compositionally identical libraries were fabricated at one time. They were annealed under different partial pressures of O_2, H_2, He, and Ar and at temperatures ranging from 1100°C to 1400°C for up to 4 hours following an extensive low temperature annealing process in air. The refractory nature of the precursors used in these materials prevented significant changes in stoichiometry due to thermal evaporation. Indeed, the effects of heat treatment under various partial gas pressures (e.g., reducing or nonreducing) were also found to be reversible, further supporting the fact that the composition of phosphor in the libraries does not change during the thermal processing. Photoluminescence images of the libraries were obtained by color photography under broad-wavelength UV irradiation using a Hg lamp (Figure 2).

As is evident from Figure 2, the thin-film combinatorial approach is very effective for comparative evaluation of the luminescent properties of phosphors. Selected lead composition phosphors are encircled, and their corresponding compositions and processing conditions are listed in Table 2. Compounds with the compositions $GdAl_{1.6}O_x:Eu^{3+}_{0.08}$, $Gd_{0.6}Sr_{0.4}Al_{1.6}O_xF_{1.8}:Eu^{2+}_{0.08}$, and $LaAl_{3.1}O_x:Eu^{2+}_{0.08}$ produced the brightest red, green, and blue photoluminescence emission, respectively. Red phosphors with the composition $LaAlO_x:Eu^{3+}_y$ ($y < 0.01$) (12) and $GdAlO_x:Eu^{3+}_y$ ($y < 0.01$) (14) have been previously reported, but they had not been optimized or characterized with regard to their QEs. The valence states of Eu, namely, $2+$ or $3+$, in the thin-film libraries were estimated from the color of emission and confirmed with near-edge x-ray absorption fine-structure spectroscopy (NEXAFS) using the microfocused synchrotron x-ray beam.

2. High-Throughput Screening of the Phosphor Libraries

The photoluminescent spectral properties of these phosphors were characterized using a customer-built spectrophotometer in reflection mode (Figure 3). A monochromatized UV light was focused onto each sample in the library (focus spot is ~0.5 mm in diameter). The fluorescent signal was collected with a reflecting mirror into a spectrograph with a CCD detector operating at -120°C (cooled with liquid N_2). The photoluminescent excitation and emission spectra of as-synthesized $GdAl_{1.6}O_x:Eu^{3+}_{0.08}$, $Gd_{0.6}Sr_{0.4}Al_{1.6}O_xF_{1.8}:Eu^{2+}_{0.08}$, and $LaAl_{3.1}O_x:Eu^{2+}_{0.08}$ thin-film phosphors are shown in Figure 4. The red emission is characteristic of the Eu^{3+} $^5D_0-^7F_2$ transition, with a major narrow emission peak centered around 613 nm. The green and blue phosphors displayed broadband emissions characteristic of the d–f electronic transitions for the activator Eu^{2+}, with shifted emission-peak positions, possibly due to the differences in covalent [nephelauxetic (1)] effects and in the crystal field strength of the two host materials on the Eu^{2+} activator.

There are difficulties in directly measuring QE of thin-film phosphors due to the lack of standard tricolor thin-film phosphors of known QE and absorption.

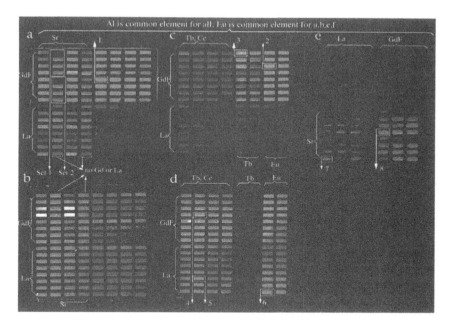

Figure 2 Photoluminescent images of a series of phosphor libraries processed under different conditions: (a) La(or GdF$_3$)$_m$(Sr$_n$)AlO$_x$: Eu$^{2+/3+}_h$, where $0.375 \leq m \leq 1$, $0.25 \leq n \leq 0.4$, $1.88 \leq y \leq 12\%$ in atomic ratio, were annealed at 1150°C in 10% H$_2$/Ar for 4 hours; (b) same as (a), but annealed at 1400°C in 40% H$_2$/He for 4 hours; (c) La(or GdF$_3$)$_m$AlO$_x$:Tb$^{3+}_y$(Ce$^{3+}_z$):Eu$^{2+/3+}_h$, where $0.32 \leq m \leq 1$, $1.29 \leq y \leq 6\%$, $0.65 \leq z \leq 4\%$, $1.29 \leq h \leq 8\%$ in atomic ratio, were annealed in air at 1150°C for 4 hours; (d) same as (c), but annealed at 1400°C in 40% H$_2$/He for 4 hours; (e) La(or GdF$_3$)$_m$(Sr$_n$)AlO$_x$: Eu$^{2+/3+}_h$, where $0.178 \leq m \leq 0.714$, $0.17 \leq n \leq 0.4$, $0.75\% \leq y \leq 16.7\%$ in atomic ratio, were annealed at 1150°C in 4% H$_2$/Ar for 4 hours. All substrates were 1 in. $\times$ 1 in., and each site is 1 mm $\times$ 2 mm; thin films were 0.4–1 μm thick. Sputtering was carried out at a base pressure of 10^{-5}–10^{-6} torr with Ar gas and with a deposition rate of 0.1–0.5 Å/s. The images were taken with a UV lamp (UVP, UVGL-25 mineralight) at a wavelength of 254 nm. The indicated sites correspond to the elemental compositions and stoichiometries of samples listed in Table 2. Set 1 and Set 2 samples in library (a) are identical sample groups.

After discovery of lead phosphor compositions from the thin-film libraries, powder forms of these compositions were synthesized using a standard solid-state synthetic method via solution precursors. Aqueous solution of all the elements involved were made from ultrahigh-purity nitrate salts (>99.99%), with the following molar concentrations: 0.5 M La(NO$_3$)$_3$, 0.5 M Gd(NO$_3$)$_3$, 0.5 M Al(NO$_3$)$_3$, 0.5 M Zn(NO$_3$)$_2$, 0.5 M Sr(NO$_3$)$_2$, 0.1 M Eu(NO$_3$)$_3$, 0.04 M Ce(NO$_3$)$_3$, and 0.04

Table 2 Lead Phosphors Discovered from Thin-Film Libraries of Fig. 2.

Sample	Synthesis conditions (°C), 4 h	Quantum efficiency (%)
1. $GdAlO_x$: $Eu^{3+}_{0.06}$ (R)	1400 air	85
2. $Gd_{0.77} Al_{1.23} O_x$: $Eu^{3+}_{0.06}$ (red)	1400 air	90
3. $GdAlO_x$:$Tb^{3+}_{0.08}$ (G)	1150 H_2/Ar	31
4. $Gd_{0.69} Al_{1.31} O_x$: $Ce^{3+}_{0.028}$:$Tb^{3+}_{0.041}$ (G)	1375 H_2/He	37
5. $La_{0.57} Al_{1.43} O_x$: $Ce^{3+}_{0.023}$:$Tb^{3+}_{0.034}$ (G)	1375 H_2/He	5
6. $La_{0.5} Al_{1.5} O_x$: $Eu^{2+}_{0.04}$ (B) film	1375 H_2/He	~60
7. $La_{0.43} Sr_{0.24} Al_{1.33} O_x$: $Eu^{2+}_{0.10}$ (G)	1375 H_2/He	60
8. $(Gd_{0.46} Sr_{0.31})Al_{1.23} O_x F_{1.38}$:$Eu^{2+}_{0.06}$ (G)	1375 H_2/Ar	99
9. $Gd_{0.59}Sr_{0.18}Al_{1.23}O_x$:$Eu^{2+}_{0.06}$ (G)	1375 H_2/Ar	90

Samples 1–5, 7, and 8 are bulk samples with compositions of the encircled samples in Fig. 2. Quantum efficiencies were measured relative to commercial tricolor phosphors of known efficiencies, after the absorption difference at the maximum excitation wavelength was corrected.

M $Tb(NO_3)_3$. F^- was from the aqueous solution of NH_4F (0.5 M). Solution precursors were mixed according to the cation mole ratio of the preceding formula.

For example, to synthesize $Gd_{0.6}Sr_{0.4}Al_{1.6}O_xF_{1.8}$:$Eu^{2+}_{0.08}$, 6 mL of $Gd(NO_3)_3$, 4 mL of $Sr(NO_3)_2$, 16 mL of $Al(NO_3)_3$, 18 mL $(NH_4)F$, and 4 mL of $Eu(NO_3)_3$ were mixed. After removing the water and drying at 200 °C, the mixture was calcined at 600 °C for 12 hours in air to decompose the nitrate into oxides. The resulting powders were then extensively ground and sintered in an ultrapure (>99.99%) 40% H_2/He mixture (with a total pressure of 2 psi in a vacuum furnace) at 1375°C for 8 hours. The resulting green powder was again extensively ground before the luminescence measurement.

The fluorescent spectra of these fine powder samples were taken. The shape of the emission and excitation spectra in thin-film and bulk powder phosphors were found to be very similar to each other, suggesting that thin-film phosphors were correctly reproduced in the powder forms. With the standard powder phosphors of known QE available, it becomes possible to measure the QE of these phosphors by a direct comparison method, as done by Bril et al. (20).

A thick layer (2 mm) of fine powder was used in measuring the spectral properties with a Hitachi F-4500 spectrophotometer. The UV absorption of incident light was measured from a diffuse reflectance measurement. Emission spectra were recorded at the maximum excitation of each phosphor with a monochromatized UV light, and the QE was calculated at the maximum-excitation edge, which is what is typically reported in QE measurements (21, 22). The wavelength dependence of the UV source and the photomultiplier tube (PMT) response were also calibrated. The integration of the emission counts over a certain wavelength

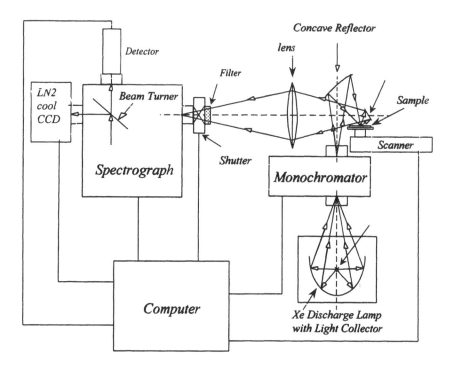

Figure 3 Schematic diagram of a customer-built scanning spectrophotometer for characterizing thin-film phosphor libraries.

range can be compared between the new and standard phosphors in order to calculate QE.

3. Luminescent Properties

A QE of >90 % was obtained for both the red phosphor $GdAl_{1.6}O_x:Eu^{3+}_{0.08}$ and the green phosphor $(Gd_{0.6}Sr_{0.4})Al_{1.6}$ $O_xF_{1.8}:Eu^{2+}_{0.08}$, using the earlier procedure, compared to the commercial phosphors $Y_2O_3:Eu^{3+}$ (red, QE = 97 %) (1, 7) and $LaPO_4:Tb^{3+}$, Ce^{3+} (green, QE = 93%) (1, 7), respectively. It was found that $Gd(La)_mSr_nAlO_x:Eu^{2+}_y$ can be effectively excited using a broad UV-visible band ranging from about 300 nm to 450 nm. The λ^{max}_{ex} of the red phosphors $Gd_{>m}AlO_x:$ Eu^{3+}_y is approximately 275 nm, and the λ^{max}_{ex} of the green phosphor $Gd(La)_mSr_nA-lO_x:Eu^{2+}_y$ is about 300 nm. The Stokes shifts of these new phosphors are smaller than those of the commercial oxide phosphors, such as $Y_2O_3:Eu^{3+}$ (red, λ^{max}_{ex} ~254 nm) and $LaPO_4:Tb^{3+}$, Ce^{3+} (green, λ^{max}_{ex} ~290 nm). In general, bulk samples qualitatively reproduced the luminescent properties of the samples from

Intensity
(a.u.)

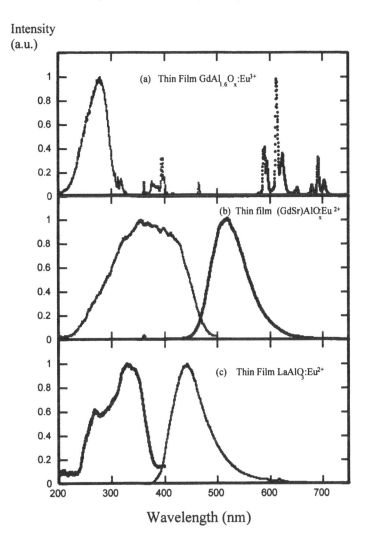

Wavelength (nm)

Figure 4 Photoluminescence emission and excitation spectra of (a) $GdAl_{1.6}O_x:Eu^{3+}_{0.08}$, (b) $Gd_{0.6}Sr_{0.4}Al_{1.6}O_xF_{1.8}:Eu^{2+}_{0.08}$, and (c) $LaAl_{3.1}O_x:Eu^{2+}_{0.08}$.

the libraries, and higher QEs were obtained for samples with stoichiometries corresponding to the thin-film library samples with brighter luminescence. The only exception was the efficient blue phosphors (with QE $\sim$60%) found in thin-film samples of $LaAl_{3.1}O_x:Eu^{2+}_{0.08}$. Bulk samples of the same composition displayed considerably reduced QEs (approximately 20%). This is probably due to the difference in the behavior and response of film and powder samples to processing under strongly reducing atmospheres.

Library studies can also provide considerable information on the relationship between compositions and luminescence properties from the fluorescent imaging and composition map of libraries. In general, for the activator Eu^{3+}, La_mAlO_x hosts displayed reduced luminescence intensities compared to $Gd_m(Sr_n)AlO_x$ hosts; and for Eu^{2+}, $La_mSr_nAlO_x$ hosts displayed reduced luminescence intensities compared to $Gd_mSr_nAlO_x$ hosts. These results are consistent with the fact that Gd^{3+} is more efficient in transferring exciton energy to activators than La^{3+} (1, 23). In each class of phosphors in the series, the luminescence brightness and the emitting color chromaticity can be adjusted by varying the stoichiometry and the doping level of the activators. We found that activators can be accommodated at high concentrations (sometimes up to 16%) in these host series without resulting in significant quenching of luminescence, suggesting that quenching by energy transfer between activators is inefficient in these hosts. Finally, we found that inclusion of fluoride ions in the synthesis of $(Gd_{0.6}Sr_{0.4})Al_{1.6}O_xF_{1.8}:Eu^{2+}_{0.08}$ increases its QE relative to the same composition without fluoride.

Another interesting observation is that high-efficiency tricolor emissions can be achieved with the same activator (Eu), in the same structure (perovskite) and same series of host $Gd(La)_m(Sr_n)AlO_x$, under identical processing conditions (the exact host compositions are $m, n = 0$ for red, $m = 0$ and $n = 1$ for green, and $m = 1$ and $n = 0$ for blue, as seen in Figure 2). Refractive oxides have high thermal and chemical stability due to their high lattice binding energy. Currently, three major refractory oxide host structures, $Y_2O_3:Eu^{3+}$, $Y_3Al_5O_{12}:Tb^{3+}$, and $BaMgAl_{10}O_{17}:Eu^{2+}$, are known for red, green, and blue, with thermal quenching temperatures around 800°C, 300°C, and 300°C, respectively. They are used in plasma displays and projection TVs, where they have to withstand harsh excitation conditions. These phosphors can host only these specific activators for their respective color generation. Tricolor phosphors are presently made in different hosts with different activators and with different processing conditions. Multistep deposition and patterning are needed to fabricate color display panels in such devices as thin-film electroluminescence displays. It is a nontrivial task to make full-color, high quality (epitaxial) multilayer thin-film display panels because of the structural incompatibility of these three host structures.

In order to obtain tricolor emissions in a single host material, ion implantation of luminescence centers had been investigated (24). Unfortunately, this process usually introduces defects and charge imbalance in phosphors, resulting in quenched luminescence. Because of this, delicate annealing processes are required following the ion implantation, leading to a high cost in fabrication of display screens. The present tricolor perovskite phosphors with the host given by $Gd(La)_m(Sr_n)AlO_x$ may provide a solution to simplify the fabrication of tricolor display panels, with high QE, high quenching temperature and stability, structural compatibility, easier synthesis, and identical processing conditions.

4. Processing Conditions Optimized by Combinatorial Approach

As expected, the processing temperature and atmosphere also significantly affect luminescent properties. It is apparent from Figure 2 that identical-composition thin-film phosphors processed at different temperatures have different luminescence properties (compare Figure 2a with Figure 2b, and 2c with 2d). A similar processing-condition dependence of luminescence properties was reproduced in bulk powder phosphors. For example, in a phosphor library study, $(Gd_{0.6}Sr_{0.4})Al_{1.6}O_xF_{1.8}:Eu^{2+}_{0.08}$ displayed a more intense green luminescence (QE $\sim 100\%$) when synthesized at 1375 °C in 10% H_2/He than when it was processed at 1150°C in 10% H_2/He (QE $\sim 42\%$). This is consistent with our x-ray diffraction studies, which indicate that the higher-intensity luminescence is directly correlated with the formation of a single-phase perovskite structure. We also note that Tb^{3+}-activated phosphors with a Ce^{3+} sensitizer in this series of compounds can be synthesized only at a relatively high temperature (1375°C) in a reducing atmosphere. Tb^{3+}-activated phosphors without Ce^{3+} are not efficient under this processing condition, and a relatively low processing temperature (1100°C) produces much brighter phosphors. In the case of a thin film blue phosphor $La_mAlO_x:Eu^{3+}_y$, processing in an H_2 atmosphere at temperatures above 1300°C is required, consistent with more complete reduction of the Eu activator. Even under these strongly reducing conditions, red phosphors were still obtained with strong luminescence and good chromaticity at the $Gd_mAlO_x:Eu^{3+}_y$ thin-film sites containing F^- (Figure 2b, d). Thus, tricolor phosphors can be synthesized with the same activator (Eu) in a perovskite host under reducing conditions (Figure 2b). Again, these results underscore the importance of including processing conditions as a parameter in the combinatorial search.

To study the processing-temperature dependence of luminescence properties in a combinatorial way, a special-purpose furnace with a linear gradient temperature zone was set up. After a lead composition red phosphor $GdAl_{1.6}O_x:Eu^{3+}_{0.08}$ was discovered with the combinatorial method, a second library, containing a column of identical composition $GdAl_{1.6}O_x:Eu^{3+}_{0.08}$, was deposited on a (100) $LaAlO_3$ single-crystal substrate. To find its optimum processing temperature, the "single-composition" phosphor library was processed under a continuously varying temperature along the column. The furnace was equipped with an in situ gradient temperature monitor in air. The gradient temperature was set up along a 5-inch distance between a hot zone (with temperature up to 1500°C) and the room temperature with the aid of a fan. With proper venting and air-flowing control, a stable linear temperature gradient up to 250°C/inch in certain regions of the furnace can be applied on the library at high temperatures. After processing two identical libraries with temperatures ranging from 900°C to 1415°C in air for 4 hours, their emission photoimages were again taken under a broadband Hg UV lamp (Figure 5). It was found that while the minimum necessary processing condition to synthesize thin films of $GdAl_{1.6}O_x:Eu^{3+}_{0.08}$ was around 1100°C, processing at 1400°C resulted in phosphors of slightly higher brightness without apparent sample evaporation. This information was used in the reproduction of powder phosphors.

From the photoemission spectra of phosphors, the Commission International del'Eclairge (CIE) color coordinates of phosphors listed in Table 2 are

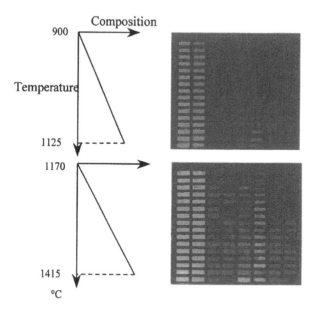

Figure 5   Photoluminescent emission images of libraries containing $Gd(La)_mAlO_x:Eu^{3+}$, (Tb^{3+}), with $GdAl_{1.6}O_x:Eu^{3+}_{0.08}$ and $GdAlO_3:Eu^{3+}_{0.04}$ in the first two columns. A linear temperature gradient from 900°C to 1125°C (a) and from 1170°C to 1415°C (b) was applied along columns of identical samples in a gradient furnace. Each library was annealed in air for 4 hours.

calculated, and they plotted in Figure 6. The phosphor $(Gd_{0.6}Sr_{0.4})Al_{1.6}O_xF_{1.8}$: $Eu_{0.08}^{2+}$ (green) has better chromaticity ($x = 0.27$, $y = 0.58$) than that of the standard Tb_{3+}-activated green phosphor, $LaPO_4:Tb^{3+}$, which is widely used in fluorescent lamps (1). It is also better than $Y_3(Al,Ga)_5O_{12}:Tb^{3+}$ ($x = 0.36$, $y = 0.55$), which is used in projection TVs (5). The new green phosphor, $(Gd_{0.6}Sr_{0.4})Al_{1.6}O_xF_{1.8}:Eu_{0.08}^{2+}$, has a high QE, a high concentration of activators (up to 16 %) without dramatic luminescent quenching, and high quenching temperatures (250–350°C). It is chemically stable and expected to have a short emission transition time ($5d-4f$ parity allowed transition). These properties are vital for its application in plasma displays. The presently used green phosphor, $Zn_2SiO_4:Mn^{2+}$, suffers from slow decay time and a correspondingly slow frame transition speed (25).

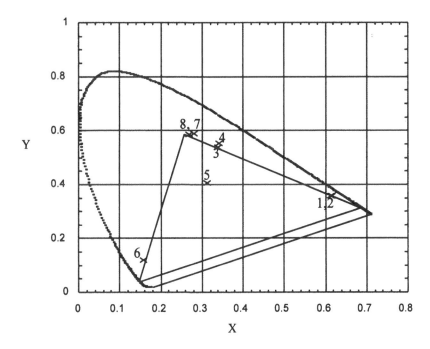

Figure 6 CIE (Commission International del'Eclairge) chromaticity diagram and color coordinates of phosphors from Table 2. The point on the curve periphery indicates monochromatic color. The triangular region covers all colors produced from the three internationally accepted optimum primitive colors for color TV applications. (From Ref. 1.)

5.   Structure and Phase Analysis

An x-ray diffraction (XRD) analysis of selected powder samples (Figure 7) shows that the phosphors in this series of compounds have the perovskite structure, with a slight ($\alpha = 90°4'$) rhombohedral distortion ($LaAlO_3$) or orthorhombic structures ($GdAlO_3$) (26, 27). The powder sample of the green phosphor, $(Gd_{0.6}Sr_{0.4})$ $Al_{1.6}O_xF_{1.8}:Eu_{0.08}^{2+}$, was found to have a minority phase (~5 %) of $SrAl_2O_4$. The synthesis of additional powder samples with stoichiometries near that of the optimal phosphor in the library produced a near-single-phase phosphor, $Gd_{0.77}Sr_{0.23}Al_{1.6}O_x: Eu_{0.08}^{2+}$, with a QE of $\sim90\%$. The Cohen's least square method was used to refine and calculate the crystal lattice parameters from the powder diffraction pattern.

Lattice constants of $GdAl_{1.6}O_x:Eu_{0.08}^{3+}$ were calculated in the following manner. It is known that $GdAl_{1.6}O_x:Eu_{0.08}^{3+}$ has an orthorhombic lattice structure. In Bragg's law, given by

$$Sin^2\theta = \frac{\lambda^2}{4a^2}h^2 + \frac{\lambda^2}{4b^2}k^2 + \frac{\lambda^2}{4c^2}l^2 \tag{1}$$

θ is the calculated diffraction angle, λ is the x-ray wavelength, (a, b, c) is the lattice constant of the orthorhombic lattice, and (h, k, l) is the Miller index of the plane of diffraction.

To obtain (a,b,c) from the experimental observed diffraction patterns, an experimentally observed diffraction angle ω' is introduced. To get the minimum difference (least square of the experimental diffraction) and the refined (calculated) diffraction, a set of (a,b,c) has to be calculated to minimize their difference (Eq. (2)).

$$\Delta = sin^2\theta' - sin^2\theta = sin^2\theta' - (\frac{\lambda^2}{4a^2}h^2 + \frac{\lambda^2}{4b^2}k^2 + \frac{\lambda^2}{4c^2}l^2) \tag{2}$$

From the error in the measurement of the diffraction angle from the instrument (Goniometer), an extra "drift error" term $D * \delta$ is introduced. Let $A = 1/4a^2$, $B = 1/4b^2$, $C = 1/4c^2$, $\alpha = h^2\lambda^2$ $\beta = k^2\lambda^2$, and $\gamma = l^2\lambda^2$. Eq. (2) now reads

$$\Delta = sin^2\theta' - (A\alpha + B\beta + C\gamma + D\delta) \tag{3}$$

By differentiating Δ against A,B,C,D, respectively, and setting them equal to zero, the following set of equations for an orthorhombic lattice is obtained:

$$\Sigma\alpha\,sin^2\theta' = \Sigma A\alpha^2 + \Sigma B\alpha\beta + \Sigma C\alpha\gamma + \Sigma D\alpha\delta$$
$$\Sigma\beta\,sin^2\theta' = \Sigma A\alpha\beta + \Sigma B\beta^2 + \Sigma C\beta\gamma + \Sigma D\beta\delta$$
$$\Sigma\gamma\,sin^2\theta' = \Sigma A\alpha\gamma + \Sigma B\beta\gamma + \Sigma C\gamma^2 + \Sigma D\gamma\delta$$
$$\Sigma\delta\,sin^2\theta' = \Sigma A\alpha\delta + \Sigma B\beta\delta + \Sigma C\gamma\delta + \Sigma D\delta^2 \tag{4}$$

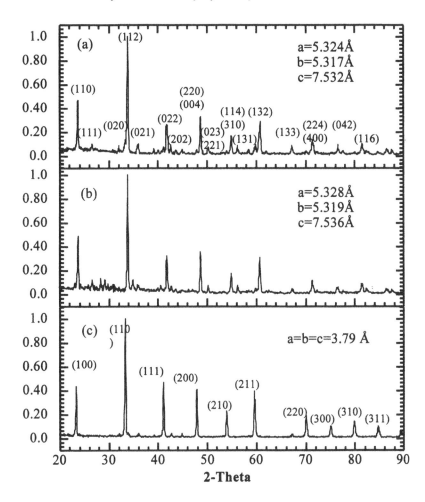

Figure 7   X-ray diffraction spectra (θ–2θ scan) of thin films of (a) $GdAl_{1.6}O_x:Eu^{3+}_{0.08}$; (b) $(Gd_{0.6}Sr_{0.4})Al_{1.6}\ O_xF_{1.8}:Eu^{2+}_{0.08}$, and (c) $LaAl_{3.1}O_x:Eu^{2+}_{0.08}$ taken with a Siemens D5000 x-ray diffractometer operated with the Cu Kα line (1.54 Å). Vertical axes indicate intensity (arbitrary units). Peaks are indexed for the perovskite structure, and their positions agree with the literature values. (From Refs. 26 and 27.)

The summation is over all the diffraction peak positions. To solve for the crystal lattice parameter of $GdAl_{1.6}O_x:Eu_{0.08}^{3+}$, its diffraction peaks were first indexed according to the ideal composition perovskites of $GdAlO_3$ (27) to obtain (h,k,l) for each peak. Ω' is obtained from the diffraction position of each peak (Figure 7a). After solving this four-equation set, the lattice constants of $GdAl_{1.6}O_x:Eu_{0.08}^{3+}$ were determined to be ($a = 5.324$ Å, $b = 5.317$ Å, $c = 7.532$ Å). Note that these values are larger than those of the stoichiometric $GdAlO_3$ (27) ($a = 5.30$, $b = 5.25$, $c = 7.44$). The lattice parameters of other phosphors in this series of compounds can also be calculated this way.

B. Other Thin-Film Phosphor Libraries Made with the Binary Masking Method

1. Primary Screening of Phosphors with Discrete-Composition Libraries

There are some arguments about the role of empirical knowledge in guiding combinatorial research. In our opinion, an educated design of a library may generate more "hits" in a certain number of experiments, but too much influence of educated design may limit the possibility of "surprises." The high throughput of phosphor synthesis and screening can be used to investigate a large number of previously unexplored compositions. Besides perovskites, we studied various other oxide hosts with RE activators. In one investigation, we concentrated on the compositions of the form $A_mB_nC_kO_x:R_y$, where A was Gd, La, or Y, B was Ta, Zr, W, Mo, or Al, and C was Zn, Mg, or Sr. R is a RE element (Tm, Eu, Tb, or Ce). m, n, and k ranged from 0 to 1, y ranged from 0.005 to 0.1, and x depended on the stoichiometries of specific compositions to maintain the charge neutrality. The sputtering targets (>99.9% purity) used here included GdF_3, La_2O_3, Y_2O_3, Ta_2O_5, Zr, WO_3, Mo, ZnO, Al_2O_3, MgO, $SrCO_3$, TmF_3, EuF_3, TbF_3, and CeF_3. Libraries with 128 different compositions or stoichiometries were deposited on 1-inch × 1-inch polished (100) $LaAlO_3$ single crystal substrates. The library design was based on the binary scheme (Figure 1a). Three identical libraries were made simultaneously and processed at different partial pressures of O_2, H_2, He, and Ar in temperatures ranging from 1100°C to 1400°C for up to 4 hours. Photoluminescence images of these libraries under broad wavelength UV irradiation (Hg lamp) were obtained by color photography (Figure 8).

2. Multiple-Color Tm-Activated Phosphor Compositions

As is seen in Figure 8, some Tm-activated sites of nominal compositions—$La_{1.2}AlTa_{0.5}O_x:Tm_y$ ($0.005 < y < 0.075$ $4.5 < x < 4.7$) (Fig. 8b), $LaZrO_x:Tm_y$($0.005 < y < 0.055$ $3 < x < 3.2$) (Figure 8a and 8c), and $LaTaAl_{1.3}MgO_x$:

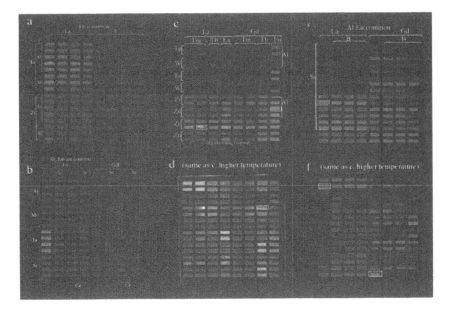

Figure 8 Photoluminescence images of a series of phosphor libraries processed under different conditions: (a) La(or Y)$_m$ (Al$_n$)Ta(or Zr)O$_x$: Tm$_y^{3+}$, where $0.375 \leq m \leq 0.75$, $0 \leq n \leq 1$, $0.005 \leq y \leq 0.075$, annealed at 1375°C in 40% H$_2$/60% He mixture for 4 hours; (b) La(or GdF$_3$)(Sr or Ce)$_n$Ta(Al, or Mo) O$_x$:Tm$_y^{3+}$, where $0.375 \leq m \leq 0.75$, $0 \leq n \leq 0.3$, $0.005 \leq y \leq 0.075$, annealed at 1375°C in 40% H$_2$/60% He mixture for 4 hours; (c) La(or GdF$_3$)$_m$(Al)$_n$(Mg)$_l$Ta(W, Zr, Zn)O$_x$: Tm$_y^{3+}$ (Tb$_z^{3+}$ or Eu$_h^{2+/3+}$), where $m = 1$, $0 \leq n \leq 1$, $0 \leq l \leq 1$, $0.005 \leq y \leq 0.055$, $z = 0.04$, $h = 0.04$, annealed in 10% H$_2$/90% Ar mixture at 1150°C for 4 hours; (d) same as (c) but annealed at 1375°C in 40% H$_2$/60% He mixture for 4 hours; (e) La(or GdF$_3$)$_l$Sr$_m$B$_n$AlO$_x$:Eu$_y$, where $0 \leq m \leq 0.4$, $0 \leq n \leq 0.1$, $l = 0.5, 1, 1.5, 2$, $0.01 \leq y \leq 0.2$. Library was annealed in 10% H$_2$/90% Ar mixture at 1150°C for 4 hours; (f) same as (e) but annealed at 1375°C in 40% H$_2$/60% He mixture for 4 hours. The numbered lead samples have the following nominal compositions: #1: La$_{1.2}$Al Ta$_{0.5}$O$_x$:Tm$_{0.055}$; #2: La$_{1.2}$Zr$_{0.8}$O$_x$:Tm$_{0.055}$; #3: LaTaAl$_{1.3}$MgO$_x$: Tm$_{0.005}$; #4: GdZnO$_x$:Eu$_{0.04}$; #5: GdTaO$_4$:Tb$_{0.04}$; #6: LaAl$_{1.3}$ Si$_{0.5}$Ox:Eu$_{0.04}$. All substrates are 1 inch $\times$ 1 inch, and each sample site is 1 mm $\times$ 2 mm; the thin films range from 0.4 μm to 1 μm in thickness. All photos were taken under the same exposure condition.

$Tm_{0.005}$ (Figure 8d)—were found to emit green, blue, and orange-red lights, respectively, under 254-nm centered broad UV spectrum excitation with a Hg lamp, after processing at 1375°C for 4 hours in ultrapure (99.99%) 40% H_2/60% He mixture (at 16.7 psi). Tm-activated phosphors are known to have complicated energy-level schemes due to the strong deviation from the Russell–Saunders coupling in the $(4f)^{12}$ configuration (28). As a consequence, the relaxation of excited states of the Tm^{3+} ions may take place via a large number of relaxation paths, giving rise to UV, visible, and IR emissions (29). Indeed, we found the Tm^{3+}-activated sites often had multiple wavelength emissions with medium to low brightness measured with the thin-film spectrophotometer.

These libraries also provide information on the influence of composition and processing conditions on the brightness and chromaticities of phosphors. For example, addition of Sr or Ce was found to be detrimental to Tm activation in $La_mTa_nAlO_x$ compounds, while Mg addition shifts emission to a long wavelength. Silicon doping (~5%) assists blue emission in the thin film La_mAlO_x:Eu^{2+}, but doping of B (up to 1%) in this compound has a very little effect. We also found that La_mZrO_x is a better host to Tm for blue luminescence than Gd_mZrO_x, and this bright green–emitting compound can only be synthesized in H_2 atmosphere at above 1350°C. These observations help us to better understand the complex interactions of hosts and activators in order to obtain improved luminescence materials.

3. $GdZn_mO_x$:Eu^{3+} Phosphor

Some bright red and green thin film phosphors activated by Eu^{3+} and Tb^{3+}, respectively, were also identified. In the library, the samples with the brightest emission of red and green had the nominal compositions of $GdZnO_x$:$Eu^{3+}_{0.04}$ and $GdTaO_4$: $Tb^{3+}_{0.04}$, respectively (shown in Figure 8c and 8d). $GdTaO_4$:Tb^{3+} is a well-known x-ray phosphor (30), but to our knowledge there had been no study of luminescence properties on compositions of $GdZn_mO_x$:Eu^{3+}. To systematically study this system, a series of bulk powder samples of nominal composition $GdZn_mO_x$:$Eu^{3+}_{0.04}$, with m ranging from 0 to 1, were synthesized using the conventional solid-state-synthesis method. Solution precursors (>99.99% purity), including aqueous 0.5M $Gd(NO_3)_3$, 0.5M $Zn(NO_3)_2$, and 0.1M $Eu(NO_3)_3$, were mixed with the mole ratio as given by the formula. The mixture was dried, calcined in air at 900°C for 12 hours to decompose nitrates to oxides, crushed, finely ground, and then sintered in air at 1400°C for 10 hours. The powders were again finely ground before the structural and optical spectral measurements.

Powder XRD of nominal composition $GdZn_mO_x$:$Eu^{3+}_{0.04}$ was measured with a Simens D500 x-ray diffractometer with the CuK_α line (1.54 Å). For $m <$ 0.3, the XRD patterns (Figure 9) show that the samples have the single phase (monoclinic) crystal structure of Gd_2O_3 (31, 32). The variation in relative peak

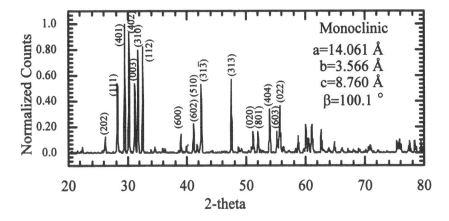

Figure 9 Powder x-ray diffraction pattern of nominal composition GdZn$_m$O$_x$: Eu$^{3+}_{0.04}$. The peaks are indexed to the monoclinic crystal structure of Gd$_2$O$_3$.

intensities compared to that of pure Gd$_2$O$_3$ indicates the incorporation of Zn^{2+} in the Gd$_2$O$_3$ lattice. For $m > 0.3$, a ZnO phase appears as indicated in the XRD patterns. This result indicates that the maximum solubility of Zn^{2+} in Gd$_2$O$_3$ is around $m = 0.3$ with the appropriate formula (Gd$_{1.54}$Zn$_{0.46}$)O$_{3-\delta}$:Eu^{3+}.

The photoluminescent excitation and emission spectra (Figure 10) of (Gd$_{2-x}$Zn$_x$)O$_{3-\delta}$: Eu$^{3+}_y$ ($0.18 < x < 1$, $0.02 < y < 0.073$) were taken with an F-4500 Hitachi spectrophotometer in the diffuse-reflection mode. The excitation spectra (Figure 10a) of (Gd$_{2-x}$Zn$_x$)O$_{3-\delta}$:Eu$^{3+}_y$ ($0.18 < x < 1$, $0.02 < y < 0.073$) ($\lambda^{max}_{ex} \approx 270$ nm) are slightly different from that of Y$_2$O$_3$:Eu^{3+} ($\lambda^{max}_{ex} \approx 254$ nm). This shift can be explained by the higher covalency of Gd$_2$O$_3$ compared to Y$_2$O$_3$ [nephelauxetic effect (1)]. The main emission peak of (Gd$_{2-x}$Zn$_x$)O$_{3-\delta}$: Eu$^{3+}_y$ (for $x > 0.18$) is located at 621 nm (Figure 10b), with the CIE chromaticity ($x = 0.656$, $y = 0.344$) better than that of Y$_2$O$_3$:Eu^{3+} ($x = 0.642$, $y = 0.358$) in color saturation. This better color saturation is apparent to the naked eye. QE (defined at corresponding λ^{max}_{ex}) and CIE coordinates were calculated by integrating emission counts from a photomultiplier tube (PMT) detector. The QE was measured relative to that of the commercial phosphor Y$_2$O$_3$:Eu^{3+} (with QE of around 97%) under identical measurement conditions and the range of integration (580–720 nm). The maximum QE ($\approx$86%) was obtained for compound (Gd$_{1.54}$Zn$_{0.46}$)O$_{3-\delta}$:Eu$^{3+}_{0.06}$ in the current synthesis condition.

The origin of the Eu^{3+} ($^5D_0$ to $^7F_2$) emission spectra shift from 610 nm to 621 nm in Y$_2$O$_3$:Eu^{3+} and (Gd$_{1.54}$Zn$_{0.46}$)O$_{3-\delta}$:Eu$^{3+}_{0.04}$, is intriguing. We found that both the undoped monoclinic Gd$_2$O$_3$:Eu^{3+} (prepared at 1400°C) and the

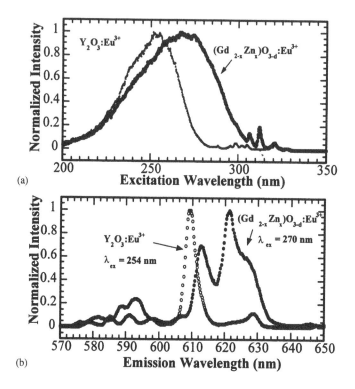

(a)

(b)

Figure 10 Photoluminescence excitation (a) and emission (b) spectra of powder samples of $(Gd_{1.54}Zn_{0.46})O_{3-\delta}$: $Eu^{3+}_{0.06}$ and Y_2O_3:Eu^{3+}. Quantum efficiencies (QEs) at corresponding λ^{max}_{ex} were measured relative to that of the commercial phosphor Y_2O_3:Eu^{3+} (QE = 97%). Thick ($\sim$2 mm) compacted powder samples were used in the measurements. The accuracy in the measurements of QEs is $\pm$5%. A set of standard phosphors was measured with this method, and their QEs thus determined agreed with the literature values.

cubic $(Gd_{1.54}Zn_{0.46})O_{3-\delta}$:$Eu^{3+}_{0.04}$(prepared at 950°C) have main emission peaks at 610 nm. We believe that the incorporation of Zn^{2+} effectively introduces oxygen vacancies. To confirm this, we prepared a monoclinic Gd_2O_3:Eu^{3+} sample under a strong reducing environment (1300°C in ultrapure 40% H_2/He mixture for 1 hour). The main emission peak of this sample is located at 621 nm, with a slightly lowered QE ($\sim$78%). We conclude that the coexistence of monoclinic structure and oxygen vacancies is the necessary condition for this shift of emission peak.

Pure monoclinic Gd_2O_3 was determined to have a symmetry group C_{2h} with a center of inversion symmetry in the unit cell (31, 32). Any loss of oxygen anion in this host structure will break the inversion symmetry and reduce the symmetry from C_{2h} to C_s. The local crystal field of this C_s symmetry should split the degenerate Stark levels of $^7F_2$ and result in the observed multiplets ($^5D_0$ to $^7F_2$) and $^5D_0$ to $^7F_0$ transitions at 580 nm in the emission spectrum of monoclinic $(Gd_{1.54}Zn_{0.46})O_{3-\delta}:Eu^{3+}_{0.04}$ (28, 33).

The finding of significant improvements in color saturation and QE resulting from Zn^{2+} addition is crucial for applications, since the reducing process of phosphors is often incompatible with device fabrication processes. The processing of this phosphor is much simpler compared to that of $YVO_4: Eu^{3+}$ (with a main emission at 621 nm), which requires an additional process to reduce the impurity phase of V_2O_5, which significantly decreases its QE (7).

4. Cathodoluminescence (CL) Properties

The CL properties of $(Gd_{1.54}Zn_{0.46})O_{3-\delta}:Eu^{3+}_{0.04}$ and a few other phosphors discovered with the combinatorial approach were measured in ultrahigh vacuum environment (up to 10^{-10} torr) (34). Samples were deposited on a glass substrate coated with indium tin oxide (ITO) using a settling/evaporation technique with isopropyl alcohol as a solvent with subsequent baking at 200°C. A low-voltage electron gun was used for testing of phosphors at energies between 200 eV and 1500 eV. It was found that $(Gd_{1.54}Zn_{0.46})O_{3-\delta}:Eu^{3+}_{0.04}$ has a brightness of 2510 cd/m² at only 7.5 keV [projection TVs usually operate with brightness of 1000 cd/m² (1)]. Its single emission peak at 626 nm ensures a better CIE color chromaticity than that of the industrial phosphor $Y_2O_3:Eu^{3+}$ (emission at 610 nm). This makes $(Gd_{1.54}Zn_{0.46})O_{3-\delta}:Eu^{3+}_{0.04}$ a potential candidate for phosphors in projection TVs and low-voltage field-emission displays.

C. Thin-Film Phosphor Libraries Made with Quaternary Masking Method

The thin-film libraries discussed thus far were made with the binary masking method. We have also fabricated a high-density phosphors library (1024 phosphors per in.²) on a Si (100) substrate using the quaternary lithographic masking process. The five quaternary masks A, B, C, D, and E (illustrated in Figure 1b) were used to generate the library, where the sequence of masking and the precursor depositions was as follows: A_1: Ga_2O_3 (355 nm); A_2: Ga_2O_3 (426 nm); A_3: SiO_2 (200 nm); A_4: SiO_2 (400 nm); B_1: Gd_2O_3 (577 nm); B_2: ZnO (105 nm); B_3: ZnO(210 nm); C_1: Gd_2O_3 (359 nm); C_2: Y_2O_3 (330 nm); and C_3: Y_2O_3 (82.5 nm); D_1: CeO_2 (3.5 nm); D_2: EuF_3 (11.3 nm); D_3: Tb_4O_7 (9.2 nm); E_1: Ag (3.8 nm); E_2: TiO_2 (6.9 nm); E_3: Mn_3O_4(5.8 nm). Here, the numbers in parentheses are the thickness of the deposited films (the notation X_i represents a deposition step with mask X rotated $(i-1) \times 90°$ counterclockwise relative to the position depicted in Figure 1b). The steps corresponding to B_4, C_4, D_4, and E_4 were omitted to include compounds consisting of less than five elements in the libraries in order to study the effects of dopants. A photograph of the as-deposited library under daylight is shown in Figure 11(a).

After extensive low temperature annealing, the library was heated at 1000°C for 2 hours in flowing Ar gas, with the ramping and lowering rates of 1°C/min. The luminescence photograph of the processed library under UV illumination with a 254-nm broadband filter is shown in Figure 11(b). For this library, in order to perform accurate quantitative analysis of the libraries, a scanning optical spectrophotometer was used to measure the excitation and emission spectra of individual luminescent samples in the library. The relative photon output of a blue-emitting site with the nominal composition $Gd_{5.2}Ga_{3.33}O_z$ was calculated to be ~75% of that of a $Zn_2SiO_4{:}Mn_{0.05}$ site included in the library (which has a QE of 70%) (7). Because of complex optical effects in thin films, QEs of thin film luminescent materials are difficult to determine in general. (In fact, the photon output is the quantity that is most relevant to applications.) If one assumes the same absorption and thin-film optical effect for the reference sample $Zn_2SiO_4{:}Mn_{0.05}$ and $Gd_{5.2}Ga_{3.33}O_z$, the QE of this sample is estimated as ~53%.

Surprisingly, it was discovered that the thin-film composition of $Gd_{5.2}Ga_{3.33}O_z$ was luminescent only on silicon substrate. This led to the suspicion

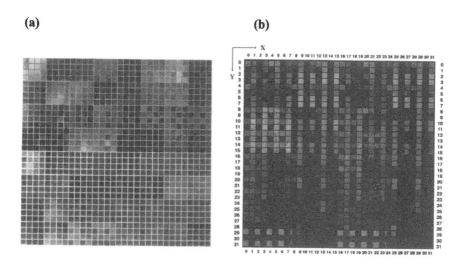

(a) **(b)**

Figure 11 (a) Photograph of an as-deposited quaternary library under ambient light. The diversity of colors in different sites stems from variations in film thickness and optical indices of refraction. (b) Luminescent image of the same library after processing under UV excitation with a 254-nm broadband filter.

that diffusion of Si into the sample was necessary for the blue photoluminescence. To identify the optimal phosphor composition, we fabricated a two-dimensional gradient library of composition $Gd_xGa_{1-x}O_z(SiO_2)_y$, where the amount ratio of Gd and G_a (x) varied from 0 to 1 in the x-direction and the amount of Si varied from 0.003 to 0.1 in the y-direction. After synthesis, $Gd_3Ga_5O_{12}:(SiO_2)_{0.08}$ was identified as the composition with the highest luminescent intensity. Although XRD of the sample showed no trace of silicate, x-ray photoemission spectroscopy (XPS) and energy-dispersive x-ray analysis showed the presence of Si in the sample. We concluded that the active phosphor composition is a composite oxide with the composition of $Gd_3Ga_5O_{12}:(SiO_2)_{0.08}$.

IV. POWDER PHOSPHOR LIBRARIES MADE WITH SOLUTION PRECURSORS AND LIQUID HANDLERS

A. Application of an Inkjet-Solution Dispenser to Make Powder Libraries for Investigating $(La_mGd_{1-m})Al_nO_x$: Eu_y^{3+}

Libraries of phosphors can be generated in thin film forms and screened with scanning or parallel detectors (15, 16, 35, 36). However, for many inorganic and organic materials, such as polymers, zeolites, and homogeneous catalysts, syntheses are best accomplished by solution-phase methods. The solution-based methods have the advantage that they allow mixing of precursors at the molecular level, thereby reducing the need for thermally induced interdiffusion of precursors required in the afore-mentioned solid-state synthesis processes (37). We developed a scanning multiple-inkjet delivery system capable of accurately and rapidly delivering nanoliter volumes of reagents. This was used to fabricate a library of phosphors based on rare earth activated refractory metal oxides.

The inkjet technology is compatible with the microscale and high speeds required to generate combinatorial libraries, and it has been used to microdispense both biochemical and chemical reagents for a variety of applications (38–41). This raises the possibility that inkjets can also be used to reproducibly deposit different compositions of solid state precursors that can be subsequently processed into high-density materials libraries. To test this idea, we fabricated a delivery system consisting of an array of single-nozzle, "drop on demand" (i.e., the ejection of droplets is on a discrete, one-drop-per-pulse basis) inkjets that use PZT piezoelectric discs as the electrical-pulse-to-pressure transducing element (Figure 12) (42–47). A reservoir filled with chemical solution is connected via Teflon tubing to the pressure chamber of the inkjet device, which was designed to be cone shaped in order to amplify the impulse pressure from the piezo disc. When an electrical pulse of fixed signal size and duration is applied to a particular piezo disc, its diaphragm bends inward and relaxes to actuate a small droplet ($\sim$50 μm in diameter) of solution through a nozzle. This discrete drop formation process and the droplet size can be monitored under a microscope with a synchronized flash lamp.

The volume of the droplet is determined by the nozzle size, the piezoelectric diaphragm dimension, the surface tension of the solution, and the waveform of

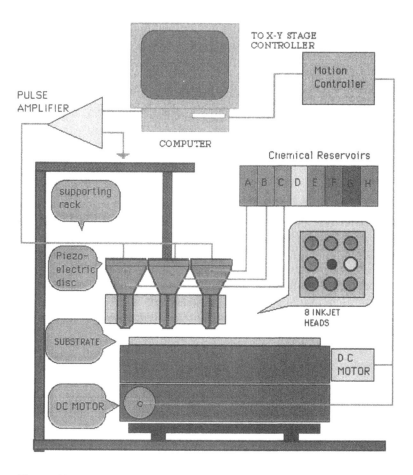

Figure 12 Schematic of scanning multihead injet delivery system. The system is in an integration of ''drop on demand,'' single-nozzle piezoelectric inkjets; operating voltage was between 20 and 150 V; pulse widths were from 20 to 200 μs; operating frequency was 100–2000 Hz. The nozzle is single-crystal sapphire with a nozzle inner diameter of 50 μm (Swiss Jewel Co., Philadelphia, PA, #0051M47). The piezo disc/diaphragm is from American Piezo Inc., Mackeyville, PA, #P120685A with #42 alloy.

the applied electrical pulse. With a fixed chemical, it is proven, both with visual observation through an optical microscope and the weight measurement with a microbalance after collection of a certain number droplets of (1000–10000), that the variation in the droplet volume from drop to drop is less than 2% over hours of continuous operation. The volume of each droplet is measured to be on the order of 1 nanoliter. Droplets are delivered sequentially to individual reaction wells (each 1-mm diameter $\times$ 1-mm depth) machined into a ceramic or a single-crystal substrate (Al_2O_3, BN, Si, Macor, or MgO) mounted on a motorized x-y stage. By coordinating the number of electrical pulses applied to each inkjet device and the stage movement by a computer, the stoichiometry of each site in the library can be controlled systematically.

The scanning inkjet dispensing system was used to optimize the elemental composition in a discovered phosphor series of the composition (La_mGd_{1-m}) Al_nO_x:Eu_y^{3+} discovered from a thin-film library. (17) Four individual inkjets were used, with each connected to one of the four liquid elemental precursors. Precursor solutions (>99.99% purity) included aqueous 0.5 M $La(NO_3)_3$, 0.5 M $Gd(NO_3)_3$, 0.5 M $Al(NO_3)_3$ and 0.1 M $Eu(NO_3)_3$ containing 15–30% ethylene glycol as a humectant.

After evaporation of solvent, the library was processed at 900°C in air for 1 hour (with less than 2°C variation across the library). Processing the library at higher temperatures or for longer times did not significantly affect the relative luminescence of individual samples. This is most likely due to the small reaction size and to the similarity in response of the samples in the library to processing conditions. To show that the compositions of the phosphors in the library corresponded to the intended stoichiometries after processing, they were subjected to atomic absorption analysis. The stoichiometry of the phosphor corresponding to the intended composition, $Gd_{0.8}Al_{1.2}O_x$: $Eu_{0.06}^{3+}$, was determined to be Gd (0.8 $\pm$ 0.03), La (<0.01), Al (1.23 $\pm$ 0.03), and Eu (0.06 $\pm$ 0.01). The luminescence image of the library was taken to identify those samples with the brightest emission (Figure 13). It is apparent that $Gd_{0.8}Al_{1.2}O_x$: $Eu_{0.06}^{3+}$ (encircled in Figure 13) yields the brightest red emission under 254-nm UV excitation. The relative QE of a bulk sample of the same composition processed under the same condition was approximately 98% in comparison to the commercial red phosphor Y_2O_3: Eu^{3+} with a known QE of 97%.

B. Optimization of the Powder Phosphor $(Y_xCe_{0.06}A_{1-x})_3(Al_yGa_{1-y})_5O_{12}$ with an Automated Positive-Displacement Solution Dispenser

Besides inkjet technology, commercially available liquid dispensing systems can also be applied in making powder phosphor libraries. This section will discuss the application of a positive-displacement liquid dispenser in making phosphor

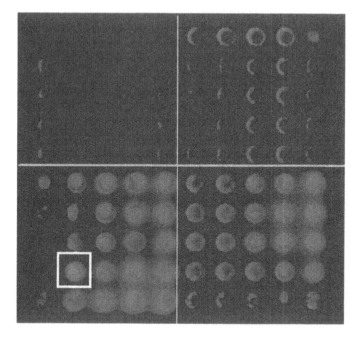

Figure 13   Photoluminescence image of a powder phosphor $(La_mGd_{1-m})Al_nO_x:Eu_y^{3+}$ library excited with a Hg UV lamp with a 254-nm broadband filter.

libraries. Figure 14 is a schematic of such a system. There are eight liquid channels and dispensers in a linear array. Neighboring dispensers are 3 mm apart, and the volume of liquid is individually controlled by a stepping motor, based on a positive-displacement mechanism for liquid deposition. Up to eight different metal precursors can be simultaneously delivered in the volume range from 10 nanoliters to 250 microliters. The liquid deposition is in a non-contact mode, which minimizes the possible cross contamination of solutions. The coefficient of variation in volume from each of the eight dispensers was measured to be less than 2%.

A sintered alumina (99.5% purity) plate with 128 (8 × 16 array) machined cells (2 mm in diameter and ~6 mm in depth) was used to hold liquid precursors for powder phosphors. The neighboring cells are 3 mm in center to center separation and registered in position to the array of liquid dispenser. By controlling the scanning motion of an X-Y-Z table (upon which a substrate is mounted) with the spatial resolution of 0.1 mm and the amount of liquid volumes delivered from dispensers, a library with 128 different solution mixing precursors of powder phosphors can be generated in a matter of minutes.

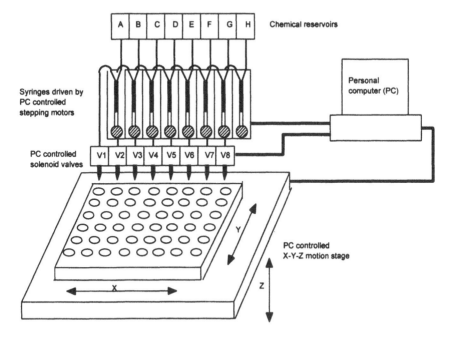

Figure 14 Schematic of an automated liquid-dispensing system for making powder phosphor libraries.

Using the automated liquid dispenser, we optimized the composition of the $Y_3Al_5O_{12}$(YAG):Ce^{3+}-based phosphors series. YAG is a phosphor host well-known for its refractory nature (7), and it has been used in projection TVs and solid-state lamps based on blue GaN LEDs (48). YAG:Ce can be effectively excited by blue photons and emit in the broad yellow band. To vary the range of emitting wavelength of this phosphor for different luminescent colors, a few substituting elements of varying concentrations are introduced into the host. The host, $Y_3Al_5O_{12}$, has a garnet structure, where the compositional Al^{3+} may be substituted with Ga and Y^{3+} can be substituted with rare earth cations such as Gd^{3+}, and Lu^{3+}. Such cross substitutions are expected to influence both the intensity and the emission color in this series of phosphors. The liquid-dispensing system was used to study the effect of cross-substitutions on the luminescent properties in this series of phosphors: $(Y_xA_{1-x})_3(Al_yGa_{1-y})_5O_{12}$:$Ce^{3+}$ (A: Gd, Lu). The compositional map of the library is shown in Table 3.

High-purity ($\geq99.9\%$) clear aqueous nitrate solutions were used in making the powder library. They were $Y(NO_3)_3$ (1 M), $Gd(NO_3)_3$ (1 M), $Lu(NO_3)_3$ (1 M), $Al(NO_3)_3$ (1 M), $Ga(NO_3)_3$ (1 M), and $Ce(NO_3)_3$ (0.5 M). Following a scheme

Table 3 Compositional Map of the Powder Phosphor Library $(Y_xCe_{0.06}A_{1-x})_3(Al_yGa_{1-y})_5O_{12}$

Lu2.625Y0.375Al5Ce0.06	Lu2.25Y0.75Al5Ce0.06	Lu1.875Y1.125Al5Ce0.06
Lu2.625Y0.375Al4.375Ga0.625Ce0.06	Lu2.25Y0.75Al4.375Ga0.625Ce0.06	Lu1.875Y1.125Al4.375Ga0.625Ce0.06
Lu2.625Y0.375Al3.75Ga1.25Ce0.06	Lu2.25Y0.75Al3.75Ga1.25Ce0.06	Lu1.875Y1.125Al3.75Ga1.25Ce0.06
Lu2.625Y0.375Al3.125Ga1.875Ce0.06	Lu2.25Y0.75Al3.125Ga1.875Ce0.06	Lu1.875Y1.125Al3.125Ga1.875Ce0.06
Lu2.625Y0.375Al2.5Ga2.5Ce0.06	Lu2.25Y0.75Al2.5Ga2.5Ce0.06	Lu1.875Y1.125Al2.5Ga2.5Ce0.06
Lu2.625Y0.375Al1.875Ga3.125Ce0.06	Lu2.25Y0.75Al1.875Ga3.125Ce0.06	Lu1.875Y1.125Al1.875Ga3.125Ce0.06
Lu2.625Y0.375Al1.25Ga3.75Ce0.06	Lu2.25Y0.75Al1.25Ga3.75Ce0.06	Lu1.875Y1.125Al1.25Ga3.75Ce0.06
Lu2.625Y0.375Al0.625Ga4.375Ce0.06	Lu2.25Y0.75Al0.625Ga4.375Ce0.06	Lu1.875Y1.125Al0.625Ga4.375Ce0.06
Gd2.625Y0.375Al5Ce0.06	Gd2.25Y0.75Al5Ce0.06	Gd1.875Y1.125Al5Ce0.06
Gd2.625Y0.375Al4.375Ga0.625Ce0.06	Gd2.25Y0.75Al4.375Ga0.625Ce0.06	Gd1.875Y1.125Al4.375Ga0.625Ce0.06
Gd2.625Y0.375Al3.75Ga1.25Ce0.06	Gd2.25Y0.75Al3.75Ga1.25Ce0.06	Gd1.875Y1.125Al3.75Ga1.25Ce0.06
Gd2.625Y0.375Al3.125Ga1.875Ce0.06	Gd2.25Y0.75Al3.125Ga1.875Ce0.06	Gd1.875Y1.125Al3.125Ga1.875Ce0.06
Gd2.625Y0.375Al2.5Ga2.5Ce0.06	Gd2.25Y0.75Al2.5Ga2.5Ce0.06	Gd1.875Y1.125Al2.5Ga2.5Ce0.06
Gd2.625Y0.375Al1.875Ga3.125Ce0.06	Gd2.25Y0.75Al1.875Ga3.125Ce0.06	Gd1.875Y1.125Al1.875Ga3.125Ce0.06
Gd2.625Y0.375Al1.25Ga3.75Ce0.06	Gd2.25Y0.75Al1.25Ga3.75Ce0.06	Gd1.875Y1.125Al1.25Ga3.75Ce0.06
Gd2.625Y0.375Al0.625Ga4.375Ce0.06	Gd2.25Y0.75Al0.625Ga4.375Ce0.06	Gd1.875Y1.125Al0.625Ga4.375Ce0.06

Table 3 Continued

Lu1.5Y1.5Al5Ce0.06	Lu1.125Y1.875Al5Ce0.06	Lu0.75Y2.25Al5Ce0.06
Lu1.5Y1.5Al4.375Ga0.625Ce0.06	Lu1.125Y1.875Al4.375Ga0.625Ce0.06	Lu0.75Y2.25Al4.375Ga0.625Ce0.06
Lu1.5Y1.5Al3.75Ga1.25Ce0.06	Lu1.125Y1.875Al3.75Ga1.25Ce0.06	Lu0.75Y2.25Al3.75Ga1.25Ce0.06
Lu1.5Y1.5Al3.125Ga1.875Ce0.06	Lu1.125Y1.875Al3.125Ga1.875Ce0.06	Lu0.75Y2.25Al3.125Ga1.875Ce0.06
Lu1.5Y1.5Al2.5Ga2.5Ce0.06	Lu1.125Y1.875Al2.5Ga2.5Ce0.06	Lu0.75Y2.25Al2.5Ga2.5Ce0.06
Lu1.5Y1.5Al1.875Ga3.125Ce0.06	Lu1.125Y1.875Al1.875Ga3.125Ce0.06	Lu0.75Y2.25Al1.875Ga3.125Ce0.06
Lu1.5Y1.5Al1.25Ga3.75Ce0.06	Lu1.125Y1.875Al1.25Ga3.75Ce0.06	Lu0.75Y2.25Al1.25Ga3.75Ce0.06
Lu1.5Y1.5Al 0.625Ga4.375Ce0.06	Lu1.125Y1.875Al0.625Ga4.375Ce0.06	Lu0.75Y2.25Al0.625Ga4.375Ce0.06
Gd1.5Y1.5Al 5Ce0.06	Gd1.125Y1.875Al5Ce0.06	Gd0.75Y2.25Al5Ce0.06
Gd1.5Y1.5Al4.375Ga0.625Ce0.06	Gd1.125Y1.875Al4.375Ga0.625Ce0.06	Gd0.75Y2.25Al4.375Ga0.625Ce0.06
Gd1.5Y1.5Al3.75Ga1.25Ce0.06	Gd1.125Y1.875Al3.75Ga1.25Ce0.06	Gd0.75Y2.25Al3.75Ga1.25Ce0.06
Gd1.5Y1.5Al3.125Ga1.875Ce0.06	Gd1.125Y1.875Al3.125Ga1.875Ce0.06	Gd0.75Y2.25Al3.125Ga1.875Ce0.06
Gd1.5Y1.5Al2.5Ga2.5Ce0.06	Gd1.125Y1.875Al2.5Ga2.5Ce0.06	Gd0.75Y2.25Al2.5Ga2.5Ce0.06
Gd1.5Y1.5Al1.875Ga3.125Ce0.06	Gd1.125Y1.875Al1.875Ga3.125Ce0.06	Gd0.75Y2.25Al1.875Ga3.125Ce0.06
Gd1.5Y1.5Al1.25Ga3.75Ce0.06	Gd1.125Y1.875Al1.25Ga3.75Ce0.06	Gd0.75Y2.25Al1.25Ga3.75Ce0.06
Gd1.5Y1.5Al0.625Ga4.375Ce0.06	Gd1.125Y1.875Al0.625Ga4.375Ce0.06	Gd0.75Y2.25Al0.625Ga4.375Ce0.06

of library design, solution precursors of the phosphor library were generated on a 1-in. × 2-inch alumina plate (with 128 sample cells). The total amount of sample in each cell is approximately 10 —mole. The general formula of the library was $(Y_{x}A_{1-x})_3(Al_yGa_{1-y})_5O_{12}:Ce_{0.06}^{3+}$(A: Gd, Lu), where $3 \geq x \geq 0.375$; $5 \geq y \geq 0.625$. The as-deposited library precursor was placed on a orbital shaker for mixing of solution precursors while placed under an infrared lamp for evaporating the solvent, with the sample surface temperature of approximately 80°C. After samples on the plate were dried, the library was placed in a reducing furnace

Table 3 Continued

Lu0.375Y2.625Al5Ce0.06	Y3Al5Ce0. 06
Lu0.375Y2.625Al4.375Ga0.625Ce0.06	Y3Al4.375Ga0.625Ce0.06
Lu0.375Y2.625Al3.75Ga1.25Ce0.06	Y3Al3.75Ga1.25Ce0.06
Lu0.375Y2.625Al3.125Ga1.875Ce0.06	Y3Al3.125Ga1.875Ce0.06
Lu0.375Y2.625Al2.5Ga2.5Ce0.06	Y3Al2.5Ga2.5Ce0.06
Lu0.375Y2.625Al1.875Ga3.125Ce0.06	Y3Al1.875Ga3.125Ce0.06
Lu0.375Y2.625Al1.25Ga3.75Ce0.06	Y3Al1.25Ga3.75Ce0.06
Lu0.375Y2.625Al0.625Ga4.375Ce0.06	Y3Al0.625Ga4.375Ce0.06
Gd0.375Y2.625Al5Ce0.06	Y3Al5Ce0.06
Gd0.375Y2.625Al4.375Ga0.625Ce0.06	Y3Al4.375Ga0.625Ce0.06
Gd0.375Y2.625Al3.75Ga1.25Ce0.06	Y3Al3.75Ga1.25Ce0.06
Gd0.375Y2.625Al3.125Ga1.875Ce0.06	Y3Al3.125Ga1.875Ce0.06
Gd0.375Y2.625Al2.5Ga2.5Ce0.06	Y3Al2.5Ga2.5Ce0.06
Gd0.375Y2.625Al1.875Ga3.125Ce0.06	Y3Al1.875Ga3.125Ce0.06
Gd0.375Y2.625Al1.25Ga3.75Ce0.06	Y3Al1.25Ga3.75Ce0.06
Gd0.375Y2.625Al0.625Ga4.375Ce0.06	Y3Al0.625Ga4.375Ce0.06

The orientation of the map is in registry to that shown in Fig. 15. Stoichiometry values are not subscripted. O_{12} is omitted from formula.

at 1400°C for 2 hours. The heating and cooling rates were controlled to be about 10°C/min.

After the thermal treatment, there was no detectable cross-contamination between neighboring samples, and no apparent alumina (substrate) diffusion effect on phosphor samples was observed. The luminescent image of the library under 365-nm UV excitation is shown in Figure 15, and the cells are in registry with the layout of the compositionalmap shown in Table 3. It is apparent that this series of phosphors can produce three color ranges of emission: orange, yellow, and green. The library shows that it is unlikely that a bright red phosphor under 365-nm excitation can be synthesized from this series of phosphors.

The following trends of composition–luminescence relationships were observed in $(Y_x Ce_{0.06} Lu_{1-x})_3 (Al_y Ga_{1-y})_5 O_{12}$:

Increasing the Ga amount decreases the emission wavelength and the emission intensity.

Increasing the Lu amount increases the emission intensity slightly and decreases the emission wavelength slightly.

For $y > 0.375$, increasing the Gd amount causes a red shift of the emission wavelength, while for $y < 0.375$, it quenches the emission intensity.

The color changes with composition variation in this series of phosphors have been reproduced in bulk quantity synthesis of powder phosphors. These phosphors are efficiently excited in broadband and at long wavelengths (up to blue), and the information from the phosphor library has since been validated and applied to solid-state lamp applications.

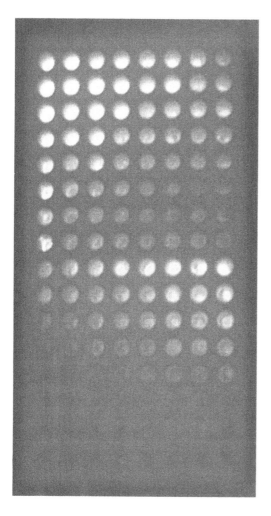

Figure 15 Photoluminescence image of a powder phosphor $(Y_xCe_{0.06}A_{1-x})_3(Al_y\text{-}Ga_{1-y})_5O_{12}$ library excited with a Hg UV lamp with a 365-nm broadband filter.

V. SUMMARY

The combinatorial approach is a fast experimental strategy in generating and studying chemical compounds for various applications. It challenges the conventional chemical synthetic approach by introducing methods of high-throughput synthesis and screening processes in developing new chemicals or materials and by miniaturizing the scale of experiments (49). This is enabled by crosslinking conventional chemistry with modern computerized robotics, microelectronic technologies, high-speed and sensitive detection techniques, and high-capacity-computer databases. Its application in phosphor research has dramatically increased the rate at which new phosphors are developed.

In general, there are four steps in the combinatorial process loop in search of new phosphors. Figure 16 illustrates the cycle of combinatorial chemistry. It starts from the design of a phosphors library, which is the collection of a large number of different chemical compositions or processes. In the design of a phosphor library, knowledge and input from a historical database will reduce the random nature in the design. The second step is high-throughput synthesis of phosphors. Phosphor materials can be used in either thin-film (e.g., thin-film electroluminescent displays) or powder forms (e.g., fluorescent lamps and CRTs), and combinatorial methods have been developed to allow the synthesis of both forms of phosphors. The third step is high-throughput luminescent and spectrum screening. For phosphors, luminescent emission intensity under certain excitation can be screened in a parallel mode, while the excitation and emission spectrum of each individual phosphor on the library can be measured with a fast sequential mode. The structure and phase information of each phosphor can be analyzed with an x-ray diffraction (XRD) method. In the fourth step, the information on composition, processing, and analysis results of the phosphor library will be fed into a large capacity database and analyzed with automated procedures. The "hit" samples that meet certain specifications will be sorted out for validation via conventional solid-state chemistry process before scaling up to industrial-scale manufacture.

Combinatorial methods have been successfully applied in searching for new phosphors, which resulted in the identification and optimization of several

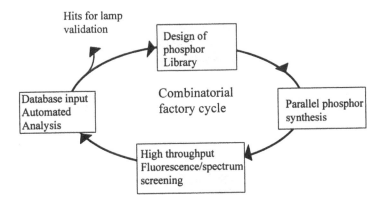

Figure 16 Combinatorial cycle in search of phosphors.

oxide phosphors with high efficiency, superior color coordinates, and good thermal and chemical stability. For example, $Gd_{0.6}Sr_{0.4}Al_{1.6}O_xF_{1.8}$: $Eu^{2+}_{0.08}$ was found to have high QE ($\sim$97%) and a better color chromaticity than those of commercial green phosphor $LaPO_4$:Tb^{3+}:Ce^{3+}. A preliminary optimized phosphor ($Gd_{1.54}Zn_{0.46}$)$O_{3-\delta}$:$Eu^{3+}_{0.06}$) has been identified with a better CIE chromaticity than the commercial phosphor Y_2O_3:Eu^{3+}, with a slightly lower photoluminescent QE $\approx$ 86%. We have shown that it is also an excellent CL phosphor. The application of combinatorial synthesis will allow a fast buildup of databases on composition–property relationships, and its high-throughput nature allows fast screening of new phosphors to match the pace of materials discovery with the needs from various electronic display technologies.

Combinatorial methods are a set of tools to enable higher-speed experiments on various material targets. The ultimate goal in combinatorial chemistry is not to develop fancy but impractical facilities, which boasts large numbers or high rates of materials generation, but to discover materials that can be validated in manufacturing processes. In that regard, it is important that the materials synthesized in the library are close in form to the bulk materials, and the screening tools should be refined toward quantification. A few corporate and institutes worldwide have embraced the combinatorial approach in searching for phosphors, for lighting or displays, and only the future will tell the impact of this technique in advancing the materials and manufacturing processes. As with any other emerging technique, it faces both technical and cultural challenges. With careful planning and validation processes, effective and high throughput experimental methods with laboratory automation and miniaturization can be developed to accelerate the solution of various sophisticated material problems.

ACKNOWLEDGMENTS

Most of the work described in this chapter is based on my Ph.D. dissertation work at the University of California, Berkeley (50). I would like to thank my former advisors, Prof. Peter Schultz of UC Berkeley and Dr. X.-D. Xiang of Lawrence Berkeley National Lab (LBNL), for the help and guidance they gave me throughout the course of this work. I would also like to thank my other collaborators at LBNL.

REFERENCES

1. Blasse, G.; Grabmaier, B.C. Luminescent Materials; Springer-Verlag: New York, 1994.
2. Justel, T.; Nikol, H.; Ronda, C. Angew. Chem. Int. Ed. 1998, 37, 3084–3103.

3. Ronda, C.R. J. Alloys Compounds. 1995, 225, 534–538.
4. Wickersheim, K.A.; Lefever, R.A. J. Electrochem. Soc. 1964, 111, 47–51.
5. Shmulovich, J. Information Display; 1989; Vol. 3, 17–19.
6. Poort, S.H.M.; Blokpoel, W.P.; Blasse, G. Chemical Materials. 1995, 7, 1547–1551.
7. Ropp, R.C. The Chemistry of Artificial Lighting Devices: Lamps, Phosphors, and Cathode Ray Tubes; Elsevier: New York, 1993.
8. Tannas, L.E., Jr. Flat-Panel Displays and CRTs; Van Nostrand Reinhold: New York, 1985.
9. Gingerich, R.G.W. Phosphors Materials. Materials Research Society, H12.1 Spring Meeting, 1996, unpublished materials.
10. Galasso, Francis S. Structure, properties, and preparation of perovskite-type compounds, 1st ed.; Pergamon Press: New York, 1969.
11. Faucher, M.; Caro, P. J. Chem. Physics. 1975, 63, 446–450.
12. Klimke, J.; Wulff, H. Fresenius J. Analyt. Chem. 1994, 245–246.
13. Bagnato, V.S. Solid State Communications. 1984, 49(1), 27–30.
14. Padua, S.J.N.; Nunes, Lao; Castro, J.C. J. Luminescence. 1989, 43, N6:379.
15. Xiang, X.-D.; Sun, X.; Briceño, G.; Lou, Y.; Wang, K.-A.; Chang, H.; Wallace-Freedman, W.G.; Chen, S.-W.; Schultz, P.G. Science. 1995, 268, 1738–1740.
16. Briceno, G.; Chang, H.; Sun, X.; Schultz, P.G.; Xiang, X.-D. Science. 1995, 270, 273–275.
17. Sun, X.; Wang, K.A.; Yoo, Y.; Wallace-Freedman, W.G.; Gao, C.; Xiang, X.D.; Schultz, P. Advanced Materials. 1998, 9, 1046.
18. Wang, J.; Yoo, Y.; Gao, C.; Takeuchi, I.; Sun, X.; Chang, H.; Xiang, X.D.; Schultz, P. Science. 1998, 279, 1712–1714.
19. Danielson, E.; Golden, J. H.; Mcfarland, E. W.; Reaves, C. M.; Weinberg, W. H.; Wu, X.-D. Nature. 1997, 389, 944.
20. Bril, A.; Hoekstra, W. Philips Res. Rep. 1961, 16, 356.
21. Stevels, A.L.N.; Schrama-de Pauw, A.D.M. J. Electrochem. Soc, 123(5), 691–697.
22. Uehara, Yasuo; Masuda, Iso; Kobuke, Yoshimasa J. Electrochem. Soc, 107(1), 1–8.
23. Kilian, H.S.; Kotte, J.F.A.K.; Blasse, G. J. Electrochem. Soc. 1987, 134, 2359–2364.
24. Maruska, H.P.; Parodos, T.; Kalkhoran, N.M.; Halverson, W. D., MRS Spring Meeting, : San Francisco, April 1994, unpublished materials.
25. Weber, L.F.; Birk, J.D. MRS bulletin, March 1996, 65–68.
26. Dalziel, J.A.W.; Welch, A.J.E. Acta Cryst. 1960, 13, 956.
27. Roth, R.S. J. Research NBS. 1957, RP 2736, 58.
28. Dieke, G. Spectra and Energy Levels of Rare Earth Ions in Crystals; Interscience: New York, 1968.
29. Nieuwesteeg, K.J.B.M.; Mutsaers, C.A.H.A. Philips J. Res. 1989, 44, 157–182.
30. Brixner, L.H.; Chen, H.Y. J. Electrochem. Soc. 1983, 130, 2435.
31. Guentert, O.J.; Mozzi, R.L. Acta Cryst. 1958, 11, 746.
32. Cromer, D.T. J. Phys. Chem. 1957, 61, 753.
33. Nieuwpoort, W. C.; Blasse, G.; Bril, A. Optical Properties of Ions in Crystals; Interscience: New York, 1967.
34. Dr. A.G. Chakhovskoi from Department of Electric Engineering, U.C. Davis did the cathodoluminescence characterization.

35. Sun, X.D.; Gao, C.; Wang, J.; Xiang, X.-D. Appl. Phys. Lett. 1997, 70, 3353–3355.
36. Sun, X.; Xiang, X.-D. Appl. Phys. Lett. 1997, 72, 525–527.
37. Stein, A.; Keller, S.W.; Mallouk, T.E. Science. 1993, 259, 1558.
38. Kimura, J.; Kawana, Y.; Kuriyama, T. Biosensors. 1988, 4, 41–51.
39. Newman, J.D.; Turner, A.P.F.; Marrazza, G. Anal. Chim. Acta. 1992, 262, 13–17.
40. Lemmo, A.V.; Fisher, J.T.; Geysen, H.M.; Rose, D.J. Anal. Chem. 1997, 69, 543.
41. Bernadini, G.L.; Rampy, B.A.; Howell, G.A.; Hayes, D.J.; Frederickson, C.J. J. Neurosci. Methods. 1991, 38, 81–88.
42. Heinz, J.; Hertz, C.H. Adv. Electronics Electron Phys. 1985, 65, 91–171.
43. Switzer, G.L. Rev. Scientific Instruments. 1991, 62, 2765–2771.
44. Ma, S.-H. US Patent 5,271,765.
45. Gendler, P.L.; Sporer, A.G.; Stremel, D.A. U.S. Patent 4,781,758.
46. Crarnahan, R.; Hou, S.L. IEEE Transactions on Industry Applications. 1977, IA-13, 95–104.
47. Taketo, N.; Kimura, Y.; Tanaka, Y. U.S. Patent 4,609,925.
48. Anonymous, Y. Electronic design. 1997, 45(10), 27.
49. Pirrung, M.C. Chem. Rev. 1997, 97, 473–488.
50. Sun, X.-D. A combinatorial approach in the discovery of advanced materials. Ph.D. dissertation; University of California, Berkeley, May 1998.

7

Combinatorial Ion Synthesis and Ion Beam Analysis of Materials Libraries

Chang-Ming Chen, Xin-Quan Liu, and Min-Qian Li
Shanghai Institute of Nuclear Research, Chinese Academy of Sciences, Shanghai, China

I. INTRODUCTION

Ion beam technology is widely used in various thin-film synthesis and thin-film surface modification. The use of ion beams for processing has many advantages. Various parameters of the ion beam, including the choice of species, its charge state, flux, energy, direction, and divergence, can all be easily quantified and controlled. An energetic ion beam can be used to introduce dopants into the surface layers of solids, a process referred to as *ion implantation*. Because ion implantation involves a beam of many ions interacting with a solid containing many atoms, the process can induce lattice damage in the target material. Chemical and physical properties of most materials can be altered by the addition of impurities or dopants. Changes in properties due to them are inherently different from changes due to radiation damage. Ion implantation is a nonequilibrium process, where the doping atoms get driven into the solid by a violent use of their excess kinetic energy. In this way, compounds that are unattainable by more conventional methods may be formed.

Ion implantation is well known for its application in large-scale manufacturing of semiconductor devices, such as metal oxide semiconductor transistors (MOSTs) and resistors (1). Its application also includes tailoring a wide range of microstructures for such diverse research goals as inducing grain refinement, providing nucleation sites, forming precipitates, and creating amorphous phases and a variety of other stoichiometric or structural alterations (2–5). Optical properties of surface layers of many materials can be altered by ion implantation via incorporation of embedded microcrystallites (5,6). By changing the refractive index, for example, it should be possible to produce all-optical switches or waveguides for optical integrated circuits (5,6). Ion implantation can be used to fabricate semiconductor and optoelectronic devices in general, ranging from doping of diamond and other wide-band-gap semiconductors to the formation of new supersaturated substitutional alloys and improved optical waveguides. Using ion implantation, ''smart'' surfaces on otherwise-inactive materials can also be fabricated. An example of this includes VO_2 precipitates embedded in Al_2O_3 single crystals as a medium suitable for optical applications such as optical data storage (7). This concept has also been extended to the fabrication of magnetic-field-sensitive nanostructured surfaces by forming magnetostrictive precipitates of such materials as Ni and RFe_2 (where R is Tm, Tb, or Sm), which are embedded in various single-crystal oxide hosts (8). Dilution of magnetic atoms in semiconductor heterostructures can also be formed to give rise to a variety of new phenomena that are strongly sensitive to magnetic field (9).

Ion beam mixing plays an important role in the formation of phases, the enhancement of interface diffusion, and the development of textures in thin-film systems. Due to the energetic impact of ions in solids, atomic displacements are produced. A consequence of this effect is the intermixing on thin-film systems due to ion irradiation. Various mechanisms, including ballistic and, in particular, chemical driving forces due to the heat of mixing of the thin-film components, can induce materials to interact at their interface. This can lead to the formation of a mixture of materials that may have new and unique properties (10–13). Stoichiometric compounds and nonequilibrium phases can be formed in this matter (13). The implantation induced intermixing techniques have been successfully used to obtain the laser wavelength shift in the implanted thin-film systems (14–16). Ion beams can also induce structural and magnetic texturing of thin films, in which diffusion phenomena can lead to grain growth and change in the structural properties of the solid (1).

One universal goal in materials research and development is to identify the surface composition of a material. From microelectronics to biotechnology, properties of nanoscaled regions can often dictate the performance of a material. A small amount of impurities or dopants can sometimes play a dominant role in determining the physical properties of a material. In one method of fabricating combinatorial thin-film libraries, multiprecursor layers are thermally annealed to induce interface diffusion and homogeneity of the composition within individual sites in the libraries. For such libraries, it is important to perform depth-resolved characterization of the profile of compositions in the individual elements of the libraries. In quantitative characterization of elemental composition, Rutherford backscattering spectrometry, (RBS) can serve as a powerful tool. Particle-induced x-ray emission (PIXE) with high beam currents with spot sizes no more than 10 micrometers in diameter is also useful for compositional analysis of small samples.

The principle of surface analysis by RBS is based on a simple physical phenomenon of elastic collisions between beam particles and atoms of a target (17,18). There are three basic factors that determine the energy spectrum in RBS. First, beam particle backscattering results in energy changes due to collision kinematics. Second, the spectrum is further altered by energy loss of particles emerging from below the target's surface. Third, the intensity of backscattered particles reflects the concentration of target atoms. A single RBS spectrum contains information about atoms in the target, their concentrations, and depth profiles.

A beam of monoenergetic ions (generally $^4He^+$ or $^1H^+$) having energies on the order of 10^6 eV can be used to nondestructively determine which atoms are present in a solid and how they are distributed. In particular, one can obtain depth profiles of different atomic constituents in the outermost microns of a sample. In general, elements comprising as little as 0.1% of a solid can be identified by ion beam analysis, and in certain cases amounts as small as 1 part per billion can also be detected. One of the advantages of the ion beam analysis is that it provides depth-resolved element distributions without the requirement of destruction of the sample by materials removal, as in the case of sputter sectioning used in Auger electron spectroscopy (AES) or secondary ion mass spectroscopy (SIMS). Moreover, backscattering spectrometry is capable of quantitative measurements without calibration standards. This technique can be further applied in the determination of crystallinity of epitaxial thin films through channeling effect measurements.

In this chapter, we discuss the use of ion implantation to synthesize combinatorial luminescent materials libraries in SiO_2 matrices, $Al_{0.35}Ga_{0.65}As/GaAs$ single quantum wells (QWs) and $Al_{0.53}Ga_{0.47}As/GaAs/AlAs/GaAs/Al_{0.53}Ga_{0.47}As$ coupled QW structures. Optical properties of the samples in the libraries are investigated under selective laser excitations or electron bombardment. We also outline applications of RBS and proton elastic scattering (PES) methods (19) to determine the depth-resolved ion distributions in the ion-implanted combinatorial samples. Furthermore, we also report applications of the cathodoluminescence (CL) technique to identify the local bonding environment, which contributes to the luminescent properties of materials libraries.

II. COMBINATORIAL ION IMPLANTATION OF SILICON-BASED MATERIALS LIBRARIES AND THEIR OPTICAL PROPERTIES

Silicon, the most common semiconductor in the microelectronic industry, has for a long time been considered unsuitable for optoelectronic applications, which remained the domain of III-V semiconductors and glass fibers. This is due mainly to the indirect bandgap of silicon, which makes it a poor emitter, and the absence of linear electro-optic effects in silicon. The enormous progress in communication technologies in recent years resulted in an increased demand for optoelectronic functions integrated with electronic circuits. Such integration would allow us to couple the information-processing capabilities of optoelectronics. In practice, silicon would be the material of choice, because of its mature processing technology and its unrivaled dominance in microelectronics. The only limiting step is the absence of efficient Si-based light sources.

A considerable amount of effort has been devoted to circumventing the physical inability of silicon to emit light. Since the discovery of light emission from porous silicon in 1990 by Canham (20), a large amount of work has been performed in studying silicon nanostructures. These efforts include investigations not only on porous silicon, but also on nanocrystals produced by several techniques as well as on silicon-insulator multilayers (21). A research group at the University of Rochester has recently reported the development of silicon-rich silicon dioxide electroluminescent devices integrated with silicon microelectronic circuitry (22). Alternative approaches to this include synthesizing nanocrystallites embedded in silicon-based hosts by ion implantation.

A. Synthesis of Materials Libraries in SiO_2 Matrices

It is well known that quantum dots embedded in a host material with a large bandgap exhibit strong size-dependent optical and electrical properties (2–4). During the last few years, researchers have been investigating fabrication of a variety of nanocrystals of metals and other compounds in SiO_2 that exhibit different optical emission properties suitable for various optoelectronic applications (3). Ion implantation is attractive for this purpose because of its compatibility with planar microelectronic technology and its ability to generate nanocrystals with well-defined concentration at precalculated depths below the host material surface. Implantation of various individual elements, such as Si (23), Ge (24), and C (25,26), into SiO_2 has been reported, and it has yielded different luminescent properties. Recently, a sequential ion implantation of Ga and N has been used to create GaN nanocrystals in sapphire (Al_2O_3) with two main luminescence bands in the nanocrystals that are identified as excitons deeply bound to surface defects (27). Our first application of the combinatorial approach in ion implantation is intended to greatly enhance the efficiency of sample synthesis and the possibility of novel nanocrystal formations in search of materials with higher-energy emissions that can meet the needs of high-density optical-storage devices.

For this study, boron-(p-type)-doped 0.5-mm-thick single-crystal silicon wafers (<100> orientation and resistivity of 5.0 (ΩV-cm) were used for the implantation. A SiO_2 layer of thickness $\sim$ 400 nm was formed by thermal oxidation of the wafer in $H_2 + O_2$ atmosphere. C, Ga, N, Pb, Sn, and Y ions were selected for implantation for their potential in forming nanocrystals of GaN, C, Sn, and SiC as well as other unexpected recombination centers in the host matrices. The energies for the implantation were 40–80 keV, and each ion dose was about $1 \times 10^{16}/cm^2$. Implantation of various elements was accomplished at room temperature using a mass-separated beam through an electromagnetic isotope separator (EMIS-SINR-02). The ion beam was electrostatically scanned in a 50-cm^2 area to homogenize the surface ion dose distribution. The ion beam was incident on the substrate through a set of physical masks, containing a primary mask with 2-mm-diameter circular openings in a 8×8 grid geometry. The primary mask was used in combination with a series of secondary masks (28). In this way, different combinations of ion species were implanted at different sites in the substrate. The generated materials library was subsequently thermally treated at 800°C for 30 min in flowing Ar atmosphere. Figure 1 shows the materials library containing 64 different samples generated by ion beam implantation with six masking combinations, imaged under a natural light. The different colors on the various sample sites originate from specific light reflections and interference from different interfaces.

B. Cathodoluminescences of Materials Libraries in SiO_2 Matrices

Cathodoluminescence (CL) is the emission of light as a result of electron bombardment. Analysis of CL provides microcharacterization of the optical and electronic properties of luminescent materials. It is a powerful tool for the depth-resolved studies of ion-implanted samples and characterization of semiconductor interfaces. We have performed nondestructive depth-resolved CL measurements on the ion-implanted SiO_2 library by varying the range of the electron penetration, which depends on the electron beam energy, in order to excite CL from different depths in the material. This also provides additional local information on the defects and impurities in the matrix (29–31). The CL measurement system capable of producing a beam current of 1–3 μA was employed for this study. The dispersed emission was detected by an HTV–R-446 photomultiplier and a WDG05-II monochromator.

Because the materials chip is exposed to an electron beam irradiation of 5 kV and 1 μA, we can easily distinguish the implanted samples from the pristine substrate. Most samples result in blue emission, and differences in color and intensity of emitting light from the samples can be observed by the naked eye. A yellow-green color from C containing implanted samples and a violet-blue color from Sn containing implanted samples were observed under 2-keV electron bombardment. Major CL peaks observed are listed in Table 1. We use the notation S(X, Y, Z) to indicate a site implanted with elements X, Y, and Z.

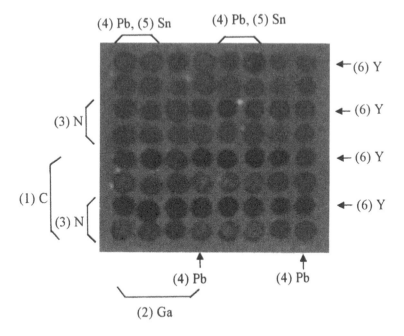

Figure 1   Image of the ion-implanted silicon-based 64 sample materials prior to annealing under a natural light. The silicon wafer is 25 mm × 25 mm, and each sample site is 2 mm in diameter. Indicated elements and the numbers in parentheses correspond to the implanted ion species and their implantation sequences, respectively. Implantation parameters for various elements are: (1) C, 50 KeV, $5 \times 10^{16}/cm^2$; (2) Ga, 80 KeV, $1 \times 10^{16}/cm^2$; (3) N, 50 KeV, $1.5 \times 10^{16}/cm^2$; (4) Pb, 80 KeV, $1.5 \times 10^{16}/cm^2$; (5) Sn, 70 KeV, $1 \times 10^{16}/cm^2$; (6) Y, 65 KeV, $1 \times 10^{16}/cm^2$.

As evident in Table 1, the blue (450 nm) luminescent band is observed in all samples. This is commonly observed in pristine SiO_2, amorphous SiO_2, and silica, and it is generally ascribed to the oxygen vacancies or the E′ center and the two-fold-correlated silicon center (24,25). The 450-nm CL band is created by the electron bombardment, and its intensity is enhanced by increasing the electron accelerating voltage to 10 keV. Implantation of Ga and N into SiO_2 is expected to form embedded GaN nanocrystals, which are of great importance in optoelectronic device applications. However, CL measurements of the S(Ga, N) sample did not show any evidence of the existence of GaN crystals in SiO_2, and no characteristic spectra for GaN were observed in this case. This conclusion was also supported by the absence of N in the same sample as determined by

Table 1 Dominant Peak Positions from Characteristic CL Spectra for Selected Samples Under 2-keV Electron Bombardment

Sample	Dominant peak position (nm)[a]		
S(C)		450	555 (0.10)
S(N)		450	
S(Ga, C)		450	570 (0.56)
S(Ga, N)		445	
S(Ga, Pb, C)		450	567 (0.60)
S(Ga, Pb, Sn, C)	415 (0.75)	450	565 (0.60)
S(Pb, Sn, N)	405 (1)	450	
S(Pb, N)	435 (0.6)	450	

[a] The data in parentheses indicate normalized emission intensities relative to ~450 nm.

PES measurements (32). This may be due to N diffusion through the film to escape into air.

All samples implanted with C ions resulted in emission of some yellow-green color. S(C) exhibits a weak additional emission peaked at 555 nm. We implanted C ions into the host, hoping to induce the formation of C-cluster, silicon carbide, or silicon oxycarbide nanoparticles. These nanocrystals are of great interest because of the shift of their luminescent emission toward higher energies compared to that of Si nanocrystals in SiO_2, with an ~680-nm peak emission. Emissions of shorter wavelength from the nanocrystal-based light-emitting diodes are expected to increase the optical storage density (3). Indeed, carbon can react with oxygen in SiO_2 to form CO and/or CO_2. We believe that reduced Si can react with carbon to form silicon carbide or silicon oxycarbide particles when there are excess carbon atoms in SiO_2. An emission band similar to this was also observed in silicon-carbon-based films prepared by plasma-enhanced chemical deposition (33).

We performed another experiment on combinatorial implantation of C into SiO_2 with a dose range 1×10^{15} to $1 \times 10^{17}/cm^2$ coupled with PL and Raman spectral analysis of the samples. The samples on the chip were separated and annealed at various temperatures for different times. The PL spectra of the implanted samples with different doses were obtained under 514.5-nm excitation. The results were similar for samples annealed under the same condition, but a stronger luminescence was present for increased implantation doses.

Figure 2 shows the PL spectra for the samples implanted with $1 \times 10^{17}/cm^2$ of C at 50 keV annealed at various temperatures. The as-implanted sample shows nearly no PL feature, and the samples annealed at 300°C exhibit a weak PL band peaked at 730 nm. For the samples annealed at 1100°C, a strong blue-

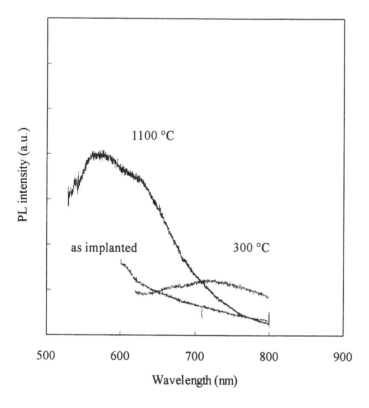

Figure 2 PL spectra of samples implanted with C at 50 keV, $1 \times 10^{17}/cm^2$, in the as-implanted state and annealed at different temperatures.

shift band peak at 560 nm with a shoulder at 620 nm is evident. The PL may originate from luminescence centers, such as nanocrystals and defects. Raman spectra measurements of the C-implanted samples show that C colloids or clusters are formed in C-implanted SiO_2, as evidenced by the well-known D and G bands in the Raman spectra from the sample implanted and annealed at 300°C for 30 min (26). The D and G bands showed peaks at 1356 and 1530 cm^{-1}, respectively, which are characteristic features of activated charcoal or glassy carbon (34). When the samples are annealed at 1100°C for 100 min, no bands from 1000 to 1700 cm^{-1} for carbon are observed, This is an indication of the absence of free carbons in SiO_2. No evident band for silicon carbide is visible, but the amount of SiC may simply be too little for detection by Raman scattering. Depth profiling for C in the samples by PES analysis shows that there is significant diffusion of carbon toward the SiO_2/Si interface and to the top surface due to the annealing (32).

In order to elucidate the mechanism of the dominant 560-nm emission in the C-implanted samples, cross-sectional transmission electron microscopy (TEM) was performed. Images showing the distribution and size of silicon carbide nanocrystals at the SiO_2/Si interface are presented in Figure 3. These nanocrystals are embedded in the Si wafer rather than in the SiO_2 and are almost uniformly oriented at the SiO_2/Si interface. Particles are in a rectangular array with dimensions of ~40 × 10 nm^2, with a spacing of ~20 nm and with their longer side parallel to the interface and shorter side normal to the interface (Figure 3a). From the HRTEM image (Figure 3b), it is evident that the interface between the nanocrystals and the Si crystalline is very sharp. This indicates that the nanocrystals are in a near-perfect epitaxial relation with the Si wafer. The electron diffraction pattern shows that all of the recrystallized nanocrystals are 3C-SiC and are aligned with the silicon matrix. Considering the fact that no silicon carbide nanocrystals are observed in SiO_2, it is assumed that high compressive stress existing in SiO_2 near the SiO_2/Si interface (35) favors formation of silicon carbide compounds at the interface. This is perhaps the cause of the PL peak at 560 nm. This self-organized growth of nanocrystals (islands) by ion implantation has the potential to be used as an alternative planar-compatible method for building dense arrays of quantum dots.

Coimplantation of C and Ga ions in SiO_2 resulted in a strong emission peak at ~570 nm. Ga can react with SiO_2 in the following way:

$$Ga + 3SiO_2 = 3\,Si + Ga_2O_3 \tag{1}$$

Ga_2O_3 is a phosphor that usually crystallizes in a highly anisotropic crystal structure (36), forming interconnected tetrahedra or octahedra in a long hollow structure. It is believed that these unique one-directional "tunnels" play a major role in contributing to its superior capacity to generate hot electrons in exciting the luminescent centers. As depicted in Figure 16a in Section IV, two peaks of the Ga concentration depth profile for sample S(Ga,C) indicate the diffusion of Ga atoms into the C concentration-peaked nearby regions after annealing. Based on this phenomenon, one may deduce that carbon implantation into the SiO_2 results in the formation of $Ga_xSi_yO_3C_x$, which may be attributed to the bonding between Ga and C atoms in SiO_2 rather than the strain induced by carbon implantation damage. Two similar Ga concentration peaks are not observed in SiO_2 coimplanted with Ga and N ions (of Figure 16a. The complex consisting of a gallium vacancy and a carbon atom may be also playing a role in the emission of ~570 nm from samples coimplanted with C and Ga ions, as was previously seen in C-implanted GaN (37).

We also conducted CL measurements on the implanted materials chip with various electron-impinging voltages. Table 1 shows that co-implantation of Sn and Pb causes an emission shift to a higher energy. Figure 4 shows a comparison of CL spectra from S(Pb, Sn, N) and the pristine substrate bombarded with elec-

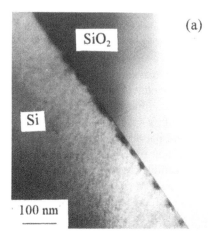

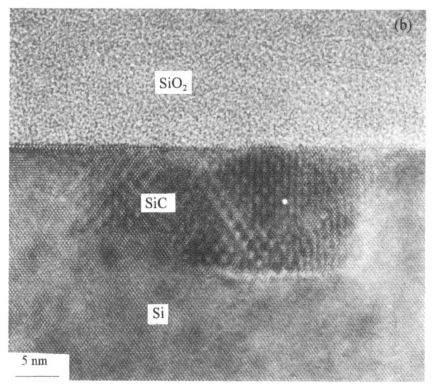

Figure 3   (a) TEM image of the region near the Si/SiO$_2$ interface in a sample implanted with C at 50 keV, 1×10^{17}/cm^2, followed by annealing at 1100°C, showing an alignment of nearly equal-spacing silicon carbide nanocrystal islands embedded in the Si wafer (b) HRTEM image of silicon carbide nanocrystals.

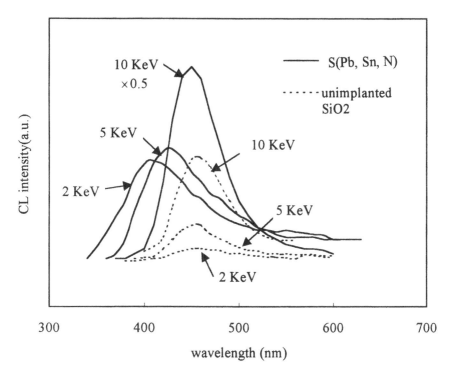

Figure 4 Cathodoluminescence spectra of S(Pb, Sn, N) and unimplanted SiO_2 at various electron energies. The spectra were measured at room temperature.

trons of various energies. The peak near 450 nm was observed in the spectra for both the implanted sample and the pristine SiO_2 substrate under the irradiation of electron beam energy of 10 keV, but the luminescence intensity of the implanted sample was four times that of pristine SiO_2. The luminescence intensity of S(Pb, Sn, N) was larger by a factor of 3.3 and 8 compared to that of the pristine substrate, measured with impinging electron energies of 5 keV and 2 keV, respectively. Figure 4 displays the depth-resolved CL spectra of S(Pb, Sn, N) measured with electron beam energies of 2 keV, 5 keV, and 10 keV. The band peak shifts to the lower-wavelength region as the beam energy decreases, with peaks at 450 nm, 422 nm and 405 nm for beam energies of 10 keV, 5 keV, and 2 keV, respectively. The shifts in the spectra as a function of electron energy reflect the depth-resolved distribution of luminescent centers inherent in the sample. This may be correlated with an inhomogeneous depth-impurity distribution in the sample.

We have calculated the energy loss of an electron penetrating into SiO_2 using a density of 2.25 g/cm^3 for SiO_2 (Figure 5). The concentration of excess electron-hole pairs created in the material is directly proportional to the energy loss as electrons penetrate the solid. Comparing Figure 4 with Figure 5, it is evident that implantation of Pb and Sn ions in SiO_2 causes an increase of pair formation and CL with increased energy. The CL peak position of S(Pb, Sn, N) at 10 keV is almost the same as that of SiO_2. This is because at 10 keV the electrons are largely excited in the SiO_2 area near the Si/SiO_2 interface. As the electron energy decreases, the region closer to the surface gets excited, and the peak band shifts to higher energies. This is possibly due to the effects of Pb- and Sn-related luminescent centers in the material. The CL-peak shift is believed to be related to the formation of new phases or defect-impurity complexes (30,38). Assuming that two-fold-coordinated silicon O–Si–O or O–Ge–O is the origin of the blue luminescence in a $Si_{1-x}Ge_xO_2$ glass (39), one may expect that O–Pb–O and O–Sn–O also contribute to the peak shifts in a similar manner. Alternatively, the nanocrystal formation may also have an effect, because Sn nanocrystals have been observed in thermally treated Sn-implanted SiO_2 (40).

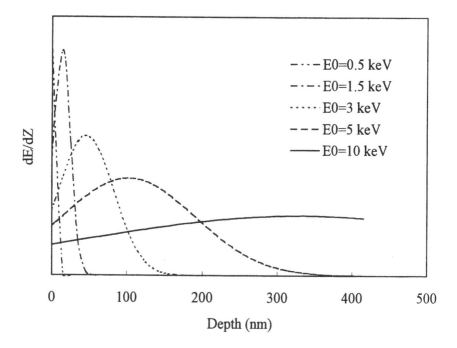

Figure 5   Calculated electron energy loss vs. depth inside SiO_2 for electrons with different energy.

As described earlier, CL measurements provide information on the redistribution of impurities and lattice damage defects during annealing of ion-implanted samples. This, in conjunction with the lateral spatial resolution available in a SEM, in principle allows three-dimensional mapping of impurities and defects in materials libraries. Furthermore, CL measurements can be effectively used in the analysis of chemical interactions at metal–semiconductor interfaces that may produce new interfacial phases. For example, pulsed laser annealing of a Cu–CdS interface can lead to an intense (relative to bandedge emission) CL peak at 1.28 eV, which can be attributed to the Cu_2S compound formation (41). Therefore, rapid detection of the formation of new semiconductor compounds in the combinatorial ion-implanted libraries by low-energy CL spectroscopy is warranted. In future, panchromatic imaging of optical properties combined with element profiling for materials libraries can be used for parallel evaluation of optical properties and, in turn, the formation of defects and novel phases.

III. FABRICATION OF MATERIALS LIBRARIES OF $AL_{0.35}GA_{0.65}AS$/GAAS SINGLE QUANTUM WELLS AND $Al_{0.53}Ga_{0.47}As$/GaAs/AlAs/GaAs/$Al_{0.53}Ga_{0.47}As$ COUPLED QUANTUM WELL STRUCTURES

Semiconductor photonic integrated circuits (PICs) are composed of many active waveguide elements, such as lasers, detectors, optical modulators, switches, and optical amplifiers. These elements can be optically coupled via a complex branching network of low-loss passive waveguides all on a semiconductor chip. Precise tailoring of the electrical and optical properties of devices of differing functionality in both vertical and lateral directions is a critical task in the development of PICs.

AlGaAs/GaAs QWs have proved to be one of the best active-region structures for optoelectronic devices such as semiconductor lasers, photo-detectors, and optical nonlinear devices. One important advantage of AlGaAs/GaAs QW structures is the application of energy bandgap engineering (EBE) by finely tuning the barrier and well parameters, including layer thicknesses and compositions. The emitting or detecting wavelength can be tuned by controlling the QW structures. The wavelength can also be tuned by postgrowth modification of the QW confinement profile using controlled intermixing across the well/barrier interfaces (41–45). Ion implantation has proved to be an effective method to realize the postgrowth tuning of the QW energy band structures by introducing defects or impurities to enhance the thermal diffusion of Al into the GaAs well layer and modifying the shape of the QW (46–51).

We have learned that an ion beam can be well controlled spatially, and, therefore, selective region wavelength tuning can be realized by controlling the ion beam position. Furthermore, the amount of tuning of the energy potential profile and the amount of shifting of the emitting or detecting wavelength can be controlled by changing the ion beam parameters. This ion implantation-induced wavelength shifting of AlGaAs/GaAs QW lasers is of particular interest, especially in wavelength-division multiplexing (WDM) applications, where multiple lasers of different wavelengths are integrated on a single chip. Laser sources of different wavelengths produce digitally modulated signal carriers, which are combined in optical multiplexes and transmitted down the fiber (51–54).

The two main methods to achieve multiwavelength emission are selective-area epitaxy (55) and postgrowth selective-region layer intermixing. The implantation-induced intermixing techniques have been successfully used to obtain laser wavelength shifting (56,57). Another promising application of postgrowth modification is to obtain a parabolic QW structure by implantation-enhanced thermal intermixing. Postgrowth thermal interdiffusion in AlGaAs/GaAs QW has been theoretically predicted (58) to be a viable alternative to the complex techniques required to fabricate parabolic QWs (59). The parabolic QWs has been predicted to have a strong third-order optical nonlinearity (60). The equally spaced eigenstate energy in a parabolic QW provides multiple pairs of resonant subbands, so the third-order susceptibility is much more enhanced than that in the case of square QWs.

Recently, the selective-area implantation-induced intermixing technique has been used to shift the infrared response wavelength of a quantum well infrared photodetector (QWIP) (61–63). The multiwavelength response QWIP structures are important for fabricating multicolor-sensitive QWIP devices, e.g., with wavelengths in the range of 3–5 μm and 8–12 μm in one chip.

The major problem associated with this method is the difficulty in precisely controlling the amount of layer intermixing, which is determined by the ion beam parameters, including species, ion energy, dose, etc. (64). Thus, finding the optimum parameters of the ion beam is essential for obtaining a certain desired amount of wavelength shift or to obtain a parabolic QW potential profile. The combinatorial method provides a quick way to search for the optimum parameters and also an easy way to get a multiwavelength-emitting-material chip (65). We have demonstrated a combinatorial implantation of As ions and protons, which are two conventionally used ion species for the modification of GaAs-related materials into AlGaAs/GaAs Single QWs.

A. Materials Libraries of Quantum Wells

1. Materials Libraries of $Al_{0.35}Ga_{0.65}As$/GaAs Single Quantum Wells

The material used for this study was an $Al_{0.35}Ga_{0.65}As$/GaAs single QW grown by molecular beam epitaxy (MBE) on a GaAs (001) substrate. As depicted in

Figure 6, a 1-μm-thick GaAs buffer layer was grown underneath the single QW structure. A 4-nm GaAs QW layer was sandwiched by two 50-nm $Al_{0.35}Ga_{0.65}As$ barrier layers. Then a 20-nm GaAs cap layer was grown. All layers were undoped and grown at 605°C. Ion implantation was accomplished at room temperature using shadow masks for spatially selective implantation. One mask made of aluminum had an array of 64×64 circular openings with a diameter of 2 mm. This mask was used together with one of four other masks that covered 4/8, 5/8, 6/8, 7/8 of the array matrix. In this experiment, the upper left 5×5 matrix was used. Four rows of openings implanted with different doses of As at an ion energy of 90 keV were obtained after the four masks were used sequentially. Figure 7 shows the elements' distribution. Implantation doses were 5×10^9 cm^{-2}, 5×10^{11} cm^{-2}, 5×10^{13} cm^{-2}, and 5×10^{15} cm^{-2} for rows labeled, respectively, As_1, As_2, As_3, and As_4 in Figure 7. The four masks were then rotated by 90° for proton implantation atenion energy of 40 keV. After four implantations with four masks, protons were implanted along four columns with doses 5×10^9 cm^{-2}, 5×10^{11} cm^{-2}, 5×10^{13} cm^{-2}, and 5×10^{15} cm^{-2} in columns labeled H_1, H_2, H_3, and H_4, respectively. The fifth column and the fifth row from the upper left corner were implanted with arsenic and protons only, respectively, and they are labeled As_i and H_i, respectively. Henceforth, we use the notation As_mH_n (m, n = 1, 2, 3, or 4) to identify cells in the upper left 4×4 matrix of the library chip. The implanted chip was subsequently annealed in a rapid thermal

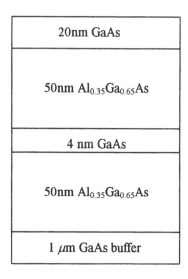

Figure 6   Schematic of the single QW structure.

H₄ H₃ H₂ H₁   Asi

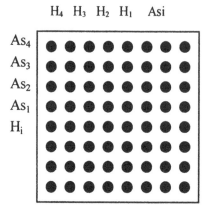

Figure 7 Schematic of the QW libraries. The labels As_1, As_2, As_3 and As_4 indicate regions with different doses of As ions, which are 5×10^9 cm^{-2}, 5×10^{11} cm^{-2}, 5×10^{13} cm^{-2}, and 5×10^{15} cm^{-2}, respectively; H_1, H_2, H_3, and H_4 represent regions with different doses of proton implantation, the doses are 5×10^9 cm^{-2}, 5×10^{11} cm^{-2}, 5×10^{13} cm^{-2}, and 5×10^{15} cm^{-2}, respectively.

annealer (RTA) at 950°C in a flowing N_2 atmosphere. During the annealing, the chip was face down on a fresh GaAs wafer, with another GaAs wafer placed on top in order to prevent excessive As loss.

2. Materials Libraries of $Al_{0.53}Ga_{0.47}As/GaAs/AlAs/GaAs/$ $Al_{0.53}Ga_{0.47}As$ Coupled Quantum Wells

An asymmetrical-coupling double quantum well has been examined intensively in recent years for the basic study of quantum tunneling effects (66–70). When the barrier between the double quantum wells is as thick as the de Broglie wavelength, interesting characteristics arise, such as abstinent transitions (71), which are not allowed in single quantum wells. For device applications, such coupled structures can be an integrated part of the active region of devices, such as QWIP and cascade laser devices (72–74). Here, we investigate the dose-dependent modification of the potential profile by proton implantation with no follow-up rapid thermal anneal at high temperatures (36).

As described in Ref. (75), protons cause little damage near the surface during implantation. This is because protons can rapidly diffuse in semiconductors, leaving fewer atoms and point defects in the QW region, compared to larger atoms such as Si, O, Be, Mg, Se, Al, and Ga. Although rapid thermal annealing can be used to simultaneously remove implanted defects and stimulate diffusion of Al from the barriers into a GaAs QW (75), it also degrades the structural quality of the QW and the performance of QW devices. Here, we investigate the dose-dependent modification of the potential profile by the combinatorial implantation of protons without rapid thermal annealing.

A GaAs/AlGaAs asymmetrical-coupling double QW was grown on a semi-insulating GaAs (100) substrate by MBE. A 1-μm GaAs buffer layer and a 50-nm $Al_{0.53}Ga_{0.47}As$ barrier were grown underneath a coupled QW structure with 3-nm and 7-nm GaAs wells separated by a 2-nm AlAs barrier followed by a 50-nm $Al_{0.53}Ga_{0.47}As$ barrier and a 20-nm GaAs cap layer. All layers were undoped and grown at 605°C, and then a coupled QW structure was generated, as shown in Figure 8. Proton implantation was carried out at room temperature at 40 keV. With the same combinatorial implantation method as that described for the implantation of single-QW chips. We fabricated four samples with different implantation doses of protons in a single wafer. The proton implantation doses of the four samples were 5×10^9 cm^{-2}, 5×10^{11} cm^{-2}, 5×10^{13} cm^{-2}, and 5×10^{15} cm^{-2}.

B. Photoluminescent Properties of the Materials Libraries of $Al_{0.35}Ga_{0.65}As$/GaAs Single Quantum Wells and $Al_{0.53}Ga_{0.47}As$/GaAs/AlAs/GaAs/$Al_{0.53}Ga_{0.47}As$ Coupled Quantum Wells

1. Photoluminescence in the Single Quantum Wells Libraries

Wavelength shifting is of great interest, especially in WDM applications. Implantation-introduced defects enhance the intermixing and induce wavelength shifts.

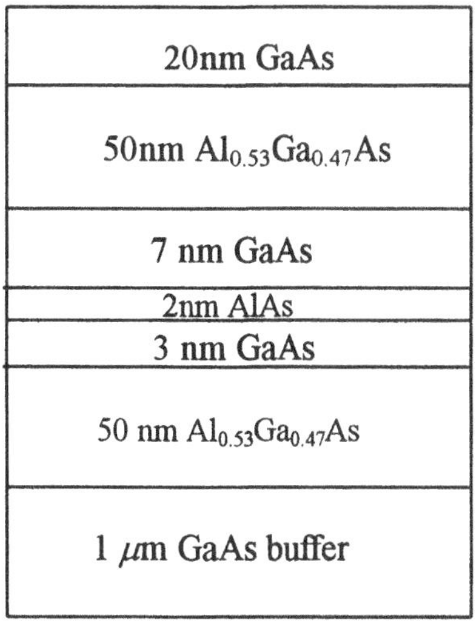

20nm GaAs
50nm $Al_{0.53}Ga_{0.47}As$
7 nm GaAs
2nm AlAs
3 nm GaAs
50 nm $Al_{0.53}Ga_{0.47}As$
1 μm GaAs buffer

Figure 8 Schematic of the coupled QW structure.

Due to the intermixing process, electron and hole energy levels can change, as shown in Figure 9. This results in the shift of the emitting wavelength. Ion implantation–induced intermixing is an attractive technique because of its ability to introduce precise defect densities by controlling the irradiation dose.

In order to elucidate the effect of ion implantation on the wavelength shifting, the microphotoluminescence (micro-PL) measurement was performed at room temperature. A 514-nm Ar$^+$ laser was used as the excitation source. Figure

Ec

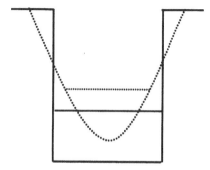

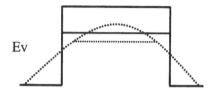

Ev

Figure 9 Schematic of intermixing process in a QW potential profile. Dashed lines represent the potential profile after intermixing; solid lines represent the as-grown QW potential profile.

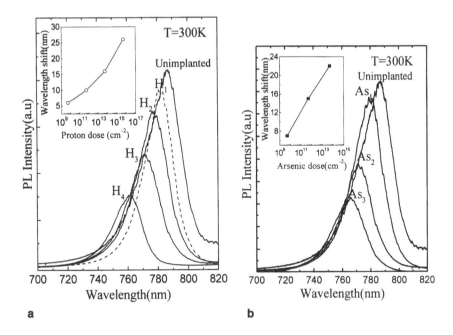

Figure 10 PL spectra for (a) proton-only-implanted and (b) Arsenic-only-implanted QW samples. Insets show corresponding dose-dependent wavelength shifts.

10a and b show the PL spectra from sites implanted with As and protons, respectively. The spectra of the unimplanted but annealed region in between the sites are also shown here as a reference. Clear wavelength shifts were observed for both As-only and proton-only sites. The insets of Figure 10a and b show implantation-dose-dependent wavelength shifts. The PL signal was quenched for the site implanted with As at a dose of 5×10^{15} cm^{-2}. The damages caused by implantation cannot be recovered by rapid thermal annealing and thus act as nonradiative centers. Four different wavelengths were observed for proton-only sites, and three different wavelength shifts were observed from the As-only sites. Maximum wavelength shifts for sites with arsenic-only and proton-only implantation are 22 nm (sites in the row As$_3$) and 26 nm (sites in the column H$_4$), respectively.

Figure 11 shows the PL spectra of the sites with mixed implantation with different doses (spectra from only the diagonal sites are shown in Figure 11). One striking feature is the observation of the weak PL signal from As$_4$H$_4$, which was implanted with the highest As and proton doses. For all of the As$_4$Hi (i = 1, 2, 3, 4) sites, PL signals can be observed, and the intensity was observed to get slightly stronger with increasing proton doses. Sixteen different wavelengths

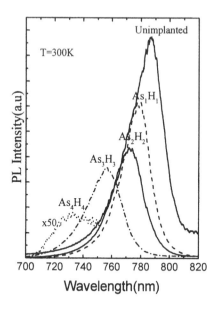

Figure 11   PL spectra for the QW samples coimplanted with As and protons.

were obtained from different doses of As and protons in various sites, labeled As_mH_n ($m, n = 1, 2, 3, 4$). The maximum wavelength shift relative to the unimplanted but annealed site is 62 nm (which corresponds to 138 meV), for As_4H_2.

Figure 12 shows the wavelength shifts of all the sites implanted with different As and proton doses. The open circles and solid squares show the implantation dose–dependent wavelength shifts of proton-only and As-only sites, for comparison. The solid triangle, star, solid diamond, and cross lines show the wavelength shifts of H_n column sites with varying arsenic doses, for $n = 1, 2, 3$, and 4, respectively. Protons were implanted after As implantation. This order seems to enhance the intermixing and shifts the wavelength more, with a maximum of 10 nm in our case. The wavelength shifts of As_mH_n sites are much larger than those of As-only or proton-only sites (As_m and H_n). The decrease of wavelength shifts seen for As_4H_3 and As_4H_4, compared with that for As_4H_2, shows the saturation effects of the wavelength shifts. The saturation effect comes from the point defects coalescing into extended defects, which will not be active for Al and Ga interdiffusion. The recovery of PL intensity of As_4H_n sites may also originate from the defect coalescence, which will not be active for trapping the carriers. This coalescence process will decrease the number of the carrier-trapping centers by forming

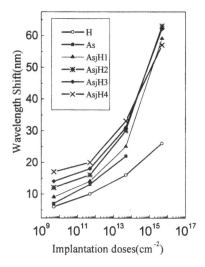

Figure 12 Wavelength shifts of the QW samples coimplanted with As ions and protons.

extended defects, and, as a result, the PL intensity is recovered slightly. The details of this process warrant further investigation.

It is well known that arsenic is often used to shift the wavelength for GaAs/AlGaAs structures. A large wavelength shift can be achieved with As ions. An As ion has a large mass, and it is effective in producing defects, which, in turn, enhance the interface interdiffusion of Ga and Al atoms. Protons have proved to be an efficient species for shifting the wavelength of GaAs/AlGaAs lasers, without much degradation (56). However, no investigation had been reported on the effect of combination of these two species. Our results show that multispecies implantation with arsenic followed by protons can be used to shift the wavelength further, i.e., from 22 nm of As_3 and 32 nm of As_3H_4, to 62 nm of As_4H_4. The PL spectra are broad, and emitting spectra often overlap with each other at room temperature. But when they are used for laser emission, their linewidths will be much narrower, and emission lines from different sites will be well resolved, For example, in Ref. (56), laser emission lines from interdiffused laser structures were demonstrated. For WDM device applications, control of wavelength shift is critical. The present combinatorial approach can be used to set up a library for investigation of wavelength shifts tuning as a function of implantation species and doses.

2. Photoluminescence in the Coupled Quantum Wells Libraries

Figure 13 displays PL spectra of different samples in the as-implanted coupled QWs libraries. The PL spectra show that transition peaks shift after implantation.

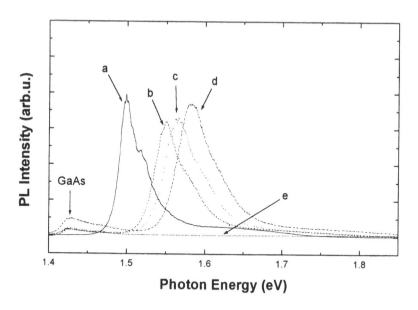

Figure 13 PL spectra of the coupled QWs implanted with different doses of proton. (a) none, (b) 5 × 10⁹ cm⁻², (c) 5 × 10¹¹ cm⁻², (d) 5 × 10¹³ cm⁻², (e) 5 × 10¹⁵ cm⁻².

The amount of wavelength shift increases with the implantation doses. When the implantation dose reaches 5×10^{15} cm^{-2}, transition peaks quench. The PL spectra were fitted using Gaussian line shapes.

An obvious blue shifting is observed at maximum 81 meV for Hh1-e1. Because rapid thermal annealing at a high temperature was not carried out, we conclude that the interdiffusion between Ga atoms in the well layers and Al atoms in the barrier layers had occurred during proton implantation. After this process, the Al composition of barriers and wells was changed, and as a result the shape of wells was modified. The energy sublevels of both electron and hole states were raised after the well shape was modified. This may be the cause of the blue shift of PL spectra.

One striking observation from this investigation is that the interdiffusion is obviously occurring, even in the as-implanted condition. Since the temperature change caused by ion implantation is estimated to be lower than 100°C during implantation, we believe that the intermixing of coupled QWs is determined by implantation-induced defects. This technique is very promising for the monolithic integration of active laser and low-loss passive waveguide circuits (76).

In summary, we selected GaAs/AlGaAs QW structures as the host material because these structures are very important for optoelectronic device applications. Since postgrowth tailoring of the QW potential profile is strongly dependent on the ion species, energy, and doses, the combinatorial method is effective in rapidly searching the parameter space in order to precisely control the potential profile. In addition, this method can be used to obtain the multiwavelength-emitting or-responding chip for laser and detector purpose. To find the optimum set of implantation parameters quickly, parallel characterizing methods, such as the parallel micro-PL system, which can characterize optical properties of multiple samples simultaneously, would be valuable.

IV. ION DEPTH-RESOLVED PROFILING OF MATERIALS LIBRARIES BY RUTHERFORD BACKSCATTERING AND PROTON ELASTIC SCATTERING

Besides screening of physical properties of materials libraries, establishing composition–property relationships in films is very important in combinatorial materials science. This is because there are many phenomena, such as interaction between various thin films and the substrate, that can affect the resulting physical properties (77). Ion beam analysis techniques, such as Rutherford backscattering (RBS) and proton elastic scattering (PES), can provide quantitative information on surface impurities and depth-resolved profiling of thin-film compositions (19,78,79). Detection of surface impurities with a heavy mass on a lower-mass substrate can be achieved using low-energy backscattering techniques. For detection of trace amounts of an impurity with low atomic mass constrained within a substrate with higher atomic mass, a nuclear reaction and a resonant scattering can be used to improve the sensitivity. These analysis techniques are nondestructive, which is very critical for ''in situ'' determination of the composition–properties relationship in materials libraries.

In our experiments, RBS analysis was performed with a 4-MeV pelletron using 2-MeV $^4He^+$ beams at a scattering angle of 170°. The energy resolution of the detector was about 16 keV, and the total dose of $^4He^+$ was 3 μC. The scheme of the library chip holder for RBS and PES measurements is shown in Figure 14. A rectangular stainless steel substrate is used to hold the materials chip. Four quartz strips, each having length equal to that of the substrate, are affixed along each side of the chip by silver paste and covered by a rectangular stainless steel frame with the same dimension as that of the quartz strip. On each side of the frame are eight holes of 1 mm diameter (with 2.5-mm spacing between them to match the sample spacing in the chip) to serve as windows for exposing the quartz surface. The materials chip is placed on top of the stainless steel substrate, with sample sites aligned with the holes located in the frame. In the

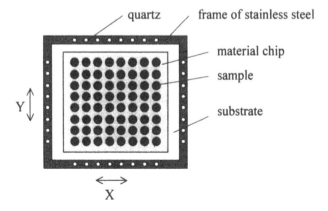

Figure 14 Schematic of the materials chip holder for RBS and PES analyses. The holder can be moved along X and Y directions to align selected sample sites with the analyzing beam.

RBS and PES measurements, a primary reference point is first located by moving the chip holder along the X and Y directions using a feedthrough manipulator, until one of the exposed quartz spots lights up due to ion bombardment emitting reddish-white light. This can be observed either by the naked eye or through a telescope. Ion beam analysis of selected samples is accomplished by moving the substrate location relative to the reference point.

Ion analysis was also used for quantitative determination of compositions and for elucidating the interaction among implanted ions within each sample. Spectra from RBS measurements of the SiO_2 materials library annealed at 800°C are partially shown in Figure 15. Figure 15a shows the spectra taken for S(Ga, N) and S(Ga, C). The Ga spectrum of S(Ga, N) shows a peak at ~36 nm from the surface, with no observable peak shift with respect to that of the sample prior to annealing (not shown in the figure). The N spectrum is estimated to have a peak at about 130 nm from the surface, as calculated by the TRIM code (80). Thus, we conclude that no GaN was formed in the region of the N peak. We believe this was due to N diffusion and escaping during the annealing process, as evidenced by the following results. Previously, GaN nanocrystal formation had been reported in Ga- and N-implanted Al_2O_3 after annealing (27).

Two peaks of the Ga spectrum from S(Ga, C) are shown in Figure 15a. They indicate that the presence of carbon induces the redistribution of Ga atoms in SiO_2 after annealing. The profile of the C-atom distribution results in a peak at about 180 nm from the surface, as estimated by the TRIM code calculation (80). This matches the location of the lower energy peak of the Ga spectrum (172 nm from the surface). Carbon-induced Ga-related compounds may have been formed in the SiO_2.

Subsequent implantation of Pb ions or coimplanting Pb and Sn ions into samples already implanted with carbon ions did not alter the Ga spectrum peak location, as shown partially in Figure 15b. We therefore conclude that the Ga redistribution is affected only by carbon implantation. Comparing the spectra of S(Pb, N) and S(Pb, Sn, N) in Figure 15c, it is evident that there is little interaction between Pb and Sn in the samples.

These observations are also confirmed by the ion concentration-depthprofiles calculated with the RUMP program (81). These are shown in Figure 16 for the spectra from Figure 15. The Ga profile of S(Ga, C) is peaked at 36 nm and 172 nm from the surface, for corresponding atom fractions of 1.5% and 0.56%, respectively Figure 16a. The profiling curves of S(Ga, Pb, C) and S(Ga, Pb, Sn, C) also each show two peaks of Ga concentration, located at nearly the same positions as in the curve from S(Ga, C). Figure 16b reveals that Sn atoms have diffused to the surface in S(Pb, Sn, N). This is believed to be the cause of the high-energy luminescent emission when bombarded with relatively lower energy electrons.

Figure 17 displays the PES spectra of S(C), S(N), and pristine SiO_2 taken with 3-MeV H^+ beams at a scattering angle of 170°C. The peak concentration of carbon atoms in S(C) is 4.12 at.%, calculated using the proton elastic-scattering cross-sectional enhancement factors given in Ref. (19). As shown in Figure 17, the spectrum of S(N) did not reveal any trace of N after the sample was annealed. This is due to the diffusion and loss of N atoms in the SiO_2 matrix during annealing, as discussed earlier. We have previously shown that for proton elastic scattering with energy > 2.5 MeV, the cross sections of light elements such as C, N, and O are enhanced, and this allows simultaneous profiling of N, C, and O over a sufficient depth range in a heavy material (19).

The RBS and PES techniques have been acknowledged as primary tools for heavy- and light-element detection in materials analysis. To a large extent the choice of technique depends on the nature of the problem. For example, low-energy backscattering is most suitable for surface studies and nuclear reactions for low-atomic-mass elements. Above 2.5 MeV, the proton-scattering cross section is influenced by nuclear force interaction, and it may be significantly enhanced compared with the Rutherford value. The two basic advantages of nuclear reactions with respect to backscattering are that they provide background-free detection of light elements on heavier substrates and that they allow perfect discrimination between two isotopes of the same element. In addition to its high sensitivity, nuclear reaction microanalysis allows a very fast determination of the absolute total amount of light elements in thin films up to 1 μm thick. This is especially useful for kinetic studies of compound formation. In the future, the PES technique should thus be extensively exploited. However, in order to quantitatively deter-

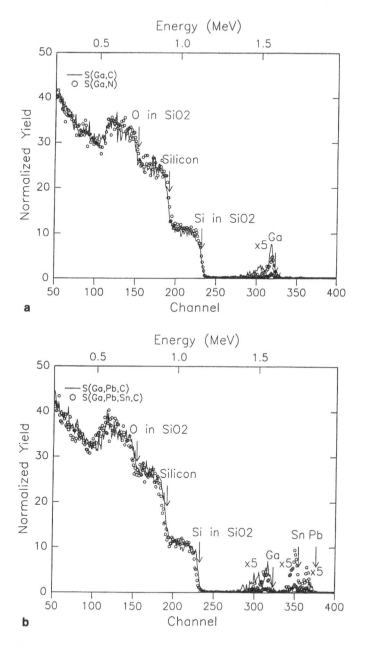

Figure 15   2.0-MeV⁴He⁺ RBS spectra of samples in a silicon-based material chip after it is annealed at 800°C for 30 min in Ar gas. (a) S(Ga, C) and S(Ga, N); (b) S(Ga, Pb, C) and S(Ga, Pb, Sn, C); (c) S(Pb, Sn, N) and S(Pb, N).

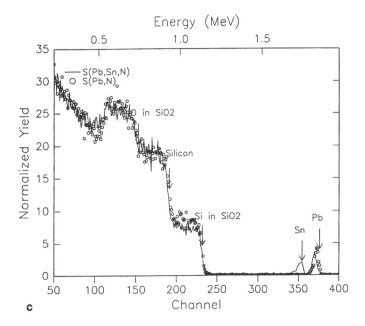

c

Figure 15

mine the light-element distribution, the proton-scattering cross section for various light elements need to be measured.

V. SUMMARY AND FUTURE DIRECTIONS

We have demonstrated panchromatic imaging of optical properties combined with element profiling on silicon-based materials libraries for effective evaluation of the formation of defects and novel phases. Samples coimplanted with C and Ga ions yielded a yellow-green CL emission, owing to the interaction between Ga and C atoms. This is corroborated by the matching of CL spectra and RBS measurements. Coimplantation of Sn and Pb ions induces an emission shift to higher-energy bands, accompanied by possible formation of Sn nanocrystals. No evidence of GaN nanocrystal formation was observed from CL and RBS measurements. Micro-PL measurement coupled with micro-Raman measurement for the samples implanted with different doses of C followed by annealing reveals a correlation between the C-related compound formation and optical properties. Depth profiling for C in the samples by PES analysis further reveals that there is significant diffusion of elementary carbon to the interface and top surface after

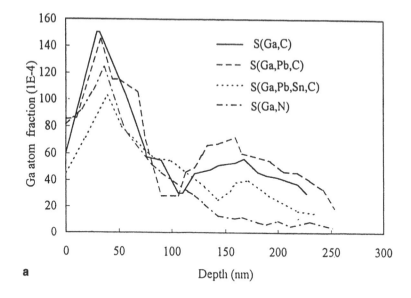

a

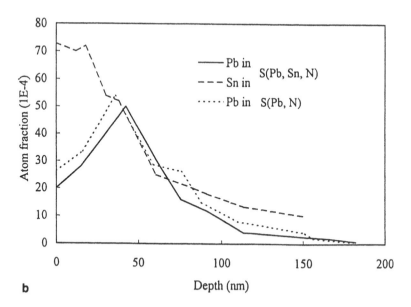

b

Figure 16 Ion concentration-depth profiles determined from the RBS measurements for samples in the silicon-based materials chip. (a) Ga profiles for S(Ga, C), S(Ga, Pb, C), S(Ga, Pb, Sn, C), and S(Ga, N); (b) profiles of Pb and Sn for S(Pb, Sn, N) are compared with the Pb profile for S(Pb, N).

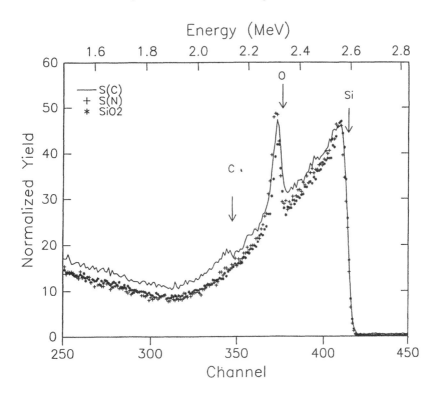

Figure 17 Proton elastic scattering spectra taken with a 3.0-MeV H$^+$ beam for SiO$_2$, S(C), and S(N).

thermal annealing. And TEM examination reveals that silicon carbide nanocrystals are formed at the SiO$_2$/Si interface. They can account for the dominant 560-nm emission in the C-implanted samples annealed at 1100°C for 100 min.

Using the combinatorial implantation method, we have farbicated multi-wavelength-emission materials libraries based on an Al$_{0.35}$Ga$_{0.65}$As/GaAs single QW and an Al$_{0.53}$Ga$_{0.47}$As/GaAs/AlAs/GaAs/Al$_{0.53}$Ga$_{0.47}$As coupled QW structures. Implantation of As ions into the Al$_{0.35}$Ga$_{0.65}$As/GaAs single QW at a dose of 5 × 10^{15} cm^{-2} induces a wavelength shift of 22 nm. For the proton-implanted samples, a maximum wavelength shift of 26 nm was observed. Of all the samples within the libraries, a maximum wavelength shift of 62 nm was obtained. More than 20 wavelengths were measured on a single-QW chip. For the Al$_{0.53}$Ga$_{0.47}$As/GaAs/AlAs/GaAs/Al$_{0.53}$Ga$_{0.47}$As coupled QW, clear wavelength shifts were obtained without high-temperature rapid thermal annealing. The combinatorial technique clearly is promising for WDM applications. Moreover, it can be utilized to set up a library to quickly find the optimum potential shape for the optical nonlinear application. This is superior to the fabrication of QW structures with special potential profiles by the expitaxy method. In addition, this is also useful for fabrication of multicolor-sensitive detectors, such as QWIP structures. Together with the mature technologies of ion implantation and semiconductor QW structures for optoelectronic devices, the present combinatorial implantation method will be highly effective in pursuing novel quantum electronic device structures.

In combinatorial materials science, ion beam techniques can play important roles in various aspects. Using ion beam implantation to synthesize nanocrystals offers a clear advantage over conventional techniques, since almost any ion can be embedded into essentially any substrate with extremely high chemical purity. Metal-doped glasses are promising candidates for applications in nonlinear integrated optics and photonics, i.e., in all optical switching device technology. In fact, glasses containing nanometer-sized metal clusters exhibit an enhanced intensity-dependent refractive index in the visible range, roughly seven orders of magnitude larger than silica glass (6,82).

Recently, we synthesized a diluted magnetic semiconductor (DMS) and nanomagnet-embedded semiconductor composite with high magnetoresistance at room temperature for the first time using the combinatorial implantation of magnetic ions into GaAs. This was accomplished using a scanning superconducting quantum interference device (SQUID) microscope and a physical property measurement system (PPMS) (83). Such DMS and hybrid ferromagnet-semiconductor structures are expected to play important roles in future spin-dependent electronics applications. There is also interest in these composite materials as strongly confined systems.

In summary, we have successfully developed combinatorial ion synthesis and ion beam analysis techniques to study ion-implanted materials libraries. We described the use of ion implantation to combinatorially synthesize luminescent materials libraries in SiO_2 matrices, $Al_{0.35}Ga_{0.65}As/GaAs$ single quantum wells (QW), and $Al_{0.53}Ga_{0.47}As/GaAs/AlAs/GaAs/Al_{0.53}Ga_{0.47}As$ coupled-QW structures. We have investigated their luminescent properties and depth-resolved distribution of elements in the libraries.

The combinatorial ion implantation can be used to rapidly dope semiconductors or form multielement composite nanocrystallites with core/shell structures in host materials, in order to obtain specific size-dependent optical and electronic properties. The ion beam analysis technique also provides a powerful tool for depth-resolved profiling of materials libraries. Apart from being a physical property evaluation tool, these techniques have a significant potential to increase the possibility of discovering of novel QW structures and new nanocrystallites embedded in hosts, and they will play a key role in ''in situ'' determining the composition–property relation of various combinatorial materials libraries. In order to quantitatively determine light-element distribution, the PES technique should be extensively exploited.

ACKNOWLEDGMENTS

The authors would like to thank all those who contributed to the work presented here, carried out at the Shanghai Institute of Nuclear Research and the Shanghai Institute of Technical Physics, the Chinese Academy of Sciences, and, in particular, Professors Dezhan Zhu and Jun Hu from the Shanghai Institute of Nuclear Research, Professor Wei Lu, Dr. Zhifeng Li, and Dr. Zhonglin Miao from the Shanghai Institute of Technical Physics. Dr. Changming Chen also thanks the financial support of the National Natural Science of China under Contract No. 19875075 and Shanghai Science and Technology Committee (Qimingxing Program) under Contract No. 00QE14035. The authors would also like to think Professor S.K. Pabi for proofreading the chapter.

REFERENCES

1. Carter, G.; Grant, W. A. Ion Implantation of Semiconductors; Edward Arnold: London, 1976.
2. White, C. W.; Budai, J. D.; Withrow, S. P.; Zhu, J. G.; Pennycook, S. J.; Zuhr, R. A.; Hembree, D. M., Jr.; Henderson, D. O.; Magruder, R. H.; Yacaman, M. J.; Mondragon, G.; Prawer, S. Encapsulated nanoscrystals and quantum dots formed by ion beam synthesis. Nucl Instrum Meth. 1997, B127, 545–552.
3. Alivisatos, A. P. Semiconductor clusters, nanocrystals, and quantum dots. Science. 1996, 271, 933–937.
4. Budai, J. D.; White, C. W.; Whithrow, S. P.; Chisholm, M. F.; Zhu, J.; Zuhr, R. A. Controlling the size, strucuture and orientation of semiconductor nanocrystals using metastable phase recrystallization. Nature. 1997, 27, 384–386.
5. Hosono, H. Chemical interaction in ion-implanted amorphous SiO_2 and application to formation and modification of nanosize colloid particles. J. Non-Cryst. Solids. 1995, 187, 457–472.
6. Chakraborty, P. Review: metal nanoclusters in glasses as non-linear photonic materials. J. Mater. Sci. 1998, 33, 2235–2249.
7. Gea, L. A.; Boatner, L. A.; Evans, H. M.; Zuhr, R. Optically active surfaces formed by ion implantation and thermal treatment. Nucl. Instrum Meth. B. 1997, 127/128, 553–556.
8. Gea, L.; Honda, S.; Boatner, L. A.; Haynes, T. E.; Sales, B. C.; Modine, F. A.; Meldrum, A.; Budai, J. D.; Beckers, L. Proceedings of the 1997 Fall Meeting of the MRS. A new approach to the fabrication of ''smart'' near-surface nanostructure composites; Gonsalves, K. E., Baraton, M. I., Chen, J. X., Akkara, J. A., eds.; Materials Research Society: San Francisco, 1997, 137–142.

9. Shi, J.; Gider, S.; Babcock, K.; Awschalom, D. D. Magnetic clusters in molecular beams, metals, and semiconductors. Science. 1996, 271, 937–941.

10. Nordund, K.; Averback, R. S. Inverse Kirkendall mixing in collision cascades. Phys. Rev. B. 1999, 59, 20–23.

11. Nordund, K.; Keinonen, J.; Ghaly, M.; Averback, S. Coherent displacement of atoms during ion irradiation. Nature. 1999, 398, 49–51.

12. Rissanen, L.; Dhar, S.; Lieb, K. P.; Engel, K.; Wenderoth, M. Ion-beam irradiation effects on NiN_3/Si bilayers. Nucl. Instrum. Methods B. 2000, 161–163, 986–991.

13. Chang, G. S.; Jung, S. M.; Song, J. H.; Kim, H. B.; Woo, J. J.; Byun, D. H.; Whang, C. N. Cohesive energy effects on the atomic transport induced by ion beam mixing. Nucl. Instrum. Meth. B. 1997, 121, 244–250.

14. Whitehead, N. J.; Gillin, W. P.; Bradley, I. V.; Weiss, B. L.; Claxton, P. A. Ion implantation induced mixing of InGaAs/InP multiquantum wells using phosphorus. Semiconductor Sci. Technol. 1990, 5, 1063–1066.

15. Tan, H. H.; Jagadish, C. Wavelength shifting in GaAs quantum well lasers by proton irradiation. Appl. Phys. Lett. 1997, 71, 2680–2682.

16. Poole, P. J.; Charbonneau, S.; Dion, M.; Aers, G. C.; Goldberg, R. D.; Mitchell, I. V. The fabrication of a broad-spectrum light-emitting diode using high-energy ion implantation. IEEE Photonics Technol Lett. 1996, 8, 1145–1147.

17. Chu, W. K.; Mayer, J. W.; Nicolet, M. A. Backscattering Spectrometry; Academic Press: London, 1978.

18. Chu, W. K.; Mayer, J. W.; Nicolet, M. A.; Buck, T. M.; Amsel, G.; Eisen, F. Principle and applications of ion beam techniques for the analysis of solids and thin films. Thin Solid Films. 1973, 17, 1–41.

19. Yang, G.; Zhu, D.; Xu, H.; Pan, H. Proton elastic scattering for light element cross-enhancement with $E_p > 1.5$ MeV. Nucl. Instrum. Meth. B. 1991, 61, 175–177.

20. Canham, L. T. Silicon quantum wire array fabrication by electrochemical and chemical dissolution of wafers. Appl. Phys. Lett. 1990, 57, 1046–1049.

21. Lu, Z. H.; Lockwood, D. J.; Baribeau, J. M. Quantum confinement and light emission in SiO_2/Si superlattices. Nature, 378, 258–260.

22. Hirschman, K. D.; Tstbeskov, L.; Duttagupta, S. P.; Fauchet, P. M. Silicon-based visible light-emitting devices intergrated into microelectronic circuits. Nature. 1996, 384, 338–341.

23. Song, H. Z.; Bao, X. M.; Li, N. S.; Zhang, J. Y. Relation between electroluminescence and photoluminescence of Si^+-implanted SiO_2. J. Appl. Phys. 1997, 82, 4029–4032.

24. Zacharias, M.; Christen, J.; Bäsing, J.; Bimberg, D. Visible luminescence from Ge nanocrystals embedded in α-$Si_{1-x}O_x$ films: correlation of optical properties and size distribution. J. Non-Cryst. Solids. 1996, 198–200, 115–118.

25. Liao, L. S.; Xiong, Z. H.; Zhou, X.; Liu, X. B.; Hou, X. Y. Photoluminescence from C^+-implanted Si_xO_y films grown on crystalline silicon. Appl. Phys. Lett. 1997, 71, 2193–2195.

26. Chen, C. M.; Liu, X. Q.; Li, Z. F.; Yu, G. Q.; Zhu, D. Z.; Hu, J.; Li, M. Q.; Lu, W. Aligned SiC nanocrystals at SiO_2/Si interface by C implantation into SiO_2 matrices. J. Vac. Sci. Technol. A. 2000, 18(5), 2591–2594.

27. Wolk, J. A.; Yu, K. M.; Bourret-Couchesne, E. D.; Johnson, E. Synthesis of GaN nanocrystals by sequential ion implantation. Appl. Phys. Lett. 1997, 70, 2268–2270.

28. Xiang, X. D.; Sun, X. D.; Briceno, G.; Lou, Y.; Wang, K. A.; Chang, H.; Wallace-Freedman, W. G.; Chen, S. W.; Schultz, P. G. A combinatorial approach to materials discovery. Science. 1995, 268, 1738–1740.

29. Yacobi, B. G.; Holt, D. B. Cathodoluminescence Microscopy of Inorganic Solids; Plenum Press: New York, 1990, 151–155.

30. Pierce, B. J.; hengehold, R. L. Depth-resolved cathodoluminescence of ion-implanted layers in zinc oxide. J. Appl. Phys. 1976, 47, 644–651.

31. Goldbergy, M.; Fitting, H. J.; Trukhin, A. Cathodoluminescence and cathodoelectroluminescence of amorphous SiO_2 films. J. Non-Cryst. Growth. 1997, 220, 69–77.

32. Chen, C. M.; Pan, H. C.; Zhu, D. Z.; Hu, J.; Li, M. Q. Cathodoluminescence and ion beam analysis of ion-implanted combinatorial libraries. Nucl. Instrum. Meth. B. 1999, 159, 81–88.

33. Rüter, D.; Rolf, S.; Bauhofer, W. Amorphous silicon-carbon–based thin films with efficient ultraviolet-excited photoluminescence and low self-absorbtivity in the emission spectral range. Appl. Phys. Lett. 1995, 67, 149–151.

34. Krautwasser, P.; Begun, G. M.; Angelini, P. Raman spectral characterization of silicon carbide nuclear fuel coatings. J. Am. Ceram. Soc. 1983, 66, 424–434.

35. Grunthaner, F. J.; Grunthaner, P. J.; Vasquez, R. P.; Lewis, B. F.; Maserjian, J. High-resolution x-ray photoelectron spectroscopy as a probe of local atomic structure: application to amorphous SiO_2 and the $Si–SiO_2$ interface. Phys. Rev. Lett. 1979, 43, 1683–1686.

36. Xiao, T.; Kitai, A. H.; Liu, G.; Nakua, A.; Barbier, J. Thin film electroluminescence in highly anisotropic oxide materials. Appl. Phys. Lett. 1998, 72, 3356–3359.

37. Ogino, T.; Aoki, M. Mechanism of yellow luminescence in GaN. Jap. J. Appl. Phys. 1980, 19, 2395–2405.

38. Koyama, H. Cathodoluminescence study of SiO_2. J. Appl. Phys. 1980, 61, 2228–2235.

39. Ginzburg, L. P.; Gordeev, A. A.; Gorchakov, A. P.; Jilinsky, A. P. Some features of the blue luminescence in $V\text{-}Si_{(1-x)}Ge_{(x)}O_2$. J. Non-Cryst. Solids. 1995, 183, 234–242.

40. Nakajima, A.; Futatsugi, T.; Horiguchi, N.; Yokoyama, N. Formation of Sn nanocrystals in thin SiO_2 film using low-energy ion implantation. Appl. Phys. Lett. 1997, 71, 3652–3654.

41. Brillson, L. J. Advances in characterizing and controlling metal–semiconductor interface. Appl. Surf. Sci. 1985, 22/23, 948–968.

42. Yuan, Shu; Kim, Yong; Jagadish, C.; Burke, P. T.; Gal, M.; Zou, J.; Cai, D. Q.; Cockayne, D. J. H.; Cohen, R. M. Novel impurity-free interdiffusion in GaAs/AlGaAs quantum wells by anodization and rapid thermal annealing. Appl. Phys. Lett. 1997, 70, 1269–1271.

43. Hughes, P. J.; Li, E. H.; Weiss, B. L. Thermal stability of AlGaAs/GaAs single quantum well structures using photoreflectance. J. Vac. Sci. Technol. 1995, B13, 2276–2283.

44. Liu, X. Q.; Li, N.; Chen, X.; Lu, W.; Xu, W. L.; Yuan, X. Z.; Li, N.; Shen, S. C.; Yuan, S.; Tan, H. H.; Jagadish, C. Wavelength tuning of GaAs/AlGaAs quantum-well infrared photodetectors by thermal interdiffusion. Jpn. J. Appl. Phys. 1999, 38, 5044–5045.

45. Deppe, D. G.; Holonyak, N. Atom diffusion and impurity-induced layer disordering in quantum well III-V semiconductor heterostructures. J. Appl. Phys. 1988, 64, R93–113.
46. Liu, X. Q.; Li, Z. F.; Chen, X. S.; Lu, W.; Shen, S. C.; Tan, H. H.; Yuan, S.; Jagadish, C. Arsenic implantation–induced intermixing effects on AlGaAs/GaAs single QW structures. Phys. Lett. A. 2000, 271, 213–216.
47. Liu, X. Q.; Lu, Wei; Chen, X. S.; Shen, S. C.; Tan, H. H.; Yuan, S.; Jagadish, C.; Johnston, M. B.; Dao, L. V.; Gal, M.; Zou, J.; Cockayne, D. J. H. Wavelength shifting of adjacent quantum wells in V-groove quantum wire structure by selective implantation and annealing. J. Appl. Phys. 2000, 87, 1566–1568.
48. Tan, H. H.; William, J. S.; Jagadish, C.; Burke, P. T.; Gal, M. Large energy shifts in GaAs-AlGaAs quantum wells by proton irradiation-induced intermixing. Appl. Phys. Lett. 1996, 68, 2401–2403.
49. Koteles, E. S.; Elman, B.; Melman, P.; Chi, J. Y.; Armiento, C. A. Partial intermixing of strained InGaAs/GaAs quantum well. Optical Quantum Electron. 1991, 23, 981–984.
50. Kash, K.; Tell, B.; Grabbe, P.; Dobisz, E. A.; Craighead, H. G.; Tamargo, M. C. Aluminum ion-implantation enhanced intermixing of GaAs-AlGaAs quantum well structures. J. Appl. Phys. 1988, 63, 190–194.
51. Elman, B.; Koteles, E.; Melman, P.; Armiento, C. GaAs/AlGaAs quantum well intermixing using shallow ion implantation and rapid thermal annealing. J. Appl. Phys. 1989, 66, 2104–2107.
52. Straus, C. Wavelength division multiplexing (WDM). J. Telesis. 1980, 7, 2–6.
53. Zhou, C.; Yang, Y. Multicast communication in a class of wide-sense non-blocking optical WDM networks; Proceedings 7th International Conference on Computer Communications and Networks, 321–328.
54. White, A. E.; Gates, J. V.; Bruce, A. J.; Cappuzzo, M. A.; Gomez, L. T.; Henry, C. H.; Laskowski, E. J.; Madsen, C. K.; Muehlner, D. J.; Shmulovich, R. S. Integrated optics for WDM. J. Lucent. Technol, IEEE/LEOS Summer Topical Meeting, 1998, II/51–52.
55. Koren, U.; Koch, T. L.; Miller, B. I.; Einsenstein, G.; Bosworth, R. H. Wavelength division multiplexing light source with integrated quantum well tunable lasers and optical amplifiers. Appl. Phys. Lett. 1989, 54, 2056–2058.
56. Tan, H. H.; Jagadish, C. Wavelength shifting in GaAs quantum well lasers by proton irradiation. Appl. Phys. Lett. 1997, 71, 2680–2682.
57. Poole, P. J.; Charbonneau, S.; Dion, M.; Aers, G. C.; Goldberg, R. D.; Mitchell, I. V. The fabrication of a broad-spectrum light-emitting diode using high-energy ion implantation. IEEE Photonics Technol. Lett. 1996, 8, 1145–1147.
58. Li, E. H. Interdiffusion as a means of fabricating parabolic quantum wells for the enhancement of the nonlinear third-order susceptibility by triple resonance. Appl. Phys. Lett. 1996, 69, 460–462.
59. Tada, K.; Nishimura, S.; Ishikawa, T. Polarization-independent optical waveguide intensity switch with parabolic quantum well. Appl. Phys. Lett. 1991, 59, 2778–2780.
60. Huang, Y.; Lien, C. The enhancement of optical third harmonic susceptibility in a parabolic quantum well by triple resonance. J. Appl. Phys. 1994, 75, 3223–3225.

61. Liu, X. Q.; Li, N.; Lu, W.; Li, N.; Yuan, X. Z.; Shen, S. C.; Tan, H. H.; Fu, L.; Jagadish, C. Wavelength tuning of GaAs/AlGaAs quantum-well infrared photo-detectors by proton implantation induced intermixing. Jpn. J. Appl. Phys. 2000, 39, 1687–1689.

62. Li, N.; Li, N.; Lu, W.; Liu, X.Q.; Yuan, X.Z.; Li, Z.F.; Dou, H.F.; Shen, S.C.; Fu, Y.; Willander, M.; Fu, L.; Tan, H.H.; Jagadish, C.; Johnston, M.B.; Gal, M. Proton implantation and rapid thermal annealing effects on GaAs/AlGaAs quantum well infrared photodetector. Superlattices Microstructures. 1999, 26, 317–24.

63. Johnston, M. B.; Gal, M.; Li, Na; Chen, Z. H.; Liu, X. Q.; Li, Ning; Lu, Wei; Shen, S. C. Interdiffused quantum-well infrared photodetectors for color sensitive arrays. Appl. Phys. Lett. 1999, 75, 923–925.

64. Hirayama, Y.; Suzuki, Y.; Okamoto, H. Ion-species dependence of interdiffuion in ion-implantation GaAs-AlAs superlattices. Jpn. J. Appl. Phys. 1985, 24, 1498–1502.

65. Liu, X. Q.; Chen, C. M.; Li, Z. F.; Lu, W.; Zhu, D. Z.; Hu, J.; Li, M. Q.; Shen, S. C. Application of combinatorial approach to fabrication of multi-wavelength emitting material Chip. Appl. Phys. Lett. 1999, 75, 2238–2240.

66. Sauer, R.; Thonke, K.; Tsang, W. T. Photoinduced space-charge buildup by asymmetric electron and hole tunneling in coupled quantum wells. Phys. Rev. Lett. 1988, 61, 609–612.

67. Livescu, G.; Fox, A. M.; Miller, D. A. B.; Sizer, T.; Knox, W. H.; Gossard, A. C.; English, J. H. Resonantly enhanced electron tunneling rates in quantum wells. Phys. Rev. Lett. 1989, 63, 438–441.

68. Roskos, H. G.; Nuss, M. C.; Shah, J.; Leo, K.; Miller, D. A. B.; Fox, A. M.; Rink, S. S.; Kohler, K. Coherent submillimeter-wave emission from charge oscillations in a double-well potential. Phys. Rev. Lett. 1992, 68, 2216–2219.

69. Suen, Y. W.; Engel, L.W.; Santos, M. B.; Shayegan, M.; Tsui, D. C. Observation of a $\gamma = 1/2$ fractional quantum Hall state in a double-layer electron system. Phys. Rev. Lett. 1992, 68, 1379–1382.

70. Takagaki, Y.; Muraki, K.; Tarucha, S. Splitting of resistance peaks and anomalous Hall plateaus in asymmetric double-quantum well structures. Phy. Rev. 1997, B56(3), 1057–1060.

71. Redinbo, G. F.; Craighead, H. G. Proton implantation intermixing of GaAs/AlGaAs quantum wells. J. Appl. Phys. 1993, 74, 3099–3102.

72. Domoto, C.; Vaccaro, P. O.; Ohtani, N. Population inversion between subbands in simple periodical GaAs/AlAs superlattices. IEEE Proceedings—Optoelectronics. 2000, 147, 225–228.

73. Sirtori, C.; Kruck, P.; Barbieri, S.; Collot, P.; Nagle, J.; Beck, M.; Faist, J.; Oesterle, U. GaAs/Al$_x$Ga$_1-x$ As quantum cascade lasers. Appl. Phys. Lett. 1998, 73, 3486–3488.

74. Faist, J.; Capasso, F.; Sivco, D. L.; Sirtori, C.; Hutchinson, A. L.; Cho, A. Y. Quantum cascade laser. Science. 1994, 264, 553–556.

75. Cibert, J.; Petroff, P. M.; Werder, D. J. Kinetics of implantation-enhanced interdiffusion of Ga and Al at GaAs-Ga$_x$Al$_1-x$As interfaces. Appl. Phys. Lett. 1986, 49(4), 223–225.

76. Werner, J.; Kapon, E.; Stoffel, N. G.; Colas, E.; Schwarz, S. A.; Schwarta, C. L.; Andreadakis, N. Integrated extent cavity GaAs/AlGaAs lasers using selective quantum well disordering. Appl. Phys. Lett. 1989, 55, 540–542.

77. Wang, J.; Yoo, Y.; Gao, C.; Takeuchi, I.; Sun, X. D.; Xiang, X. D.; Schultz, P. G. Identification of a blue photoluminescent composite material from a combinatorial library. Science. 1998, 279, 1712–1714.

78. Barradas, N. P.; Jeynes, C.; Mironov, O. A.; Phillips, P. J.; Parker, E. H. C. High depth resolution Rutherford backscattering analysis of Si–Si$_{0.78}$Ge$_{0.22}$/(001) Si super-lattices. Nucl. Instrum. Meth. B. 1998, 139, 238–241.

79. Link, F.; Baumann, H.; Bethge, K.; Klewe-Nebenius, H.; Bruns, M. C and N depth profiles of SiCN layers determined with nuclear reaction analysis and AES. Nucl. Instrum. Meth. B. 1998, 139, 268–272.

80. Biersack, J. P.; Haggark, L. G. A Monte Carlo computer program for the transport of energetic ions in amorphous targets. Nucl. Instrum. Methods B. 1980, 174, 257–269.

81. Doolittle, L. R. Algorithms for the rapid simulation for Rutherford backscattering spectra. Nucl. Instrum. Meth. 1985, B9, 344–351.

82. Bertoncello, R.; Gross, S.; Trivillin, F.; Cattaruzza, E.; Mattei, G.; Caccavale, F.; Mazzoldi, P.; Battaglin, G.; Baolio, S. Mutually reactive elements in a glass host matrix: Ag and S ion implantation in silica. J. Mater. Res. 1999, 14, 2449–2457.

83. Chen, C. M. To be published.

8

Mapping of Physical Properties: Composition Phase Diagrams of Complex Materials Systems Using Continuous Composition Materials Chips

Young K. Yoo and Xiao-Dong Xiang
Intematix Corporation, Moraga, California, U.S.A.

I. INTRODUCTION

Understanding complex materials systems is a major challenge of 21st century science. Historically the scope of materials research has been limited mostly to single-element or binary systems. Discovery of high-temperature superconductors in 1986 greatly accelerated the exploration into complex materials systems. With rapidly growing demands for better functional materials from high-technology industries, complex materials are increasingly receiving more attention in the materials research community. As a consequence, the basic research efforts in condensed-matter physics are also driven toward more complex aspects of phenomena in condensed-matter systems.

One such example is the highly correlated electronic systems. In these systems, the strength of the electron–electron interactions is comparable to the kinetic energy of the electrons. The effects of strong interelectron interactions give rise to remarkably rich physics, producing phenomena such as the fractional quantum Hall effect, colossal magnetoresistivity, and high-temperature superconductivity. Modeling of these systems has proven difficult, since the strong electron–electron interactions reduce the applicability of field theory methods; the quasi-particle approximation is no longer accurate. The experimental challenges are equally daunting. In the past, discovery of materials exhibiting these phenomena was very often serendipitous. It took more than 70 years for scientists to discover high-temperature superconductivity in cuprate systems since the initial discovery of superconductivity in 1911. While the original discovery involved simple one-element metal systems, high-temperature superconductivity was first found in multicomponent (consisting of more than four elements) oxide systems. Obviously the discovery process gets even more difficult and time-consuming as the materials consist of more elements. In addition, conventional studies are usually performed on individual samples of discrete compositions, and they often miss the subtle details and important correlations between different compositions.

If this situation is to change, effective and systematic methods to explore complex systems are needed. This effort is important for the theoretical understanding of complex systems. If we take the case of high-temperature superconductivity, nothing in the history of science drew such an intense and concentrated global research effort to solve a problem. The euphoria just after the discovery is still fresh in our memory, but one wonders after 15 years of intense global efforts if we are going to reach a point of understanding the underlying microscopic mechanism any time soon. Lack of reliable, thorough, and systematic experimental data to guide the theoretical efforts is partially attributed to this.

Since 1995 we have carried out a systematic research effort to develop methods and techniques that promise to dramatically change the situation. In our initial efforts, the discovery aspect of the exploration was emphasized. The experimental approach (mainly discrete composition material chip technique) was designed to explore the multicomponent-phase space as broadly as possible. In the next two sections, we describe these studies. It soon became apparent that this approach has very limited use in studying complex physics phenomena. We then adopted a more thorough and systematic approach, i.e., the continuous phase mapping of physical property using an epitaxial thin-film CPD growth technique combined with various high-throughput and high-resolution screening tools. We discuss our efforts in this direction in Section IV.

II. SUPERCONDUCTORS

Superconductors are among the most fascinating materials, both in terms of underlying physics and in their potential applications. The discovery of copper oxide high-T_c superconductors is one of the most important advances in materials science of the last century. More than 10 years of extensive research efforts by thousands of scientists worldwide following the initial discovery by Bednorz and Müller represent one of the most intensive material explorations in modern history (1). The discovery has raised the hope for large-scale applications of superconductivity in many areas, ranging from energy transport/storage and high-performance electronics to communication technology and NMR tomography. Even more futuristic applications, such as levitating trains, magnetohydronamic propulsion in ships, and uses in fusion reactors, were also proposed. Although many groups are trying to solve the material engineering problems that exist for the known superconductors, the discovery of new materials with higher critical temperatures (T_c), higher upper critical fields, higher critical currents, and better material properties for fabrication are clearly needed. Such materials might find immediate application in power transmission cables and wireless communication components, applications currently being pursued by companies such as Conductus, American Superconductors, and Superconducting Technologies, Inc.

A simplistic analysis of the critical temperature of the known superconductors shows a clear increase in T_c with increasing chemical complexity, as illustrated in Figure 1 (2). One explanation for this observation is that more complex structures offer more tunability in crystal chemistry for optimizing certain properties. However, among the known 24,000 inorganic phases estimated by J. C. Phillips in 1989, only 8000 are ternary compounds, compared with 16,000 binary compounds (3). When compared with roughly 100,000 ternary compounds that one might reasonably make, given only one stoichiometry of each, it becomes clear that a large number of compositions are yet to be examined for superconductivity. When one considers compounds consisting of a large number of elements, such as the quaternary compound $YBa_2Cu_3O_{7-\delta}$, the number of possibilities increases even further. Although these multicomponent compounds have the highest critical temperatures, not too much work has been reported on such systems. In fact, the quaternary compound $YBa_2Cu_3O_{7-\delta}$ is not only the first quaternary superconductor with T_C exceeding liquid N_2 temperature, but also the first quaternary metal oxide ever found (3).

Because of the overwhelming number of compositions that must be searched to find new classes of superconductors, most of the previous efforts were limited to compositions and structures related to known compounds. Elements are replaced one at a time by a time-consuming trial-and-error process. Most of the attention until 1986 was focused on intermetallic compounds such as Nb_3Ge. Few groups had considered oxides as reasonable candidates until the serendipitous discovery of the $La_{2-x}Sr_xCuO_4$ superconductor. Shortly thereafter it was found that $YBa_2Cu_3O_{7-\delta}$ has a T_c of 93 K (4). However, this superconductor, in contrary expectations, had a new structure as well. Although the Bi-, Tl-, and Hg-containing copper oxide superconductors have relatively similar structures, it took several years of intensive effort to find them. Indeed, relatively little effort has been spent searching, for example, multicomponent bismuth oxides for superconductivity, in spite of the report of a $Ba_{1-x}K_xBiO_{3-\delta}$ superconductor with a T_c above 30 K (5).

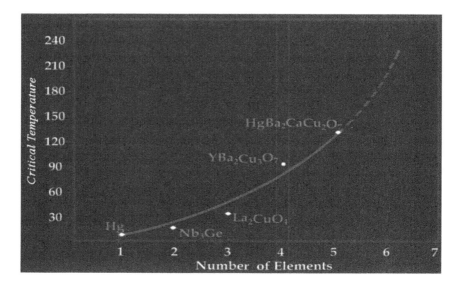

Figure 1   Relationship between T_c and the number of elements in a superconductor.

The search is further complicated by the relatively poor state of synthetic methods (in comparison to organic chemistry, where virtually any structure can be rationally synthesized) and the lack of a firm theoretical basis for correlating superconductivity with structure or even for the possibility of forming certain structures. Therefore, a combinatorial approach involving the synthesis and screening of a large phase space of solid-state materials could dramatically affect the search for new classes of superconductors with enhanced properties. Of course, empirical observations, structural knowledge, theory, and intuition should all be used to guide the design of the combinatorial materials chips.

A. Early Work: A Proof of Principle Using Discrete Material Chips

We conducted our early study in 1994 to demonstrate the feasibility of using the combinatorial approach to search for copper oxide superconductors (6). In this study, we tried to answer this question: If the technique were available 10 years ago, could we have used it to discover $YBa_2Cu_3O_{7-x}$ and $Bi_2Sr_2Ca_2Cu_3O_{10}$ superconductors after the initial discovery of $La_{1-x}Sr_xCuO_4$?

Two factors were considered initially in developing our early techniques. First, the methods were devised to rapidly generate a large number of spatially addressable arrays of solid-state materials on a single substrate in a form conducive to screening techniques and thereby to speed up both the synthesis and the characterization process. Second the methods should produce not only thermodynamically stable but also metastable phases. This consideration has led us to choose thin-film deposition/synthesis methods to generate a large collection of different compositions.

Discrete materials chip were generated by depositing precursors through a series of binary masks using RF magnetron sputtering at room temperature. The molecular ratio of deposited precursors was kept at a simple 1:1 to avoid using most after-the-fact knowledge being used about stoichiometry in the design. The film deposition thickness was monitored with a crystal microbalance and calibrated independently with a profilometer. The uniformity of the deposited films varied approximately 5% over a 2-inch-diameter area. MgO single crystals with a^{100} oriented surface were used as substrates. Libraries were generated using a binary masking scheme (Figure 2). Layers of precursors were sputtered through the binary masks step by step. In a binary synthesis, a total of 2^n compounds are formed for a given number of masking steps (n). The resulting array contains all combinations that can be formed by deleting one or more steps from the entire deposition/masking sequence.

Initially, a materials chip consisting of 16 different members was generated using the following deposition sequence: 1, Bi_2O_3, M1; 2, PbO, M2; 3, CuO, M0; 4, CaO, M3; 5, $SrCO_3$, M4. The first entry designates the deposition step, the second the element, and the third the mask. Approximately 660 Å of CuO

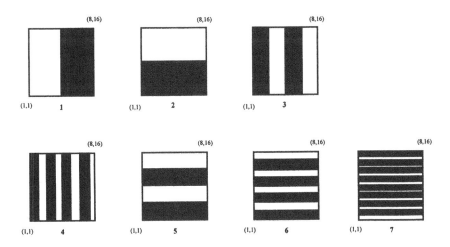

Figure 2 Binary masks used for library synthesis. Numbers in the lower left-hand and upper right-hand corners indicate the position of each member in the library. M0 corresponds to no secondary mask. (Reproduced from Ref. 6.)

was deposited, and the molar stoichiometry of each element relative to Cu was approximately 1:1. The chip was then sintered at 840°C in air. Following this process, the resistance of each site was measured using an in-line four-probe configuration. Two films were found to be superconducting, and the resistance of each was measured as a function of temperature down to 4.2 K in a liquid helium cryostat. The critical temperatures (T_C) of the BiCuCaSrO and BiPbCuCa-SrO films were 80 K and 90 K, respectively, as shown in Figure 3a. The BiPbCu-CaSrO sample has a metallic resistivity from room temperature to about 100 K, whereas the resistivity of the BiCuCaSrO film increases slowly with decreasing temperature down to the critical temperature. The critical temperatures as well as the x-ray diffraction patterns of these samples are consistent with those reported in the literature for $Bi_2Sr_2Ca_1Cu_2O_8$.

A 128-member chip was then generated to further examine the effects of stoichiometry and deposition sequence on the properties of the BiSrCaCuO films (Figure 4). The library was generated as follows: 1, Bi_2O_3, M0; 2, Bi_2O_3, M1; 3, CuO, M0; 4, CuO, M2; 5, CuO, M3; 6, $SrCO_3$, M0; 7, $SrCO_3$, M5; 8, CaO, M6; 9, CuO, M4; 10, CaO, M7. The molar stoichiometry for each layer was 1:1 relative to Bi (which was deposited as a 300-Å layer), with the exception of steps 3 and 5, in which the Cu:Bi ratio was 0.5:1. Samples were sintered and measured as described earlier. In general, films with low resistivities, which resulted in hundreds of ohms, showed metallic behavior with onset Tc of 80–90 K. Films with excess Ca and Cu (Bi:Sr:Ca:Cu ratios of 2:2:4:4 and 2:2:4:5) showed a 110 K phase, consistent with the formation of Bi_2Sr_2 $Ca_2Cu_3O_{10}$. A number of films with identical stoichiometries but different deposition sequences displayed different resistivity-versus-temperature profiles (e.g., BiCuSrCaCuCaO and BiCuCuSrCaCaO in Figure 3b and 3c), suggesting that different phases may be accessible by controlling the sequence in which the layers diffuse.

The fact that libraries are generated by sequentially depositing precursors as thin films, followed by solid-state reaction, distinguishes this approach from conventional bulk synthesis and standard thin-film fabrication methods. Special processing is needed for proper interdiffusion between precursor layers and subsequent crystalline-phase formation. For example, in order to synthesize the $YBa_2Cu_3O_{7-x}$ superconductor from a film generated by sequentially depositing BaF_2, Y_2O_3 and CuO in a 1:2:3 molar ratio (sl400 Å of CuO), it was necessary to anneal samples at low temperature (200–400°C) for an extensive period prior to a high-temperature (840°C) sintering. The diffusion process at low temperatures facilitates the formation of a homogeneous amorphous intermediate without nucleation of stable lower order phases.

This protocol was used to reproducibly synthesize $YBa_2Cu_3O_{7-x}$ superconducting films down to 200-μm × 200-μm size with a metallic behavior and a T_c of ~90 K (Figure 3d). A 128-member chip fabricated from $BaCO_3$, Y_2O_3, Bi_2O_3, CaO, $SrCO_3$, and CuO was then generated to determine the compatibility

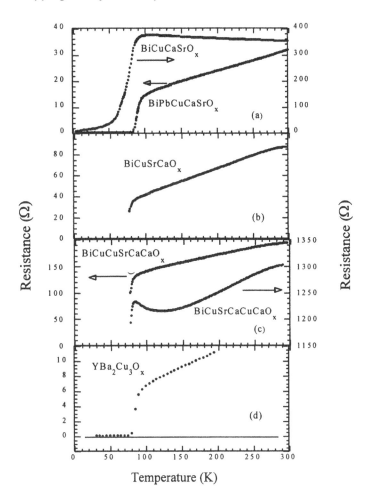

Figure 3 Resistivity-vs.-temperature plots for (a) BiPbCuCaSrO, BiCuCaSrO; (b) BiCuSrCaO; (c) Bi CuSrCaCuCaO, BiCuCuSrCaCaOn; (d) YBa$_2$Cu$_3$O$_x$. (Reproduced from Ref. 6.)

of different families of copper oxide superconductors with one common processing condition. After annealing at low temperature (200–400°C) and high-temperature sintering at 840°C, sites containing BiSrCaCuO$_x$ and YBaCuO$_x$ were all found to be superconducting. Even though these sites do not have the correct stoichiometry of superconducting compounds and the sintering temperature was not optimal for individual superconducting compound synthesis, the study did successfully result in superconductor phases.

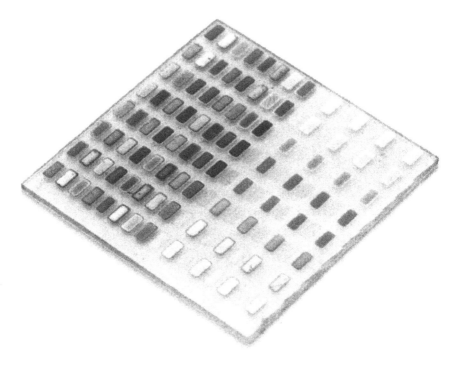

Figure 4   The 128-member binary library prior to sintering. Dimension of each discrete sample is 1×2 mm^2. The color of each sample is the natural color of reflected light from a white light source. (Reproduced from Ref. 6.)

Although the proof of concept of our technique was first demonstrated in copper oxide superconductors, using the approach to realistically search for new superconductors still represents a challenging task. We have few guidelines to follow as to where to search. We also have to implement sensitive screening techniques at cryogenic temperatures. The scanning evanescent microwave probe, the magneto-optical imaging system, and scanning SQUIDS at low temperatures to perform noncontact electrical and Meissner effect measurements are some of the techniques that may prove to be useful for this purpose.

B.   Search Strategy

We believe that theoretical and empirical guidance would be useful to help assist future exploration for superconductors. Here, we discuss only a few observations based on early knowledge in the field. We believe that more input from experts and theorists would be very beneficial for this effort.

1.   Structural Chemistry

At the time of this writing, all of the known superconductors with T_c above 40 K are copper oxides, with the exception of recently discovered MgB2 (7). These compounds all contain CuO_2 planes separated by intermediate layers. It is widely believed that it is the CuO_2 planes that carry the superconducting current and the intermediate layers that stabilize the lattice structure and act as a charge reservoir. Empirically it also seems that the number of CuO_2 planes that are closely coupled together and the charge concentration in these planes dominate the critical temperatures. It is therefore desirable to examine multicomponent structures with more closely coupled CuO_2 layers and different intermediate layers that can both stabilize the structure and supply the right amount of charge to the CuO_2 layers.

   Knowledge of structural chemistry can be used very effectively to design libraries and search for such structures. For example, most, if not all, multicomponent copper oxide superconductors can be built up based on some basic building principles. It has been pointed out that all known copper oxide superconductors can actually be derived by stacking different amounts and sequences of rock salt and perovskite-like layers of metals and oxygen (8,9). An $A_2BO_4 = (AO)(ABO_3)$ structure of the first high-T_c copper oxide structure, La_2CuO_4, is made up of a perovskite ABO_3 building block and a rock salt AO building block. If we consider the perovskite structure to be built up from CuO_2 and AO with the stacking sequence of CuO_2—AO—CuO_2—AO—, the La_2CuO_4 structure is built up with the stacking sequence of (CuO_2)—(AO)—(AO)—CuO_2—, (where the subscript c indicates that these layers have shifted by a half unit cell in a lateral direction), a general formula $A_{n+1}B_nO_{3n+1-\delta}$ can be used to guide the search for more complicated layered structures. $La_2CaCu_2O_6$ ($n = 2$) structure is an example of this building principle. Upon properly doping this structure, a T_c of about 60 K is observed (10). $Tl_mBa_2Ca_{n-1}Cu_nO_{m+2}$ is another structural family containing several high-T_c copper oxide superconductors (three with the highest T_c all belong to this family). In this family, Tl has been replaced with Bi and Hg, and Ba with Sr, and structures with $m = 1, 2$ and $n = 1, 2, 3, 4$ have been identified. There are also others (11).

   The same idea may be used to search for novel superconducting bismuth oxides beyond known compounds. Copper oxides and bismuth oxides can both form basic perovskite structures. A bismuth oxide ($ABiO_3$) can superconduct at above 30 K, as does the copper oxide A_2BO_4 structure, even though a metallic perovskite copper oxide $LaCuO_{3-\delta}$ is not a superconductor. Bi, like Cu, can have mixed valences. The structural and elemental replacement principles used in copper oxides may be adopted to construct multicomponent layered Bi oxides.

Many more transition metals form simple perovskite structures; an incomplete list includes Sc, Ti, V, Cr, Mn, Fe, Co, Ni, Mo, Nb, and Pb. However, high-order perovskite-related structures of these elements have not been explored. It may be possible to use similar construction rules to search for more complicated layered structures. It is interesting to note that these structures have many other important properties, such as ferroelectricity and magnetoresistance.

2. Empirical Observations

Villars and Phillips have noted a correlation between T_c and Villar's coordinators—the average valence electron number, orbital radii differences, and metallic electronegativity differences (11). They have mapped 600 low-T_c ($1 \text{ K} < T_c < 10 \text{ K}$) and 70 high-$T_c$ ($>10 \text{ K}$) superconductors using Villars coordinators. For low T_c superconductors the map is scattered. However, for high-T_c superconductors, the map reveals three clustered islands, which fill less than 1% of the three-dimensional volume. In island A, one finds intermetallic compounds of the A15 family as well as other complex intermetallic structures. Island B is largely dominated by the NbN family, although it also contains borides and carbides. Chevrel sulfides, copper oxides, and bismuth oxides, three of the most important classes of high-T_c superconductors, are all found in island C. This correlation between Villars coordinators and high-T_c superconductivity can be used to guide a combinatorial search. Villars and Phillips have compiled promising new candidates for high-T_c superconductors based on their analysis of known structures, as shown in Figure 5. They have also compiled a list of 1500 existing ternary oxide parent compounds that are in island C. Combinatorial material chips can be used to search this collection of compounds as well as to search for new islands.

Other systems may also be interesting to explore, including: cluster structures (such as Chevrel phases), CuCl-, SnO-, CdS-, etc., -based structures. Our strategy again will be to try to build up multicomponent layered structures with simple building blocks.

In all previously discussed cases, after a candidate parent structure is chosen, very often proper doping is crucial to induce metallic conductivity and therefore possibly superconductivity. A combinatorial strategy may be very effective in this regard. Many different dopants and dopant levels can be examined in one material chip screening.

III. MAGNETORESISTIVE MATERIALS

Magnetic memory remains a key technology basis for information technology. Currently, magnetic sensor technology is one of the enabling technologies in the $100 billion magnetic memory industry. An increase in the sensitivity of the magnetic sensing head will dramatically increase the areal density (bits/unit area)

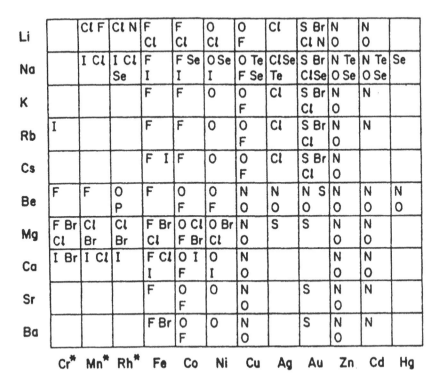

Figure 5 Tableau of promising pseudoternary (quaternary) candidates for high-T_c superconductors with compositions paralleling $YBa_2Cu_3O_2$. The compositions have the general formula $([2s +f]/3)_3(p,d)_{3p7}$, where f is Y, La, or a rare earth. The s elements are listed in the ordinate. On the abeissas are the p, d elements. Because of similar coordinates, the Mn* column includes Re, Ru, and Os, the Rh* column includes Ir, Pd, and Pt, and the Cr* column includes Mo, W, O, S, Se, and Te. (Reproduced from Ref. 11.)

of information storage. An important area is magnetic random access memory devices (MRAM). A low-cost, high-density, fast, and low-energy-consuming MRAM is very important in many existing and new applications.

Magnetoresistance (MR) refers to the change in the electrical resistance of a material in the presence of an applied magnetic field. Two figures of merit are important for applications. The first is the sensitivity, $(\Delta R/R_0)/\Delta H$, where R_0 is the zero field resistance. This is the slope of the curve of magnetoresistance, $\Delta R/R_0$, vs. magnetic field H at low field. The second is the maximum dynamic range MR ratio $(\Delta R/R_0)_{max}$, which is taken as the MR value before saturation at higher field. High sensitivity at low field (<100 gauss) is especially important in magnetic recording applications. The magnetic recording industry first replaced the old-fashioned inductive head with ferromagnetic MR heads to give a 50% improvement on performance. Permalloy, a mixture of 80% nickel and 20% iron, was used by the industry with values of $(\Delta R/R_0)/\Delta H \sim 0.5\%/Oe$ and $(\Delta R/R_0)$max of 3% (12).

The discovery and development of a new class of MR materials, known as giant magnetoresistance (GMR) materials, shows much better performance than permalloy. The successive layers of ferromagnetic $3d$ transition metals are indirectly exchange coupled through spacer layers, which can be any of the nonferromagnetic $3d$, $4d$, and $5d$ transition metals. The ferromagnetic layers are magnetically coupled, with their magnetization vector either parallel (ferromagnetic coupling) or antiparallel (antiferromagnetic coupling) relative to each other. The exchange coupling oscillates with the thickness of the nonmagnetic spacer layer, and the strength varies with the d-band filling of the spacer. For samples having antiparallel magnetic alignment (antiferromagnetic coupling), the application of a magnetic field overcomes the antiferromagnetic coupling and aligns the magnetic moments of the ferromagnetic layers in the same direction. Since the electrons scatter less at the interfaces when the magnetic moments of the layers are aligned, the electric resistance decreases. This negative MR effect $(\Delta R/R_0)_{max}$ ranges from 5% to as large as 67% in superlattice GMR materials. $(\Delta R/R_0)_{max}$ $\sim 67\%$ with $(\Delta R/R_0)/\Delta H \sim 0.001/Oe$ (at low temperature) was found in Fe/Cr superlattices (13) and $(\Delta R/R_0)_{max} \sim 17\%$ with $(\Delta R/R_0)/\Delta H$ 0.4%/Oe (at room temperature) in NiFeCo/Cu superlattices (14).

Several years ago GMR sensors moved out of the laboratory, and they are now integrated into commercial product as ''read head'' for magnetic disk drives (15). Here, a GMR sensor is used to read the state of magnetic bits. Because the capacity of disk drives continues to grow rapidly as they shrink in size, GMR read sensors become increasingly important. The data in disk drives are written in a tiny region of magnetic strip made of a thin-film magnetic material. The information (bits of 1 and 0) is read by sensing the magnetic fields just above these magnetic strips on the disk. As the density increases, these regions get smaller and the magnetic fields that must be detected to read the data become weaker. Magnetic reading heads that employ the GMR effect are the best technology currently available for detecting the fields from these tiny magnetic strips. It is expected that the GMR read sensor will allow disk capacity to reach at least 10 gigabits per square inch. At this density, a typical 3.5-inch disk drive can store 120 billion bits. Magnetoresistive-based MRAMs are also being developed that may dramatically change the situation in RAM technology (16).

There is another class of materials that has manifested an even more dramatic MR effect, which has been dubbed as colossal magnetoresistance (CMR). The MR effect in these materials is intrinsic and arises from a complex mechanism still unresolved. The first class of CMR materials, perovskite manganites in

$(La_{1-x}Ca_x)$ MnO_3, was initially synthesized by Van Santen and Jonker (17). It undergoes the metal–insulator (MI) transition at the Curie point, which is the reverse of the usual MI transition in semiconductors: It becomes a metallic phase at low temperatures and insulating at high temperatures. The MR effect can be extremely high, and the resistance changes by many orders of magnitude. The $(\Delta R/R_0)/\Delta H$ achieved in $(La_{2/3}Ca_{1/3})MnO_3$ is 0.05%/Oe with $(\Delta R/R_0)_{max}$ of 99.9% (18).

The coexistence of metallic conductivity and ferromagnetic coupling in these materials has been explained in terms of a double exchange (DE) mechanism (19–21). Crystal fields split the Mn $3d$ orbitals into two e_g orbitals that strongly hybridized with the oxygen p orbitals and three lower-energy localized t_{2g} orbitals. The DE mechanism requires a mixed-valence state of Mn^{3+} and Mn^{4+} cations in which the e_g electrons hop from a Mn^{3+} to a Mn^{4+} via the oxygen orbitals. The electron-hopping rate in these materials (t_{ij}) is $t_0 \cos(\theta_{ij}/2)$ according to the double exchange theory, where θ_{ij} is the angle between local spin directions of electrons on neighboring sites and t_0 is the hopping rate in a perfect ferromagnetic state. The application of an external field may align the local spins and thereby increase electron conduction or even induce an insulator-to-metal transition.

The CMR effect occurs when the Curie temperature of the materials is shifted toward a higher temperature by application of magnetic field, thus manifesting the MR near the Curie-temperature region. This is expected because the application of magnetic field is more amenable to the spin alignment in spite of thermal fluctuation, and this is translated to the shift in the Curie temperature to a higher temperature than without magnetic field. Once the spins are aligned, more electron carriers are available by double exchange interaction. Although existing simple theories can qualitatively explain the CMR effect, they cannot describe the quantitative values of MR observed in some materials and many other complex phenomena in this class of compounds. These phenomena may arise from the complex interplay among several competing/cooperating effects, both magnetic and electronic; spin or charge fluctuations, the Jahn–Teller effect, charge ordering, orbital ordering, polarons, dimensionality, electron and spin correlation.

A. Discovery of Colossal Magnetoresistance in Cobalt Oxides Using Combinatorial Material Chips

Soon after the first proof of principle in superconductors, we used materials chips to identify a class of cobalt oxide magnetoresistive materials of the form $(La_{1-x}Sr_x)CoO_3$ (22). Prior to this study, large magnetoresistances were found only in Mn-based perovskites, $(La, R)_{1-x}A_xMnO_{3-d}$, where R = rare earth and A = Ca, Sr, Ba. The question arises whether these effects are unique to Mn-based perovskite oxides or can be found as an intrinsic property for other materials as well. An analysis of the effects of spin configuration and electronic structure on the magnetoresistive properties of the other transition metal–based compounds can help elucidate the underlying mechanism of CMR effect. Moreover, the discovery of diverse classes of the CMR materials may help significantly in efforts to optimize these materials for eventual device applications.

Using materials chips, we searched through simple perovskite ABO_3 and related A_2BO_4 or $A_{n+1}B_nO_{3n+1}$ (with $n \geq 1$) structures, where A = (La, Y, rare earth)$^{+3}$ partially substituted with (Ca, Sr, Ba, Pb, Cd)$^{+2}$ and B = (Mn, V, Co, Ni, Cr, Fe). A 128-member materials chip was generated by combining sequential RF sputtering deposition of thin films with a series of physical masking steps designed to produce Y-, La-, Ba-, Sr-, Ca-, and Co-containing films (1 mm by 2 mm in size) with varying compositions and stoichiometries. Polished (100) $LaAlO_3$ single crystals were used as substrates, and La_2O_3, Y_2O_3, $BaCO_3$, $SrCO_3$, CaO, and Co were used as sputtering targets. Two identical chips, L1 and L2, were generated simultaneously and then thermally treated under different annealing and sintering procedures (up to 900°C) in O_2 or air. A map, which indicates the composition and stoichiometry of each sample, is illustrated in Figure 6. The resistivity of each sample as a function of magnetic field, perpendicular to the probing current, and temperature were measured using the four-probe contact method with a computer-controlled multichannel switching system. A liquid helium cryogenic system with a superconducting 12-tesla (T) magnet was used to perform variable temperature and field measurements.

A number of compounds in the library showed a significant MR effect (>5%); they are indicated in Figure 6 by solid circles. Three compounds that exhibit a large GMR effect have been identified: $La_xM_yCoO_\delta$, where M = Ca, Sr, Ba. The temperature dependence of the resistance and normalized MR of a representative sample L2 (13, 2) under different fields are shown in Figure 7. Here the number in parentheses indicates the sample coordinates in the library. In contrast to the behavior of Mn oxide MR materials (18,23), we found that the MR effect increases as the size of the alkaline earth ion increases.

The MR effects of the samples in library L1 are larger than those of L2, presumably due to differences in oxidation resulting from slightly different thermal treatments. The normalized magnetoresistance of representative samples as function of magnetic field at a fixed temperature (60 K) are shown in Figure 8. The largest MR ratio measured in this library was 72%, obtained for sample L1 (15, 2) at $T = 7$ K and $H = 10$ T. This value is comparable to those measured for films generated in a similar fashion in a Mn-based materials chip. As in the manganese oxides, optimization of composition, stoichiometry, substrate, and synthetic conditions may lead to increase in the MR ratio. In our libraries, Y-containing compounds show much smaller (<5%) MR effects.

Three bulk samples with stoichiometry $La_{0.67}(Ba,Sr,Ca)_{0.33}CoO_\delta$ were then synthesized (sintered at 1400°C in air) for further structural study. X-ray diffrac-

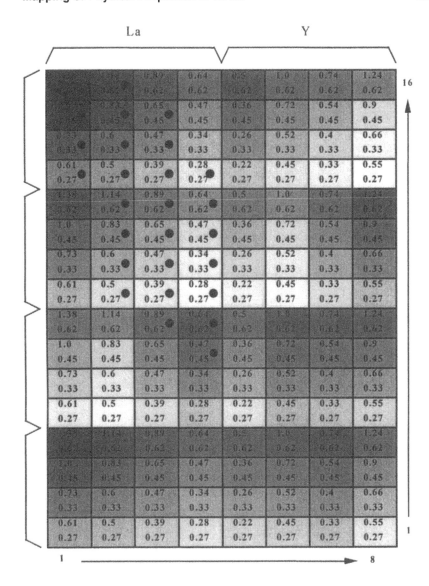

Figure 6 Map of compositions and stoichiometries ($Ln_xM_yCoO_{3-\delta}$, where $Ln = La$ and Y and $M = Ba$, Sr, Ca, and Pb) of thin-film samples in libraries L1 and L2. Samples are labeled by index (row number, column number) in the text and figure legend. The first number in each box indicates x and the second y. Solid circles indicate the samples that show significant MR effects ($>5\%$). (Reproduced from Ref. 22.)

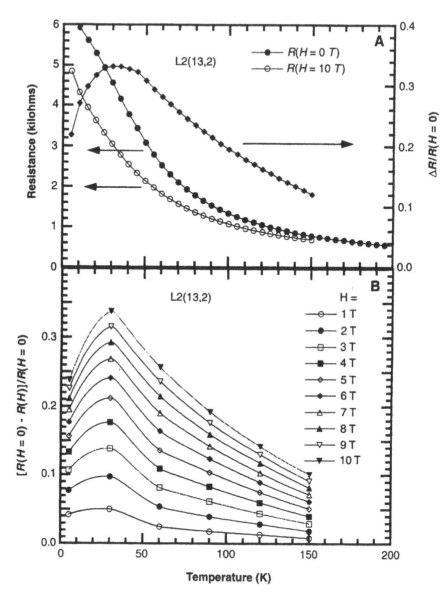

Figure 7 (A) Resistance of sample L2 (13, 2) under 0 T and 10 T and the MR ratio ($H = 10$ T) as a function of temperature; (B) MR ratios of the same sample for different magnetic fields as a function of temperature. The solid lines are guides to the eye. (Reproduced from Ref. 22.)

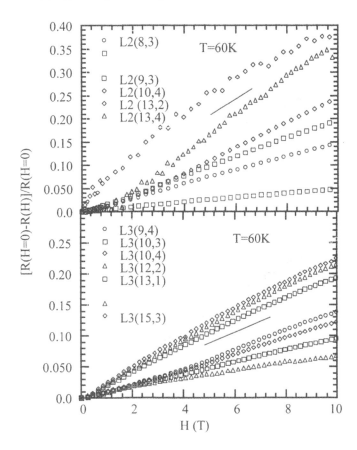

Figure 8 MR ratios of representative samples in L1 and L2 as a function of magnetic field. (Reproduced from Ref. 22.)

tion patterns show that the crystal structure is basically a cubic perovskite with the lattice constant a = 3.846 Å, 3.836 Å, and 3.810 Å for the Ba, Sr, and Ca compounds, respectively. Minor splittings of the intensity peaks are attributed to rhombohedral distortion from the cubic perovskite structure (24).

A bulk sample of stoichiometry $La_{0.58}Sr_{0.41}CoO_8$ was synthesized, and its magnetization was measured with a SQUID magnetometer (Quantum Design). The sample MR as function of magnetic field and the sample magnetization under a 1-T magnetic field as a function of temperature were measured (Figure 9). A gradual ferromagnetic transition starts at $\sim$200 K and saturates below 50 K. The MR ratio of this bulk sample (60%) is significantly higher than that of the corresponding thin-film sample on L1 (30%). The x-ray analysis of this sample confirmed the cubic perovskite structure with $a = 3.82$ Å.

This study has resulted in the discovery of a new family of Co-containing magnetoresistive materials. Magnetoresistance was found to increase as the size of the alkaline ion increased, in contrast to Mn-containing compounds, in which the magnetoresistive effect increases as the size of the alkaline earth ion decreases.

Following this study, CMR effects have also been observed in other, related compound families—the pyrochlores (e.g., $Tl_2Mn_2O_7$)(25), the spinels (e.g., ACr_2Ch_4, where A = Fe, Cu, Cd) (26), and the layered manganese oxides, so-called Ruddlesden–Popper (RP) phases (27–29). Unlike perovskites, the pyrochlores and the spinels have no mixed valence, and their A-site cation (T1 or A) can contribute to Fermi-level states with large deviations of the metal–anion–metal bond angle from 180°. In RP phases $X_{n+1}Mn_nO_{3n+1}$, where X is a lanthanide, strontium, or mixture, n-layer-thick perovskite blocks are stacked, with each block separated by a rock salt $(Sr,Ln)_2O_2$ layer that decouples the blocks electrically and magnetically. For $n = 1$, the K_2NiF_4 structure is realized,

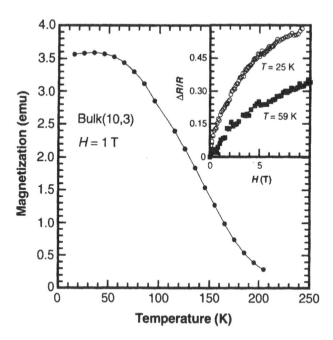

Figure 9   Magnetization of the bulk sample $La_{0.58}Sr_{0.41}CoO_8$ under 1-T field as a function of temperature. The solid line is a guide to the eye. *Inset:* MR ratios of the sample at different temperatures as a function of magnetic field. (Reproduced from Ref. 22.)

which is similar to that of high-T_c cuprates. When $n = \infty$, it becomes the cubic perovskite. While the $n = 1$ RP phase shows an insulating behavior for all x, a large MR has been observed in the $n = 2$ RP phases. For $n = 2$, it forms the so-called double-layer system, with a two-dimensional character and a reduction from 6 to 5 in the number of nearest-neighbor Mn cations around a particular transition metal site. The reduction in the number of nearest neighbors is expected to produce an anisotropic reduction in the width of the energy bands that are derived (largely) from the Mn $3d$ orbitals. This will modify the electrical conductivity and magnetic behavior of these $n = 2$ RP materials. All these discoveries and lessons learned can now be used to guide future exploration using the combinatorial materials chip approach.

IV. CONTINUOUS PHASE DIAGRAMS

Mapping out phase diagrams has been the core subject of materials science and condensed-matter physics. In materials science, phase diagrams are most often referred to the structural phase diagram displaying the formation of different crystal structures as functions of composition and temeprature. If other physical properties are mapped out together with the structural information, the phase diagrams can be made to describe structure-property relationships, the central issue in materials science. Also important are phase diagrams mapping fundamental physical properties as functions of parameters such as doping concentration and temperature. In these studies, physical phase transitions rather than structural phase transitions are often emphasized.

Traditionally, phase mapping of structural and physical properties as a function of materials composition was accomplished by the synthesis and analysis of samples, with discrete compositions prepared and characterized one at a time. This process consumes extensive time and manpower. It also has very limited ability to provide detailed mapping, due to intrinsic difficulties in producing identical growth conditions for all samples and detecting subtle difference in their physical properties. The lack of systematic and comprehensive studies often results in missing opportunities for important discoveries.

Long before the recent activity in combinatorial materials science, there were efforts among materials scientists and condensed-matter physicists to partially address this problem. For example, diffusion couples and triples have been used to map the structural phase diagram of binary and ternary systems (30). This approach relies on the existence of a small diffused region (on the order of tens of microns) at the interface of multicomponent systems after heat treatment. In this approach, the phase points being mapped out in a phase diagram are not evenly distributed in the cross section of the interface region. The key to the success of this approach is the ability of analytical tools to determine accurately the local composition and crystal structures with high spatial resolution (ideally much less than 1 micron). Electron-microscope-based techniques are often used for this application (31). However, tools with a similar resolution are not always available to characterize many other physical properties. As a consequence, the size of the approach remains limited.

The "composition spread" technique represents a different approach, which generates a map of varied composition in a large two-dimensional space. The earlier effort to use this approach can be traced back to Kennedy's work in 1965 (32). Due to a number of technical difficulties, the method had been used only to grow amorphous or polycrystalline samples. Another difficulty in this early effort was the lack of advanced screening tools. Recently, van Dover and his colleagues at Bell Lab have improved the technique and applied it to solving important industrial materials problems (33). Usually, "composition spread" techniques rely on the geometric arrangement of sputtering targets and the substrate to generate composition gradients. As a result, the gradients are not linear and can vary, often with different deposition conditions. Time-consuming chemical analysis and coordinate transformation are usually required to generate a map (or a phase diagram) conforming to standard binary or ternary representation.

We had initially developed a discrete combinatorial materials chip technique for discovery and optimization of functional materials. The technique has dramatically increased the speed of materials research by allowing access to tens of thousands of new compounds on a single chip instead of one at a time. While this method allows the rapid survey of multiple element systems, its ability to provide continuous surveys of physical properties upon variation of one or two parameters is limited. For instance, detailed mapping of different structural phases is not easily accomplished by the discrete method. We have adopted a technique that generates a continuous phase diagram (CPD) in an epitaxial thin-film format (34–37). The technique first generates a linear gradient profile of a single element or a composite by moving a shadow mask during deposition of each individual precursor. By rotating the substrate after each deposition, binary or ternary phase diagram profiles can be generated. Subsequent low-temperature diffusion and high-temperature phase formation give rise to CPDs in a format of high quality, most often epitaxial crystalline films. This approach can produce a well-defined (usually in a linear fashion for direct phase diagram mapping) local composition in an area large enough for physical property characterization. On the other hand, a CPD is small enough to match available substrates of any experiment's choice, often critical for growing high-quality crystalline films. These features are essential for reliable phase diagram mapping, which makes the CPD approach a significant improvement over the discrete materials chip technique. It could reveal, in a systematic way, the scientific phenomena of a given materials system likely missed by any conventional approach.

Numerous studies have been conducted in the past to understand the relationship between the charge dopant level or ionic radii and a variety of long-range order thermodynamic phase transitions at different temperatures. Two of the well-known studies were conducted in high-temperature superconducting cuprates (38) and colossal magnetoresistive manganites (39–43). The importance of CPD experiments is the ability to map global and systematic information on the behavior of these complex systems as a function of charge-filling and lattice parameters.

A. Continuous Phase Diagrams of Perovskite Manganites

Manganese oxides belong to a class of highly correlated electronic systems where long-range Coulomb interaction between electrons give rise to many-body correlations with a strong interplay between electronic and magnetic order. They are of technological interest, owing to the large magnetoresistance and/or spin polarization observed, and of scientific interest, owing to their complex behavior as a function of composition and temperature.

$Re_{1-x}A_xMnO_3$, where Re is a rare earth element and A is a divalent alkaline earth element, has a perovskite structure with two interpenetrating cubic lattices, one consisting of Mn and O atoms, and the other of A and Re, as shown in Figure 10. It has $(1 - x)$ electrons in each unit cell. Hopping between Mn ions, Mn^{3+} and Mn^{4+}, along p orbitals of oxygen atoms is possible via double exchange interaction (19–21). Magnetically, there are several competing interactions (Figure 11). For adjacent Mn^{3+} or Mn^{4+} ions, interaction is generally antiferromagnetic via superexchange. Between Mn^{3+} and Mn^{4+} ions via oxygen, ferromagnetism is favored, because the electron kinetic energy can be lowered when these two ions share electrons through aligning their spins (double exchange interaction). The actual phase diagrams of perovskite manganese oxides are far more complex, due to strong electronic and magnetic coupling.

A central feature of this system is that it is a doped Mott insulator, obtained by chemically adding charge carriers to a highly correlated antiferromagnetic insulator. The behavior of manganese oxides is strongly dependent on charge filling (doping) and the lattice parameter, which control the overlap of electron wave functions. Since quantum mechanical variables, such as the long-range Coulomb interaction and exchange energy, dictate different ground states of the correlated system, the best way to study this complex material is to tune its Hamiltonian variables systematically. Experimentally we can achieve this by controlling doping and bandwidth. For example, in bandwidth control, hydrostatic or strain due to substrates can be used to apply global pressure to the system, while chemical substitution of different-size atoms can be used to apply local pressure. In adjusting charge filling, either the oxygen amount from the stoichio-

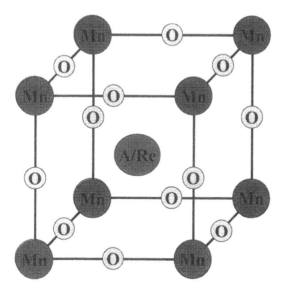

Figure 10   The structure of perovskite manganese oxides.

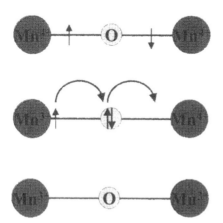

Figure 11   Magnetic interactions in manganites.

metric sample can be altered or chemically different valence ions can be substituted.

The stability of the perovskite manganites against chemical substitutions provides an opportunity to study the effect of varying the parameter of a highly correlated electronic/magnetic system. In this study we control the charge filling n of the Mott insulator by continuous charge doping over the entire range and the hopping integral t by introducing substrate-induced anisotropic strain effect in epitaxial films and varying the average ionic radius of the A site.

1. Synthesis

The synthetic methods in this study had to be tried and developed before we could grow high-quality epitaxial films, important for many applications. Different from conventional methods, in which stoichiometric targets are used to deposit and grow films on a heated substrate, our films are made from layers of different precursor films deposited at room temperature. Subsequently they are annealed at low temperature for interdiffusion of precursor layers and fired at high temperature to form crystalline phases. This method uses the fact that there exists a window of thickness and temperature within which the interdiffusion process is dominant over the nucleation process at the precursor interfaces that may prevent the formation of the desired crystalline phase. The layering sequence of different precursors is selected so that such interdiffusion among different layers can occur more efficiently. Forming an amorphous intermediate from multilayered precursors first is crucial for subsequent epitaxial crystalline film growth, since the nucleation is likely to occur only at the interface of the single-crystal substrate and the amorphous film. If the single-crystal substrates are chosen to lattice match the desired phases, high-quality epitaxial films can be formed.

To control film stoichiometry, we deposited a gradient thickness of precursor layers using in situ high-precision shutters, followed by ex situ postannealing. This technique can easily generate precisely controlled stoichiometric profiles within a very small area.

Thirty-six continuous phase diagrams (CPDs) of perovskite manganites $(Re_{1-x}A_xMnO_3)$ in an epitaxial thin-film form were made, where Re = Y, La, Ce, Pr, Nd, Sm, Eu, Gd, Tb, Er, Tm, and Yb, A = Ca, Sr and Ba, and x is varied continuously from 0 to 1. We made three sets of 36 different phase spreads on six 15-mm by 15-mm substrates of three different single crystals, (100) LaAlO₃, (100) SrTiO₃, and (110) NdGaO₃. As illustrated in Figure 12, a gradient of Re oxide is deposited at the bottom, where Y_2O_3, La_2O_3, CeO_2, Pr_6O_{11}, Nd_2O_3, Sm_2O_3, Eu_2O_3, Gd_2O_3, Tb_4O_7, Er_2O_3, Tm_2O_3, and Yb_2O_3 were used for the targets. For example, moving the horizontal shutter at a constant speed from one edge of the substrate to the other, defined by two vertical shutters during deposition (Figure 12), generated a linear thickness gradient, from $0 = \text{Å}$ to 651-Å thickness for Y_2O_3. After finishing six strips of Re oxide gradients on a substrate, gradients of Mn_3O_4 (0 Å to 453 Å) and $AMnO_3$ (0 Å to 900 Å, 1018 Å, 1134 Å for $CaMnO_3$, $SrMnO_3$, and $BaMnO_3$, respectively) were deposited as a middle layer and a top layer similarly. All the precursor films in this study were deposited at room temperature by a pulsed laser deposition (PLD) system. The forward expanding plume in high vacuum, coupled with scanning of the laser beam across the 2-in × 2-in. targets during deposition, resulted in a deposited thickness uniformity of better than 1.5% over a 15-mm × 15-mm area and therefore ensured accuracy of the deposited stoichiometry. Following deposition, the sample was annealed at 200°C for several days before it was annealed at 400°C for 30 hours followed by 2 hours' sintering at 1000°C. We believe that this low-temperature annealing is necessary to allow homogeneous mixing of precursors into an amorphous intermediate before crystallization at higher temperatures. Once the synthe-

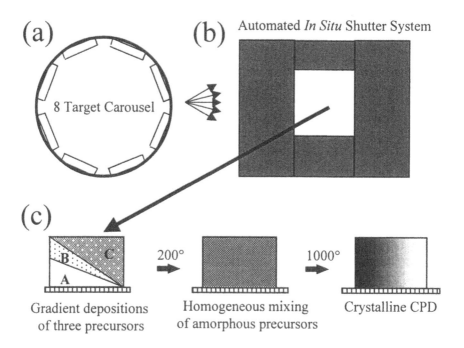

Figure 12   (a) Schematic of the eight-target carousel. This allows uninterrupted depositions without breaking the vacuum. (b) Vertical shutters are used to define the width of each phase strip on a substrate while precise gradient profiles of three precursors are deposited with horizontal shutters moving across the phase strip at constant speed. (c) Schematic steps of CPD fabrication.

sis conditions are found, fabrication of these libraries takes only a small amount of time.

In order to study the crystalline quality of the samples, we made individual samples of various compositions selected from the phase diagrams under the same fabrication and processing conditions. In Figure 13, a $\theta/2\theta$ XRD pattern for $Nd_{0.7}Sr_{0.3}MnO_3$ on a $LaAlO_3$ substrate indicates that the film is highly oriented along the c-axis. Even on a logarithmic scale, we can hardly see any other impurity phases. A ϕ scan of the (101) planes indicates that the film is in-plane aligned with the substrate. The narrow width of the peaks indicates the high-quality crystalline formation of the films. We confirmed similar epitaxial growth in many different compounds with different hole doping levels.

Figure 14a is a resistivity-vs-temperature plot of selected manganite samples. The transition temperature and resistivity scales are comparable to those from sample made with conventional thin film fabrication techniques. Figure 14b shows resistivity-vs-temperature plots of different samples at zero magnetic fields. The residual resistivity ratios are the lowest among films different groups have measured, suggesting that the films are of high quality.

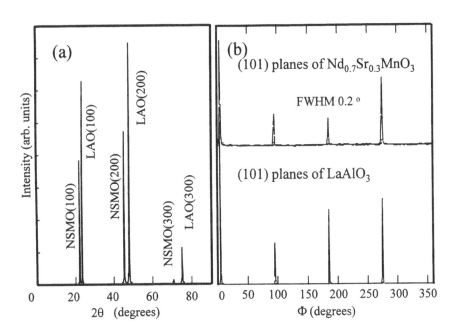

Figure 13 (a) The $\theta/2\theta$ XRD pattern and the ϕ scan of the (101) plane of a $Nd_{0.7}Sr_{0.3}MnO_3$ (NSMO) thin film made from three precursors of Nd_2O_3, Mn_3O_4, and $SrMnO_3$ on (001) $LaAlO_3$.

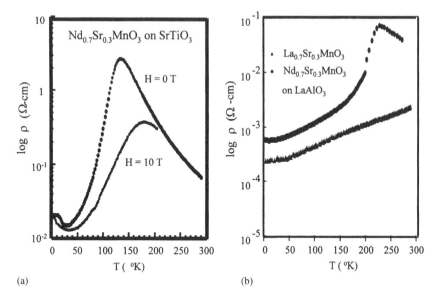

Figure 14 Resistivity-vs.-temperature plots of perovskite manganite samples grown by combinatorial synthesis: (a) $Nd_{0.7}Sr_{0.3}MnO_3$ with $H = 10$ and 0 T (b) $Nd_{0.7}Sr_{0.3}MnO_3$ and $La_7Sr_{0.3}MnO_3$ with $H = 0$.

2. Characterization

To map the electronic properties of the CPDs, we chose probes with two very different frequency scales (by 10^6): visible light and microwave. The colors roughly indicate the electronic bandwidths of the compounds. In order to map the electrical impedance at microwave frequencies, we used a scanning evanescent microwave probe (EMP) operating at 2.2 GHz (44–46). The microscope tracks the change in its resonance frequency (f_r) and the quality factor (Q) as the sample is scanned in close proximity to its probing tip. The change in f_r carries the capacitive information, and the change in Q carries the electrical loss information of the sample. The detailed description of the microscope is given in another chapter.

   To calibrate the data obtained from microwave probes, we compared the dc resistivity of a number of films to the microwave results. The proximity of a sample to the tip changes the complex impedance of the probe, changing f_r and Q. In this study, the model analysis is based on a first-order approximation, which is valid when the conductivity is low. The shift in resonant frequency is a monotonic increasing function of the conductivity. Each value of the shift in quality factor corresponds to two values of conductor, and it approaches zero as the sample becomes insulating or highly conducting. The shift in quality factor has been calibrated against the measured dc conductivity and is consistent with the theoretical analysis. To prevent shorting of the probe tip with a conducting sample, a uniform thin layer of polymer coating was applied on the sample surface. Figure 15 provides a conversion function from the microwave probe data to dc conductivity.

To map the magnetic phases at low temperature, we employed a scanning superconducting quantum interference device microscope (SSQM) (47). A miniature SQUID ring with a diameter of 10 µm was scanned over the sample surface. In this geometry, the SSQM senses a local magnetic field perpendicular to the surface, B_Z. Since the magnetic moment of a ferromagnetic phase is in plane, the SSQM is limited to the measurement of a field leaking from magnetic domains in the material (Figure 16). An oscillating signal indicates the presence of magnetic domains. The periodicity (or width d) is correlated with the domain size, while the amplitude (ΔB_z) is related to the presence of ferromagnetic ordering. The SSQM measurements were performed without an external magnetic field.

3. Results

Figure 17 is a room temperature CCD photoreflection image of 36 CPDs on three different substrates. Figure 18 shows the detailed measurements of the room-temperature microwave impedance of a CPD of $Er_{1-x}Ca_xMnO_3$. First, we observed many clear boundaries in both optical and microwave impedance images. The relevant energy scale of transitions probed by visible photons is high com-

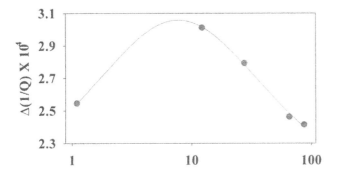

Figure 15 Microwave loss, $\Delta(1/Q)$, versus dc conductivity ($\Omega^{-1} - cm^{-1}$). Solid circles are obtained from the measured microwave and corresponding dc responses. The line is the theoretical fit. (Reproduced from Ref. 35.)

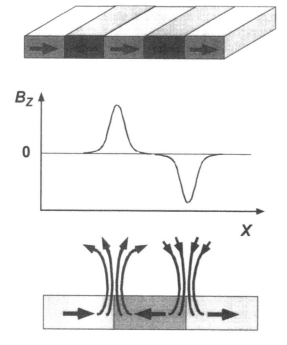

Figure 16 Schematics of magnetic domains measured by SSQM in a thin film with perpendicular magnetization.

pared to kT, thermal energy at room temperature, indicating that the phenomenon probed here is likely to be electronic in origin. Microwave loss peaks at these boundaries. X-ray data show no evidence of different structural phases besides perovskite throughout these CPDs. The existence of these boundaries suggests that complex electronic orderings (with fundamentally different physical properties) occur as a function of x.

Second, a very narrow insulating (or more accurately semiconducting) strip within a highly conducting region around 7/8 doping points on the CPDs of $Tm_{1-x}Ca_xMnO_3$ on $NdGaO_3$ substrate and $Yb_{1-x}Ca_xMnO_3$ on $SrTiO_3$ was observed at the commensurate charge-filling point $x = 7/8$. The strip is so narrow in phase width that it would have been very difficult to find it using the conventional practice of studying samples of discrete composition.

In Figure 19, different doping ranges from the CPD of $Tm_{1-x}Ca_xMnO_3$ are selected for temperature-dependent comparisons of blue light (6.2–7.8 $\times$ 10^{14} Hz) reflection images. In strips A and C, as the temperature is raised, the boundary (shown as a peak in intensity) broadens or disappears, indicating the

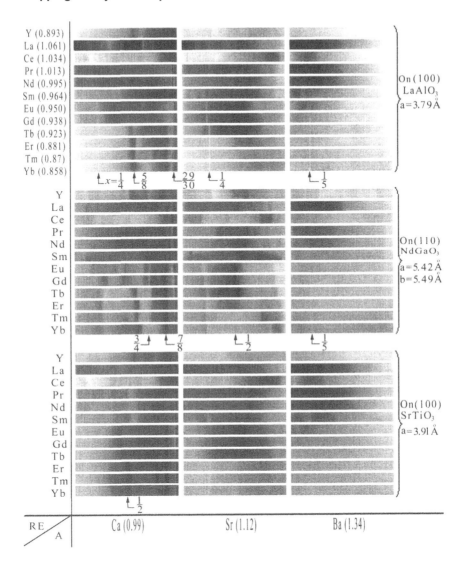

Figure 17 Room-temperature charge-coupled device (CCD) photograph (photoreflec-tion image) of 36 $Re_{1-x}A_xMnO_3$ CPDs on LaAlO3, NdGaO3, and SrTiO$_3$ substrates under white light ($4.2-7.8 \times 10^{14}$ Hz). The optical images were taken using a monochrome CCD camera. The parentheses indicate the crystal ionic radii of the elements. (From Ref. 70.) The commensurate doping points of singular phases are denoted. The effective unit cell of (110) NdGaO$_3$ has $a' = b' = 3.86$ Å ($a' = b' = \sqrt{a^2 + b^2}/2$.

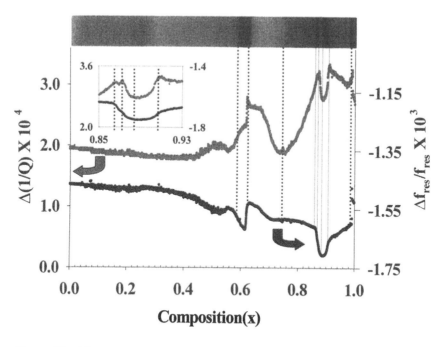

Figure 18   Microwave response of the $Er_{1-x}Ca_xMnO_3$ CPD on $NdGaO_3$. Microwave loss, $\Delta(1/Q)$, and frequency shift, $\Delta f_{res}/f_{res}$, are measured versus composition (x). The dashed lines are used to indicate phase boundaries. The inset is a zoomed portion near the commensurate singularity. (Reproduced from Ref. 35.)

effect of thermal fluctuation. In strip B, as the temperature is lowered, the suppression of intensity at the middle portion becomes more pronounced. This tends to sharpen the phase boundary toward the right side (less amount of doping) of the dark conducting phase. The dip in conducting phase at around $x = 1/2$, also corresponding to a dip in microwave conductivity measured by EMP, tends to disappear at low temperatures. In other manganite systems, such as $Nd_{1-x}Sr_xMnO_3$, a similar phase near the 1/2 doping point had been found to have two or more intrinsic competing phases (phase separation) with charge-ordering phase dominating as the temperature is lowered (48). This indicates that a more homogeneous charge-ordering phase dominates at 77 K, since an insulating and more homogeneous phase should decrease the reflected light. The gradual change in ionic radii induces unexpected abrupt changes in phase patterns. This effect can be observed in substitutions of different rare earth and divalent alkaline earth elements. We see relatively smooth phase patterns and the absence of sharp transition boundaries in the Sr-doped system. This trend is even more pronounced in the Ba-doped system. The clear effect of large band overlap due to a smaller ionic size of Ca is observed in the rich and diverse phase patterns in the Ca-doped system. The effect of substrate-induced stress can also be observed.

Figure 20 presents the results of the microwave probe scan and the SSQM scan in the CPD sample of $La_{1-x}Ca_xMnO_3$ (36). Figure 20a is the CPD of $La_{1-x}Ca_xMnO_3$; Figure 20b shows line scan profiles of microwave loss and frequency shift for $La_{1-x}Ca_xMnO_3$ film as a function of Ca concentration x measured with a scanning evanescent microwave microscope at room temperature. Figure 20c gives line scan profiles of a perpendicular magnetic field above $La_{1-x}Ca_xMnO_3$ film as a function of Ca concentration x measured with a scanning SQUID microscope at 3 K without external magnetic field. Figure 20d shows the magnetic domain structures of $La_{1-x}Ca_xMnO_3$ film at 7 K.

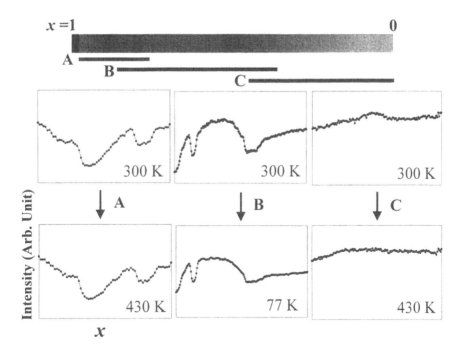

Figure 19 Room-temperature CCD color photograph of $Tm_{1-x}Ca_xMnO_3$ CPD on $SrTiO_3$ under white light. Three doping regions are selected for blue light (6.2–7.8 × 10^{14} Hz) photoreflection at three different temperatures. The optical images were taken using a monochrome CCD camera and a blue filter. Incident white light at 60° incidence was used to illuminate the sample, and scattered light at 30° incidence was measured as a function of temperature from 77–430 K. (Reproduced from Ref. 35.)

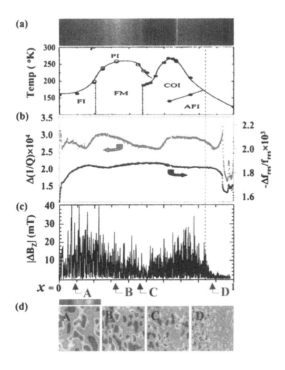

Figure 20 (a) Room-temperature CCD color photograph (photoreflection image) of the CPD of $La_{1-x}Ca_xMnO_3$ on $LaAlO_3$ substrate under white light (4.2–7.8 $\times$ 10^{14} Hz) and the magnetic phase diagram from a single-crystal study (53). The various states are: paramagnetic insulator (PI), ferromagnetic insulator (FI), ferromagnetic metal (FM), charge-ordered insulator (COI), antiferromagnetic insulator (AFI). (b) Line scan profiles of microwave loss and frequency shift for $La_{1-x}Ca_xMnO_3$ film as a function of Ca concentration x measured with a scanning evanescent probe at room temperature. $\Delta(1/Q)$ increases with increasing microwave loss and $-\Delta f_r/f_r$ increases with increasing tip-sample capacitance. (c) Line scan profiles of a perpendicular magnetic field above $La_{1-x}Ca_xMnO_3$ film as a function of Ca concentration x measured with a scanning SQUID microscope at 3 K without external magnetic field. (d) Magnetic domain structures of $La_{1-x}Ca_xMnO_3$ film at 7 K. A color bar indicates the measured perpendicular magnetic field, B_Z. Scanned areas are 300 $\times$ 300 μm^2 in all images. The Ca content x increases linearly with the horizontal distance in each image. Domain structures of ferromagnetic films are probed with in-plane magnetization. In the present thin-film sample, magnetic moments lie in plane. Thus, SSQM looks at the magnetic field flowing in or out of domain boundaries. Red and purple regions in the SSQM images correspond to the domain boundaries. The nominal values of x from left to right in each image and ΔB_Z are: (A) 0.078–0.099 and 52.8 μT, (B) 0.330–0.351 and 59.3 μT, (C) 0.449–0.470 and 48.3 μT, (D) 0.868–0.888 and 17.2 μT, respectively. (Reproduced from Ref. 36.)

The optical and microwave properties at room temperature (well above the onset of magnetic order) are clearly correlated with the magnetic properties at low temperatures. Starting at $x = 0$, the material is insulating. As doping x is slightly increased from 0, there is an abrupt increase in conductivity. In $0 < x \leq 0.2$, d, *domain size*, and ΔB_z are relatively large and increasing with x, which is consistent with the FI (ferromagnetic insulator) phase found in previous studies (49,50). A broad dip in $\Delta(1/Q)$ centered at $x \sim 0.15$ indicates an increase in conductivity. For $0.2 < x < 0.5$, d and ΔB_z are relatively large but decreasing with x, which is also consistent with results from a single-crystal study, suggesting a ferromagnetic phase approaching a charge-ordered insulating phase. A second broad peak in $\Delta(1/Q)$ centered at $x \sim 0.35$ corresponds to a decrease in conductivity. At $x = 0.5$, ΔB_z is substantially suppressed, and an abrupt change is visible in the optical image as well. This abrupt change corresponds closely to the 1/2 charge ordering observed in TEM studies (51). For $x > 0.5$, ΔB_z almost recovers its value, while d decreases monotonically. We observe a broad plateau in $\Delta(1/Q)$ centered at $x \sim 0.75$.

At $x = 0.67$, we observe a sharp boundary (white colored in Figure 20a) which, we believe, is the onset of the 2/3 Jahn–Teller type of stripe phase (52). For $x > 0.82$ ($\sim 4/5$), $\Delta(1/Q)$ decreases while f_r decreases, indicating an increase in conductivity. Simultaneously, ΔB_z reaches a minimum. At around $x = 0.85$, ΔB_z is abruptly reduced to ~ 10 μT (note that the observed ΔB_z for $x = 0.88$ is one order of magnitude larger than the noise level). At $x = 0.93$, the optical image changes abruptly. The changes in $\Delta(1/Q)$ and f_r indicate an abrupt transition to an insulator.

For $x = 0.97$, a narrow highly conducting region appears while adjacent regions are highly insulating. Such a drastic change in conductivity over a small range in doping cannot be explained by orbital orderings, and it has not been observed before in doped Mott insulators. This phenomenon may be common to many highly correlated systems. To study this type of phenomenon, CPDs are apparently a powerful tool, since they allow the precise exploration of minute changes in composition. The transport property of this region has been measured by a four-point probe measurement. Figure 21 shows the conductivity versus temperature. As the temperature decreases, the conductivity initially decreases slowly and then rapidly below 100 K. The room-temperature resistivity is less than 1 Ω-cm, and no magnetoresistance was observed.

Two-dimensional images shown in Figure 20d of selected compositions indicate the presence of phase separation in the compounds. Around $x = 0.1$ (Figure 20d A) the domain size is as large as 50 μm, and domains have almost identical widths, implying that a homogeneous ferromagnetic state is established in this region. At $x \sim 0.33$ (Figure 20d B) smaller domains dominate. The domain width is more widely distributed in Figure 20d B compared to Figure 20d A. In this case, the charge-ordered phase is believed to be embedded around the region

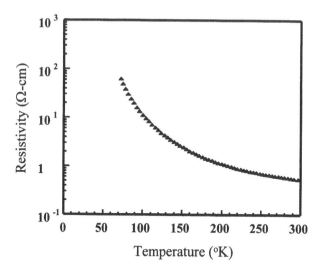

Figure 21 Resistivity vs. temperature in zero field for $La_{0.03}Ca_{0.97}MnO_3$ film. (Reproduced from Ref. 36.)

of the image, where magnetic field is weak, resulting in the appearance of domains with smaller B_Z contrast. At x around 0.46 (Figure 20d C) the domains become smaller, and, more interestingly, nonmagnetic domains of about 100 μm coexist, as can be seen in the right-hand side of the image. This nonmagnetic domain may represent an antiferromagnetic insulator (AFI) with charge ordering or, at least, an AFI-dominant region, which includes ferromagnetic metal (FM) droplets smaller than the spatial resolution of SSQM, ~5 μm. It is noted that the size of the nonmagnetic region seen from Figure 20d C is much larger than the length scale of phase separation previously reported in this compound (50). Such a large nonmagnetic region has not been seen for x greater than 0.5. However, ferromagnetic ordering seems to be in a shorter range as x increases, considering the decrease of the domain size with increasing x. Nevertheless, there are several regions with weak B_Z contrast, suggesting the coexistence of a charge-ordered phase. For $0.51 < x < 0.88$, the domain size gradually decreases. Figure 20d D includes a boundary at the center of the image, across which the amplitude of B_Z is substantially suppressed. Even in the region for $x > 0.88$, however, finer domains are still visible. This may originate from slightly canted magnetic moments in the AFI phase.

We also chose the $Nd_{1-x}Sr_xMnO_3$ system for detailed measurements (Figure 22). Previous intensive studies have revealed a rich phase diagram in this system, exhibiting a variety of magnetic and metal/insulator transitions (53,54).

We observe that the optical and microwave properties at room temperature (well above the onset of magnetic order) are clearly correlated with the magnetic properties at low temperature. First, we note that the dark conducting phase between x = 0.5 and 0.6 of the CPD clearly matches A-type antiferromagnetic (in-plane ferromagnetic and out-of-plane antiferromagnetic) metallic (A-AFM) phase from single-crystal studies. A narrow white strip near $x = 0.5$ coincides with CE-type (alternate in-plane antiferromagnetic ordering of the Mn^{3+} and Mn^{4+} ions and antiferromagnetic along the z-axis) charge/spin order phase (55). In $Nd_{1-x}Sr_xMnO_3$ there should be no long-range order phase transitions (symmetry breaking in order parameters of low-energy excitations) at room temperature. The fact that the room-temperature electronic boundaries match the low-temperature magnetic boundaries indicates that the electronic ground state (which changes abruptly with composition) dictates the magnetic order at low temperatures. Under white light, the A-AFM phase appears dark, suggesting that it is metallic. f_r decreases sharply upon entering this phase, indicating that the conductivity increases. The sample becomes conductive for $x > 0.5$ (A-AFM phase at low temperatures) and remains conductive until $x = 0.8$. The dark phase has a width of 0.052 in x. For $x > 0.8$, the sample gradually becomes insulating. At $x = 0.66$, there are two small peaks that coincide with the boundary between A-type AFM and C-type (ferromagnetic in the z-direction, antiferromagnetic in the x-y plane) AF from single-crystal studies (51). The microwave evidence suggests the onset of high-energy electronic order at room temperature.

In the ferromagnetic insulating (FI) and ferromagnetic metallic (FM) regions with $0 \leq x \leq 0.5$, ΔB_z are relatively large, as expected. For ferromagnetic materials, we expect to see large domains with small magnetic fields and narrow domain boundaries where the flux leaks out and can be detected by the SSQM. However, since the resolution of the SSQM is limited to 10 μm, if domain boundaries are less than 10 μm they will appear wider. The observed domains have almost identical widths, implying that a homogeneous ferromagnetic state is established in this region. ΔB_z rapidly decreases as x (composition of Sr) approaches 0.5 from the FM side. An abrupt change is visible in the optical image as well. This abrupt change corresponds closely to the 1/2 charge ordering in single-crystal studies (54). For $0.5 < x < 0.6$, ΔB_Z is mostly suppressed, as expected for A-type AFM phase, but slightly recovers toward $x = 0.6$, the C-type AF region. In the C-type AF phase, the magnetic moments perpendicular to the plane were thought to be slightly tilted (53). We expect that such canting of the magnetic moments in the antiferromagnetic order should induce a nonzero in-plane magnetization M. This situation is essentially the same as that of FM with in-plane moments, although its M value is much smaller. It is quite natural to see in-plane domains in the C-type AF region. Figure 22c shows how the SSQM scan of the CPD changes with temperature. The magnetic profile across

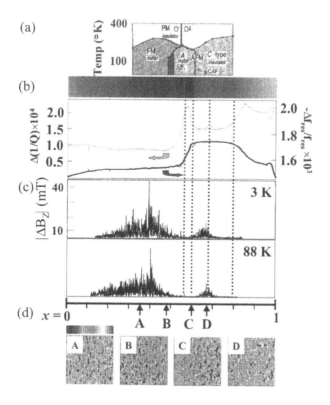

Figure 22 (a) Room-temperature CCD photograph (photoreflection image) of the CPD of $Nd_{1-x}Sr_xMnO_3$ on $LaAlO_3$ substrate under white light (4.2–7.8 × 10^{14} Hz) and the magnetic phase diagram from a single-crystal study (53). The various magnetic and orbital states are: PM: paramagnetic; FM: ferromagnetic; AFM: antiferromagnetic; COI: charge-ordered insulator; CE: Ce-type charge/spin order; A: A-type antiferromagnetic order; C: C-type antiferromagnetic order; CAF: canted antiferromagnetic order. (b) Line scan profiles of microwave loss and frequency shift for $Nd_{1-x}Sr_xMnO_3$ film as a function of Sr concentration x measured with a scanning evanescent probe at room temperature. $\Delta(1/Q)$ increases with increasing microwave loss and $-\Delta f_r/f_r$ increases with increasing tip-sample capacitance. (c)Line scan profiles of perpendicular magnetic field above $Nd_{1-x}Sr_xMnO_3$ film as a function of Sr concentration x measured with a scanning SQUID microscope at 3 K and 88 K without external magnetic field. (d) Magnetic domain structures of $Nd_{1-x}Sr_xMnO_3$ film at 3 K. A color bar indicates the measured perpendicular magnetic field, B_Z. Scanned areas are 300 × 300 μm^2 in all images. The Sr content x increases linearly with the horizontal distance in each image. Domain structures of ferromagnetic films are probed with in-plane magnetization. In the present thin-film sample, magnetic moments lie in plane. Thus, SSQM looks at the magnetic field flowing in or out of domain boundaries. Red and purple regions (different shades in black and white) in the SSQM images correspond to the domain boundaries.

the film is essentially temperature-independent below 100 K. This result is consistent with the magnetic phase diagram obtained on bulk samples.

As can be seen in Figure 22d, the ΔB_Z values are widely scattered. Namely, we observe very intense B signals from some narrow regions, which are drawn in gray and black in the figures. The domains in the present CPD of $Nd_{1-x}Sr_xMnO_3$ are very fine, on the order of 10 μm, which is close to the spatial resolution of the SSQM. Furthermore, their size is almost independent of x.

To verify the identification of the $0.5 < x < 0.6$ phase as an antiferromagnetic metal, we measured the $\rho-T$ and $M-T$ curves for $Nd_{0.47}Sr_{0.53}MnO_3$ (Figure 23). The resistivity was measured by four-point probe. The thin film was cut into a rectangular shape of 5 mm × 2 mm × 2000 Å, and electrical contacts were made with silver paint. Magnetization M was measured under a field of $\mu_0 B = 0.05$ T at 5 K using a superconducting quantum interference device magnetometer. The Curie temperature T_C was determined from the inflection point of the $M-T$ curve.

The $\rho-T$ curve is rather complicated. The ρ value drops slightly below $T_C = 230$ K and then jumps slightly below $T_N = 200$ K. This resistivity jump was suggested to accompany a ferromagnetic–antiferromagnetic transition (52). With further decrease in temperature, the $\rho-T$ curve shows another slight rise around 150 K, suggesting an antiferromagnetic–charge ordering transition. The temperature dependence of the resistivity of this sample is consistent with the measurement done on a single-crystal sample (52). In this region of doping, the charge-ordering insulating (COI) ground state competes with the AFM state (53). The charge-ordering transition ($T_{CO} = 150$ K) is supposed to give a steep rise in

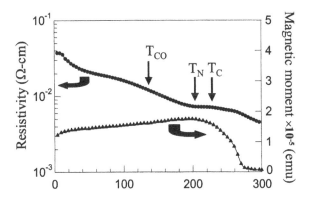

Figure 23 Temperature dependence of the resistivity at zero field and the Mn magnetic moment of $Nd_{0.47}Sr_{0.53}MnO_3$. The dc magnetization was measured in external fields of 500 gauss. The arrows denote transition temperatures.

resistivity. However, at this doping, the resistivity in the COI state is significantly reduced because of the A-type canting and the resulting enhanced itinerancy of the e_g carriers. The $M–T$ curve shows that the magnetization increases until it hits T_N, which is consistent with a FM–AFM transition. At low temperatures, there seems to be a frustrated COI phase due to lattice strain that prevents the magnetization from being completely suppressed in the AFM state.

Although there are similarities between previous phase diagram studies (mainly focused on low-temperature long-range order phases) and the inferred CPD data, the very existence of electronic phase transitions and their nature had not been observed and studied before. While some of the boundaries exist as smooth crossover to another phase, some of boundaries are quite abrupt in both electronic and optical properties. The discovery and study of these boundaries with conventional methods and approaches would be very difficult, if not impossible, due to the intrinsic nature of conventional sample synthesis.

4. Discussion

Orbital Ordering in Doped Mott Insulators.   Upon variation of the carrier density, manganese oxides pass through a variety of magnetically ordered phases at low temperatures. These states include:

F: F-type ferromagnetism, in which neighboring manganese ion spins are aligned

A: A-type antiferromagnetism, in which neighboring manganese ion spins are ferromagnetic in plane and antiferromagnetic out of plane

C: C-type antiferromagnetism, in which neighboring manganese ion spins are antiferromagnetic in plane and ferromagnetic out of plane

G: G-type antiferromagnetism, in which neighboring manganese ion spins are antiferromagnetic in three dimensions

The complexity of the various magnetic and electronic phases can be understood as arising naturally from different orbital-ordered states in this system (55,56). Orbital ordering results from the strong spin-orbital coupling in this system. The coupling of neighboring orbitals depends on a combination of direct Coulomb and exchange interactions. The orbital coupling determines the overlap of neighboring orbitals and thereby determines the exchange interactions between neighboring spins. The d orbitals of the manganese ions are split into e_g and t_{2g} orbitals (Figure 24). The two e_g orbitals, in the absence of a Jahn–Teller distortion, are degenerate. The system can be represented by a simplified Hamiltonian of the form

$$H = \sum_{ij}\left[J_{ij}\left(T_i, T_j\right) S_i \bullet S_j + K_{ij}\left(T_i, T_j\right)\right]$$

(1)

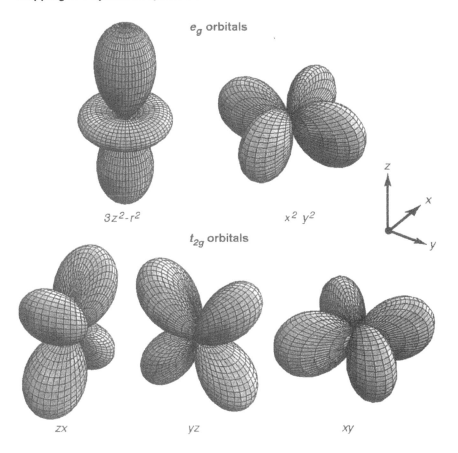

Figure 24 e_g orbitals ($3z^2 - r^2$, $x^2 - y^2$) and t_{2g} orbitals (zx, yz, xy) in manganese oxides. (Reproduced from Ref. 56.)

where T is a pseudo-spin operator denoting the occupancy of the e_g orbitals, T = 1/2 for $d_{x^2-y^2}$ and T = $-1/2$ for $d_{3z^2-r^2}$, J_{ij} is the exchange interaction between neighboring manganese ions, and K_{ij} is the interaction between neighboring orbitals. The system has been extensively modeled, and a variety of orbital-ordered phases are shown to arise.

Maezono et al. has modeled the orbital order of $La_{1-x}Ca_xMnO_3$ as a function of x using a generalized Hubbard model (56). In addition to the orbital term, they include a Jahn–Teller distortion, which is needed to stabilize some of the observed states. They find, depending on the bandwidth and doping of the system, a wide variety of orbital-ordered states:

$d_{x^2-y^2}$ and $d_{3z^2-r^2}$: generally associated with F-type ferromagnetism

$d_{x^2-y^2}$: generally associated with A-type antiferromagnetism

$d_{3z^2-r^2}$: generally associated with C-type antiferromagnetism

$d_{x^2-y^2}$ alternating with $d_{3z^2-r^2}$: associated either with F-type ferromagnetism or G-type antiferromagnetism

Figure 25 shows the schematics of orbital orderings coupled with possible spin orders. Through its exchange interaction, a given orbital ordering limits possible spin orders. Coupled with spin ordering, orbital ordering will influence the conductivity of the system. Generally, as we go from F-type to A- or C-type, conductivity will decrease, since the electron conduction is available in only one plane. In G, conductivity will further decrease, since it is antiferromagnetic in all directions and electron conduction likewise is not favored.

Typically, the orbital-ordered state is stable at temperatures greater than those for magnetic order. The modification of the exchange interaction induced in these orbital ordered phases then determines the spin ordering of the material. The present observation of phase boundaries suggests the existence of different ground states of electronic self-organization due to corresponding orbital states mediated by long-range Coulomb interaction. With the transfer interaction heavily depending on the doping level, various orbital-ordered and-disordered states may exist, with concomitant spin (and occasionally charge) orders.

For $La_{1-x}Ca_xMnO_3$, extensive modeling has been performed using a mean-field approximation by Maezono et al. (56). They find four types of spin order:

G-type: $x \sim 0$, $x > 0.9$ (three-dimensional antiferromagnetic ordering)

A-type: $0 < x < 0.1, 0.25 < x < 0.45$ (ferromagnetic in x-y plane, antiferromagnetic in the z direction)

F-type: $0.1 < x < 0.25$ (ferromagnetic)

C-type: $0.45 < x < 0.9$ (ferromagnetic in the z direction, antiferromagnetic in the x-y plane)

Figure 26 is the CPD of $La_{1-x}Ca_xMnO_3$ aligned with a theoretical phase diagram, which describes the orbital orderings in these systems as a function of composition using parameters appropriate for $La_{1-x}Ca_xMnO_3$. It shows the calculated orbital and spin order at various compositions. The correlation between the room-temperature resistivity and the low-temperature magnetic behavior is consistent with the observation of remnant orbital states. While the existence of residual orbital ordering in these materials is not unexpected, the abruptness of some transitions versus charge filling is remarkable.

For x slightly greater than 0, an abrupt increase in microwave conductivity, evidenced by curvature change in the microwave signal, correlates with the pre-

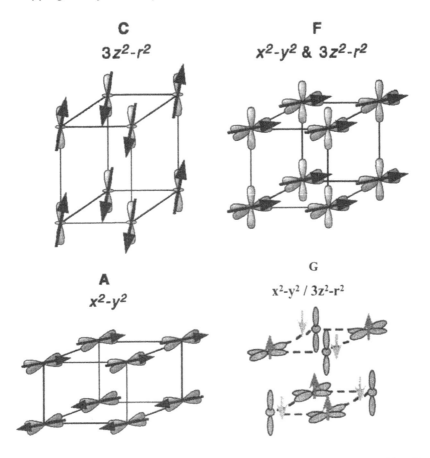

Figure 25 Orbital and corresponding spin orders in perovskite manganese oxides. (Reproduced from Ref. 56.)

dicted transition from G-type order (three-dimensional antiferromagnetic ordering) to A-type order (ferromagnetic in x-y plane, antiferromagnetic in the z direction). As x increases, a broad dip in microwave loss, $\Delta(1/Q)$, occurs around $x \sim 0.15$, indicating a further increase in conductivity that again correlates with the predicted transition from A-type to F-type orbital order (three-dimensional ferromagnetic ordering). A second broad peak in $\Delta(1/Q)$ around $x \sim 0.35$ corresponds to a decrease in conductivity, indicating that the system is returning back to A-type ordering. For large x, $\Delta(1/Q)$ exhibits a broad plateau around $x \sim 0.75$, which may be associated with a transition to a C-type orbital ordering (ferromagnetic in the z direction, antiferromagnetic in the x-y plane). At $x = 0.93$, both microwave

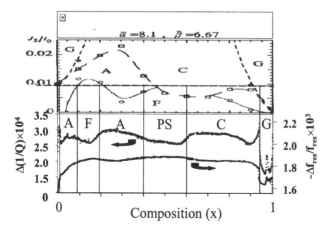

Figure 26 CPD of $La_{1-x}Ca_xMnO_3$ and calculated orbital spin orders versus charge filling with the ratio of the bandwidth to the exchange interaction indicated. (Reproduced from Ref. 56; parameters were chosen to reproduce $La_{1-x}Ca_xMnO_3$.)

loss, $\Delta(1/Q)$, and resonance frequency, f_{res}, exhibit abrupt changes, indicating that the system is returning to being a G-type insulator. Good agreements with the orbital ordering model are found throughout the entire perovskite manganites system.

The coincidence between electronic phase boundaries at room temperature and the long-range order phase boundaries at low temperatures is clearly revealed by studies of three different probes on CPD (Figures 20 and Figures 22). These electronic boundaries, we observe, coincide with those boundaries of orbital orderings predicted by Maezono et al. This correlation between the room-temperature resistivity and the low-temperature magnetic behavior is consistent with the observation of remnant orbital states. The very existence of optical phase boundaries, which originate from the electron orbital orders that persist at high temperatures, is evidence of anisotropy in the sample. The appearance of anisotropy as optical phase boundaries is only possible when the sample is grown as an epitaxial thin film.

B. Anisotropic Charge Orders and Stripes in Doped Mott Insulators

The doped Mott insulators are strongly correlated electronic systems, and the doped carriers tend to self-organize into highly anisotropic patterns (a stripe phase, for example). It is believed that the dynamical nature of this self-organization in doped Mott insulators plays a crucial role in the origins of some heavily researched phenomena, namely, colossal magnetoresistance in manganites and high-T_C superconductivity in cuprates. Recently, two competing phenomena, phase separation and charge stripes, have been observed in these highly correlated systems (49–51,57,58). Upon doping into the parent Mott insulator compounds, the charge carriers tend to separate (at a large scale) into phases of different electronic natures, and this is called the *phase separation* (59). The long-range Coulomb interaction tends to frustrate this phase separation and favors an anisotropic long-range order of charge or spin (a stripe phase, for example) (60). Note that different charge or spin-ordered phases are the ground states, which can be obtained only by solving many-body problems of electrons instead of conventional quasi-particles.

Neutron and X-ray scattering experiments can suggest the evidence of the stripe formation (57,58). In cuprates, the stripes are believed to be 1/4 filled and inserted between antiferromagnetic "banks", with a transverse periodicity corresponding to the doping level (Figure 27). In manganese oxides, the situation is rather complicated by a strong coupling of magnetic and electronic states. Here, $Mn^{4+}O_6$ charge stripes are inserted between Jahn–Teller-distorted $Mn^{3+}O_6$ "parent banks" (also antiferromagnetically paired, as in cuprates), with a periodicity corresponding to the carrier doping (Figure 28).

It must be noted that there is a definite distinction between the charge stripe and the charge density wave (CDW). In the case of CDW, due to the electron–phonon interaction, electrons form a modulated structure to lower the electron energy at the expense of lattice distortion, namely, Fermi surface nesting. Such a strong lattice coupling carries a large effective mass. The interchain interaction is usually greater than a zero-point motion, which results in a phase lock between 1D CDWs. The impurity will pin the phase-locked CDWs, favoring an insulating state. The stripe structures are believed to originate from the electron–electron correlation instead of the electron–lattice interaction (61–63). Therefore their effective mass can be low. In this case, the zero-point fluctuation can be comparable to or larger than the interchain interaction.

Kivelson and his collaborators proposed the stripe structures called *electronic liquid crystals* (64). Charge stripes, insulating charge density waves, tend to prevent superconductivity rather than promote it. Kivelson et al. circumvent this problem by allowing fluctuating stripes, which tend to break the phase locking between adjacent stripes and induce conduction. They predict a variety of stripe phases with differing degrees of symmetry. Depending on the importance of fluctuations, various stripe phases form (Figure 29). In the absence of fluctuation, we end up with a Wigner crystal, where 2D modulated electrons are locked in phase. If the fluctuation is large enough, the stripes stop being phase locked to each other. This smectic phase, or electronic liquid crystal phase, induces a translational symmetry in one direction. In this case, electrons can travel along the stripe but not in the transverse direction. Recent experiments in cuprates seem

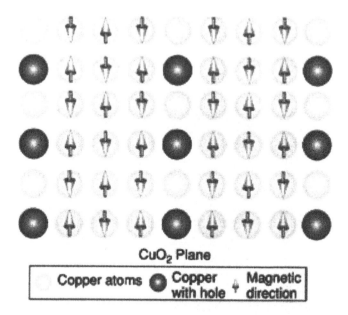

Figure 27 Schematic of charge stripes in cuprates illustrating that the stripes are believed to be 1/4 filled and inserted between antiferromagnetic "banks," with transverse periodicity corresponding to the doping level. (Reproduced from Ref. 71.)

to suggest that this kind of anisotropic transport does exist in the system. As fluctuations increase further, the stripes bend, and the resulting nematic phase has translation and reflection symmetries. Finally, the liquid phase breaks no spatial symmetry and, in the absence of disorder, is believed to be a conductor or a superconductor.

In manganese oxides the active orbital degree of freedom induces the highly anisotropic electron-transfer interaction, which gives rise to complex spin-orbital coupling effects (including stripe phases) (55,56). Similarly, the electronic boundaries we observe in the CPD sample of $Er_{1-x}Ca_xMnO_3$ can be associated with those boundaries of orbital orderings predicted by Maezono et al. with suitable antiferromagnetic interaction level. However, the CPD of $La_{1-x}Ca_xMnO_3$ revealed an unexpected narrow semiconducting region that has not been previously observed. Further studies of this narrow region is required. If the (La, Ca) ions are appreciably ordered, hopping may be enhanced if the local environment of some manganese ions is ordered in such a way as to match their energy levels;

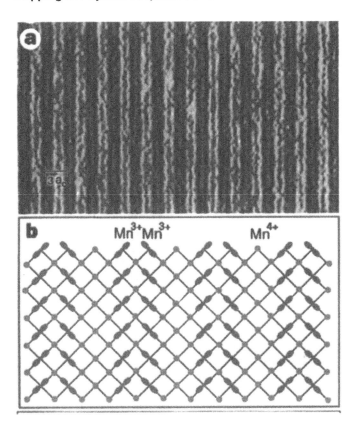

Figure 28 Charge-ordered stripes seen in $La_{0.33}Ca_{0.67}MnO_3$. (a) An electron diffraction image shows the overall structure of the stripe phase. (b) The schematic shows that $Mn^{4+}O_6$ charge stripes are inserted between the Jahn–Teller-distorted $Mn^{3+}O_6$ "parent banks" (also antiferromagnetically paired, as in cuprates), with periodicity corresponding to the carrier doping. (Reproduced from Ref. 72.)

i.e., energy splitting between adjacent Mn^{3+} and Mn^{4+} ions may approach zero, allowing increased hopping. This may also happen in the presence of charge ordering. Here, the Mn^{3+} and Mn^{4+} ions provide the local distortions. For compositions outside the conductive strip, the ordering would be broken, and the material would rapidly become insulating. Another possibility is that two directions of charge ordering compete with each other at the conductive strip, and one dominates on each side.

Furthermore, the singular phases around $x = 7/8$ in the CPD of $Er_{1-x}Ca_x$-MnO_3 on a $NdGaO_3$ substrate and $Yb_{1-x}Ca_xMnO_3$ on a $SrTiO_3$ substrate are

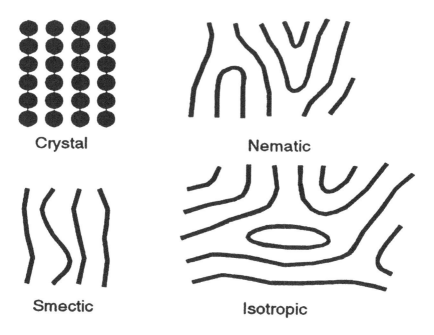

Figure 29 Various phases of stripe order, depending on the degree of fluctuation. Heavy lines represent liquid-like stripes, along which the electrons can flow, whereas the filled circles represent pinned, density-wave order along the stripes. (Reproduced from Ref. 64.)

intriguing (Figure 18). The existence of such singularity is not predicted by the phase diagram calculation of orbital orderings. A possible scenario for the current observation of phase boundaries can be the electronic self-organizations of different types. Note that around $x = 1/2$ and 7/8, we would expect commensurate locking of orbital (charge) orderings to the lattice, which would be an insulator. This commensurate singularity point is related to the static charge (or spin) orderings observed in recent experimental and theoretical studies in both copper oxides (57,58,60–66) and manganites (50,51,56,67). We expect to see a small structural anomaly associated with the singularity, owing to the strong coupling of charge ordering with the lattice at the commensurate point. If we accept this model, the highly conducting adjacent regions then suggest the existence of incommensurate, fluctuating orbital orderings. Therefore, the current observation provides preliminary evidence of the recently proposed liquid crystal phase (smectic phase) of orbital orderings in this doping range, widely believed to exist in copper oxides (64). The smeared-out critical divergence in microwave loss at the boundaries observed in this study is also consistent with this possibility. The thermodynamic nature of the transition is clearly shown by the temperature-dependent broadening or disappearance of the phase boundaries at higher temperatures. However, we believe that more studies are needed to confirm this suggestion.

CONCLUSION

The rich phase diagrams found in various manganese oxides reflect complex interactions, which determine electronic ground state. Phase mapping in condensed-matter physics is accomplished by growing single-crystal samples and measuring their physical properties at different temperatures. The famous phase diagram of $La_{1-x}Ca_xMnO_3$ took several years to construct. Technologically, knowing the location of the phase boundaries is critical in any development and optimization of materials. Obviously if we can produce a proper probe to manifest these phase boundaries and their physical nature, the immediate benefit to physical understanding of its material system and possibly the very mechanism of phenomena of interest is obvious. The impact of the current study is clear in what we can continuously and spatially realize the doping dependence of complex materials.

The observation that transitions between different electronic states at room temperature seem to drive the low-temperature magnetic order is intriguing. In a similar system of high-T_c superconductors, one also finds a rich phase diagram versus doping and temperature. It would be interesting to see whether such phenomena occur in the high-temperature superconductors. Further studies in this direction may elucidate the current debate on whether short-range Coulomb interactions lead to an exchange interaction pairing mechanism (68,69) or long-range Coulomb interactions lead to dynamic charge stripes and a smectic phase (58,64). If the same kind of doping-dependent electronic boundaries are also present in high=T_c cuprates such as $La_{2-x}Cu_xO_4$, the detailed study of the system at low temperatures can be carried out and reveal suspected quantum critical behavior at the critical doping point, such as $x = 1/8$.

The extension of the current CPD technique to other material systems may be carried out. The effectiveness of CPD technique here suggests that mapping the global properties of a condensed matter system, once thought as an intractable problem, is now possible to scientists.

REFERENCES

1. Bednorz, J. G.; Müller, K. A. Z. Phy. 1986, B 64, 189.
2. Scheel, H. J. MRS Bulletin. 1994, 19, 26.

3. Phillips, J. C. Physics of High-T_c Superconductors; Academic Press: New York, 1989.

4. Wu, M. K.; Ashburn, J. R.; Torng, C. J.; Hor, P. H.; Meng, R. L.; Gao, L.; Huang, Z. J.; Wang, Y. Q.; Chu, C. W. Phys. Rev. Lett. 1987, 58, 908.

5. Cava, R. J.; Batlogg, B.; Krajewski, JJ; Farrow, R; Rupp, L.W. Jr.; White, A.E.; Short, K.; Peck, W.F.; Kometan, T. Nature. 1990, 345, 602.

6. Xiang, X. D.; Sun, X.; Briceño, G.; Lou, Y.; Wang, K.-A.; Chang, H.; Wallace-Freedman, W. G.; Chen, S.-W.; Schultz, P. G. Science. 1995, 268, 1738–1740.

7. Nagamatsu, J.; Nakagawa, N.; Muranaka, T.; Zenitani, Y.; Akimitsu, J. Nature. 2001, 410, 63.

8. Mattheiss, L. F.; Gyorgy, E. M.; Johnson, D. W. Phys. Rev. 1988, B 37, 3745.

9. Cava, R. J.; Batlogg, B.; Krajewski, J.J.; Farrow, R.; Rupp, L.W. Jr.; White, A.E.; Short, K.; Peck, W.F.; Kometan, T. Nature. 1988, 332, 814.

10. See, for example; Santoro, A.; Beech, F.; Marezio, M.; Cava, R. J. Physica C. 1988, 156, 693.

11. Villars, P.; Phillips, J. C. Phys. Rev. B. 1988, 37, 2345.

12. Miyazaki, T.; Ajima, T.; Sato, F. J. Magn. Mater. 1989, 81, 86.

13. Schad, R.; Potter, C.D.; Belien, P.; Verbanck, G.; Moshchalkov, V.V.; Bruyseraede, Y.R. Appl. Phys. Lett. 1994, 64, 3500.

14. Jimbo, M.; Tsunashima, S.; Kanda, T.; Goto, S.; Uchiyama, S. J. Appl. Phys. 1993, 74, 3341.

15. Wall Street Journal. 10 November 1997, B8.

16. Dax, M. Semicond. Int. 1997, 20, 84.

17. van Sante, J. H.; Jonker, G. H. Physica. 1950, 16, 599.

18. Jin, S.; Tiefel, T.H.; McCormack, M.; Fastnacht, R.A.; Ramesh, R.; Chen, L.H. Science. 1994, 264, 413.

19. Zener, C. Phys. Rev. 1951, 82, 403.

20. Anderson, P. W.; Hasegawa, H. Phys. Rev. 1955, 100, 675.

21. de Gennes, P. G. Phys. Rev. 1960, 118, 141.

22. Briceno, G.; Chang, H.; Sun, X.; Schultz, P. G.; Xiang, X-D Science. 1995, 270, 273.

23. Jin, S.; O'Bryan, H. M.; Tiefel, T. H.; McCormack, M.; Rhodes, W. W. Appl. Phys. Lett. 1995, 66, 382.

24. Askham, F.; Fankuchen, I.; Ward, R. J. Amer. Chem. Soc. 1950, 72, 3799.

25. Shimakawa, Y.; Kubo, Y.; Manako, T. Nature. 1996, 379, 53.

26. Ramirez, A. P.; Cava, R. J.; Krajewski, J. Nature. 1997, 387, 268.

27. Moritomo, Y.; Tomioka, Y.; Asamitsu, A.; Tokura, Y.; Matsui, Y. Phys. Rev. 1995, B 51, 3297.

28. Moritomo, Y.; Asamitsu, A.; Kuwahara, H.; Tokura, Y. Nature. 1996, 380, 141.

29. Kumura, T.; Tomioka, Y.; Kuwahara, H.; Asamitsu, A.; Tamura, M.; Tokura, Y. Science. 1996, 274, 1698.

30. Hasebe, M.; Nishizawa, T. Application of Phase Diagrams in Metallugy and Ceramics, NBS Special Pub. 496: Washington, DC; 1978; Vol. 2, 911–954.

31. Romig, A. D., Jr Bull. Alloy Phase Diagr. 1987, 8, 308.

32. Kennedy, K.; Stefansky, T.; Davy, G.; Zackay, C. F.; Parker, E. R. J. Appl. Phys. 1965, 36, 3808.

33. van Dover, R. B.; Schneemeyer, L. F.; Fleming, R. M. Nature. 1998, 392, 162.

34. Chang, H.; Tacheuchi, I.; Xiang, X.-D. Applied Physics Lett. 1999, 74, 1165.

35. Yoo, Y.K.; Duewer, F.; Yang, H.; Yi, D.; Xiang, X.-D. Nature. 2000, 406, 704.

36. Duewer, F.; Yoo, Y.K.; Fukumura, T.; Yang, H.; Yi, D.; Hasegawa, T.; Kawasaki, M.; Koinuma, H.; Xiang, X.-D. Phy. Rev. 2001, B 63, 224421.

37. Yoo, Y. K.; Ohnishi, T.; Wang, G.; Duewer, F. W.; Xiang, X. D.; Chu, Y.-S; Mancini, D. D.; Li, Y.-Q.; O'Handley, R. C. Intermetallics. 2001, 9, 541.

38. Birgeneau, R. J.; Shirane, G. ; Ginsberg, D. M., ed. In: Physical Properties of High-Temperature Supercondcutors; World Scientific: Singapore, 1989; Vol. 1, and references therein, 152.

39. Schiffer, P.; Ramirez, A. P.; Bao, W.; Cheong, S.-W. Phys. Rev. Lett. 1995, 75, 3336.

40. Ramirez, A. P. J. Phys. Condens. Matter. 1997, 9, 8171.

41. Asamitsu, A.; Moritomo, Y.; Tomioka, Y.; Arima, T.; Tokura, Y. Nature. 1995, 373, 407.

42. Urushibara, A.; Moritomo, Y.; Arima, T.; Asamitsu, A.; Kido, G.; Tokura, Y. Phys. Rev. B. 1995, 51, 14103.

43. Tomioka, Y.; Asamitsu, A.; Kuwahara, H.; Moritomo, Y.; Tokura, Y. Phys. Rev. 1996, B 53, R1689.

44. Gao., C.; Wei., T.; Duewer, F.; Lu, Y.; Xiang, X. D. Appl. Phys. Lett. 1997, 71, 1872.

45. Gao, C.; Xiang, X.-D. Rev. of Sci. Inst. 1998, 69, 3846.

46. Gao, C.; Duewer, F.; Xiang, X. D. Appl. Phys. Lett. 1999, 75, 3005.

47. Morooka, T.; Nakayama, S.; Odawara, A.; Ikeda, M.; Tanaka, S.; Chinone, K. IEEE Trans. Appl. Supercond. 1999, 5, 3491.

48. Moreo, A.; Yunoki, S.; Dagotto, E. Science. 1999, 283, 2034.

49. Ramirez, A. P.; Cheong, S. W.; Schiffer, P. J. Appl. Phys. 1997, 81, 5337.

50. Uehara, M.; Mori, S.; Chen, C. H.; Cheong, S.-W. Nature. 1999, 399, 560.

51. Chen, C. H.; Cheong, S.-W. Phys. Rev. Lett. 1996, 76, 4042.

52. Mori, S.; Chen, C. H.; Cheong, S.-W. Nature. 1998, 392, 473.

53. Moritomo, Y.; Akimot, T.; Nakamura, A.; Ohoyama, K.; Ohashi, M. Phys. Rev. B. 1998, 58, 5544.

54. Kajimoto, R.; Yoshizawa, H.; Kawano, H.; Kuwahara, H.; Tokura, Y.; Ohoyama, K.; Ohashi, M. Phys. Rev. B. 1999, 60, 9506.

55. Kuwahara, H.; Tomioka, Y.; Asamitsu, A.; Moritomo, Y.; Tokura, Y. Science. 1995, 270, 961.

56. Maezono, R.; Ishihara, S.; Nagaosa, N. Phys. Rev. B. 1998, 58, 11583.

57. Tokura, Y.; Nagaosa, N. Science. 2000, 288, 462.

58. Tranquada, J. M.; Sternlieb, B. J.; Axe, J. D.; Nakamura, Y.; Uchida, S. Nature. 1995, 375, 561.

59. Tranquada, J. M.; Axe, J. D.; Ichikawa, N.; Moodenbaugh, A. R.; Nakamura, Y.; Uchida, S. S. Phys. Rev. Lett. 1997, 78, 338.

60. Emery, V. J.; Kivelson, S. A. Physica C. 1993, 209, 597.

61. Emery, V. J.; Kivelson, S. A. Physica C. 1994, 235–240, 189.

62. Emery, V. J.; Kivelson, S. A. Physica C. 1996, 263, 44.

63. Salkola, M. I.; Emery, V. J.; Kivelson, S. A. Phys. Rev. Lett. 1996, 77, 155.
64. Kivelson, S. A.; Fradkin, E.; Emery, V. J. Nature. 1998, 393, 550.
65. Noda, T.; Eisaki, H.; Uchida, S. Science. 1999, 286, 265.
66. Zhou, X. J.; Bogdanov, P.; Kellar, S. A.; Noda, T.; Eisaki, H.; Uchida, S.; Hussain, Z.; Shen, Z.-X. Science. 1999, 286, 268.
67. Kuwahara, H.; Tomioka, Y.; Moritomo, Y.; Asamitsu, A.; Moritomo, Y.; Tokura, Y. Science. 1996, 272, 80.
68. Scalapino, D. J. Science. 1999, 284, 1282.
69. Dai, P.; Mook, H. A.; Hayden, S. M.; Aeppli, G.; Perring, T. G.; Hunt, R. D.; Dogan, F. Science. 1999, 284, 1344.
70. Lide, D. R. CRC Handbook of Chemistry and Physics, 79th; CRC Press: Cleveland: OH, 1997, Chap. 12.
71. Service, R. F. Science. 1999, 283, 1106.
72. Littlewood, P. Nature. 1999, 399, 529.

9

Temperature-Dependent Materials Research with Micromachined Array Platforms

Steve Semancik
National Institute of Standards and Technology, Gaithersburg, Maryland, U.S.A.

I. INTRODUCTION

A. Opportunities Provided by Microfabrication

Combinatorial methodology, particularly as a general field of science outside of pharmaceutical exploration, is in a relatively early stage of development. As a "new" research area, it offers considerable challenges, but it also promises to have tremendous impact in the years ahead. Many technologies are being examined, refined, and adapted for use in combinatorial studies. The recent explosion of microfabrication methods is providing enabling technology for development in a number of related areas, including microsensors, microfluidics, and lab-on-a-chip technology, as well as for combinatorial materials science. Microscale research platforms, such as those that can now be produced by micromachining and other microfabrication methods, offer inherent features that can make them particularly beneficial for certain types of combinatorial materials research. Small microelectromechanical systems (MEMS) structures of ~100 μm are now readily constructed, and the structures can be easily replicated to produce arrays of tens, hundreds, or thousands of elements. Multilevel processing, borrowed from decades of development in the electronics industry, also makes it possible to include varied types of functionality needed for combinatorial studies on micromachined platforms. For example, heaters, thermometry components, electrodes, and control and interrogation electronics, including multiplexers, can be fully or partially integrated on a chip. When standard design rules are employed, such as complementary metal oxide semiconductor (CMOS) methods, the platforms can be made especially compatible with other instrumentation and data-handling hardware. Arrays of microplatforms allow parallel experimentation on multiple discrete samples integrated within a test specimen small enough to be easily mounted into test systems and characterization equipment. However, the small sample sizes associated with the use of these platforms do challenge analytical instrumentation to provide adequate signals, in relatively short times, for both property characterization and performance evaluation.

B.   Temperature-Dependent Research

Temperature is certainly one of the most pervasive factors in materials fabrication. General phenomena such as adsorption, desorption, diffusion, reaction, phase transformation, and structural rearrangement exhibit temperature-dependent behavior. Elevated temperatures also allow reactive species to be produced, for example, in thermally activated chemical vapor deposition (CVD), or to control surface diffusion rates to determine the type of microstructure one will achieve in film growth. Temperature cycling is critical in postdeposition conditioning of materials (sintering, curing). Obviously, it is also a critical parameter to be controlled in characterizing materials properties, such as electrical conductivity, and in examining materials performance (for example, product yields from various catalyst compositions).

In this chapter, the role that temperature-controllable microplatforms can play in combinatorial discovery is illustrated using "microhotplate" arrays that have been developed at the National Institute of Standards and Technology (NIST). The microhotplate design to be discussed was originally created as a platform for low-power, temperature-controlled gas microsensors, but in an array format it can also offer functionality required for efficiently fabricating and evaluating microsamples of varied classes of materials (1,2). Their utility can therefore impact optimization studies for a range of technologies.

As its name implies, the microhotplate makes it possible to control and measure temperatures of small areas, typically 100 μm $\times$ 100 μm, electronically. Each microhotplate element includes functional components for heating (and cooling back to ambient or a purposely reduced temperature), thermometry, and electrical probing of deposited films. Within an array, the elements are thermally isolated, and *individually addressable*, so that independent and simultaneous control and on-chip interrogation can be exercised. The arrays are well suited to temperature-dependent materials studies of discrete microsamples. We have used multielement arrays containing 16, 36, and 48 microhotplate elements as research tools for efficiently examining microsample formation and performance in varied types of temperature-dependent chemical processes (3). Figure 1 shows a micrograph of a 48-element array used in such studies.

The microhotplate arrays described here offer a matrix of temperature-controlled, discrete sites for materials processing/performance correlations applicable to a spectrum of technological areas, including, for example, catalysis, chemical sensing, and electronic materials optimization. We describe the range of methods we have used in locally depositing oxide, metal, and polymeric films onto microhotplates (including thermally activated chemical vapor deposition and addressable potential control for electrodeposition). While on-chip characterization is possible in some cases (say, for measuring microsample electrical resistance and temperature), characterization of the properties and performances for array microsamples by external probes is necessary in other cases. Obviously, success in combinatorial experimentation with such platforms requires properly adapted sample fabrication and characterization methods.

This chapter employs examples of microarray studies of oxide gas–sensing materials to demonstrate the capabilities of the micromachined technology. Gas sensing is the primary research effort within our group, but it also happens to be an area that is inherently combinatorial in nature, because different types of sensing materials must be developed and optimized for varied applications. Therefore, it provides an effective illustration of temperature-dependent materials pro-

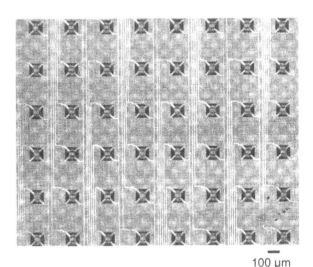

100 μm

Figure 1 Micrograph of a 48-element microhotplate array showing suspended devices and electrical connection lines.

cessing, microsample characterization by external techniques, and on-chip performance evaluation involving parallel microarray methodology. The chapter also considers issues relevant to the adaptation of sample deposition methods and both external and on-chip characterization techniques for the more generalized application of microarray methods. Such adaptation allows investigations in other technical areas and presents a basis for discussing both the limitations and the potential of the microarray approach for more general combinatorial materials research. (The discrete nature of the microarray elements are emphasized in this chapter, but it should be noted that the microfabrication techniques described also permit construction of microscale structures with purposely designed temperature gradients.) Once adapted to microplatform studies, temperature-dependent research in areas as diverse as biosystems, catalysis, electronics, polymers, photovoltaics, and sensing can be expected to be performed with significantly increased efficiency.

II. MICROHOTPLATE PLATFORMS

A. The Basic Microhotplate Structure

The temperature-controllable structure used as a base element in our conductometric gas microsensor prototypes and microarray platforms is called a *microhotplate* (1). (Conductometric microsensors measure adsorption-induced changes in the electrical conductivity of sensing films.) The micrograph shown in Figure 2a is a single suspended microhotplate that has been produced by micromachining the Si on which the device was fabricated. The most obvious features are the suspended plate, the four support beams, and the etch pit. The device shown also has four top electrodes that are used for characterizing the electrical characteristics (conductance) of deposited sensing films. The layer structure within the suspended plate, shown in Figure 2b, indicates other functionality within the basic microhotplate element. A serpentine polysilicon element embedded in SiO_2 serves as a resistive heater. Above the heater—separated by an insulating layer of SiO_2—is a metal plate that serves to conduct heat (generated by applying a current through the polysilicon heater below) more evenly across the device. The temperature coefficient of resistance of this metal plate can be calibrated to provide temperature measurements. Similarly, one may extract a temperature from the measured polysilicon resistance. In either case the measured signal can be used in a control loop for setting a temperature-versus-time schedule. To provide for measurement of the electrical characteristics of deposited films, the surface electrodes directly contact films while being isolated from the heat distribution layer by a SiO_2 layer.

a)

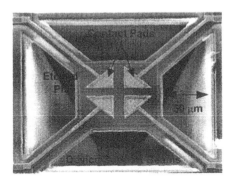

b)

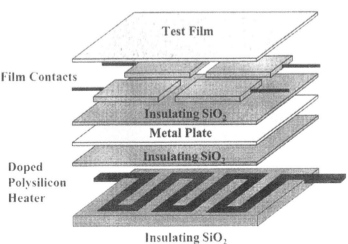

Figure 2 (a) Micrograph of a basic microhotplate device developed for conductometric gas sensing (triangular contacts). (b) Schematic of the microhotplate layer structure (shown with square contacts and including the registered position of an experimental test film).

B. Characteristics

The microhotplate elements, as well as other structures with related microheater configurations, have characteristics that are particularly well suited for developing low-power chemical microsensors and for performing temperature-dependent materials research studies on microscale samples. The lateral size of the suspended hotplate can easily be varied from 25 μm to 200 μm. The low mass of the suspended platform ($\sim$0.25 μg for a 100-μm device) allows each microhotplate to be thermally isolated (significant temperature gradients exist along the suspension legs). Local heating is made possible by flowing appropriate current levels through the polysilicon resistive heater. The accessible temperature range is 20–500°C (with heat distribution/thermometry plate) and 20–750°C (without the plate). Maximum device temperatures are largely governed by the available type(s) of metallization at a given foundry where device fabrication is performed. For Al metallization, the maximum microhotplate temperature is approximately 500°C, and metal postprocessing is needed to produce more acceptable film contact composition (such as Ir over a Cr adhesion layer deposited over the foundry Al). When other metallizations, such as W and Ti-W, are available at the foundry, devices can be heated as high as 750°C. Time constants are on the order of 1–5 ms, and heating rates can be in excess of 10^{6}°C/s, so pulsed temperature operation is not only convenient, but can be realized over a wide frequency range; when properly designed pulsed electronics are used, heating time constants can be reduced to approximately 10 μs. The top-surface microelectrodes contact films deposited on the microhotplates and can be used to measure electrical characteristics of the films during or after growth (to monitor deposition processes and sense gases, for example). Elements within an array can be individually addressed for temperature control and film electrical measurements.

C. Fabricating Multielement Arrays Through Flexible Design Methods

The use of computer-aided design in developing mask sets for foundry runs greatly simplifies the fabrication of microhotplate device variations and microarrays in which a given microhotplate design is replicated to produce multielement arrays. Proper spacing of the elements and configuration of connecting electrical lines allow array elements to be thermally isolated from each other and individually addressable. These aspects are particularly important in applying the arrays in parallel experimentation on materials processing, property examination, and performance evaluation.

In fabricating devices in wafer runs at a silicon foundry, we now design die ($\sim$15 mm $\times$ 15 mm) that have approximately 25–30 different sectors, each including array elements related to a basic microhotplate. An example of a recent die design is shown in Figure 3; this layout is step-repeated in a 6-inch wafer run to produce approximately 75 such die per wafer. The variety of platforms that can be included allows one to develop and use a range of microscale research tools and device prototypes. For example, in our chemical microsensor work, we typically use four-element arrays of differing element size and differing electrode geometry; the geometry of contacts can easily be adjusted in the mask layout. Figure 4 indicates some of the contacts we have employed on microhotplate elements. It is noted that interdigitated electrodes have been valuable in making electrical measurements on oxide and polymeric films of high resistance, while other two-contact and four-contact designs are suitable for more conductive films. The illustrated die in Figure 3 also contains 16-element arrays (Figure 5a), 48-element arrays (see Figure 1), and special 340-element arrays designed for mass spectroscopic measurements on adsorbate transients (Figure 5b) (3).

Other modifications of the basic microhotplate structure and its configuration have been introduced within the layout in Figure 3 to achieve related devices, such as microcalorimeters, preconcentrators, and catalytic filtering arrays, as well as to achieve specific required characteristics (smaller or bigger element size, faster thermal response, etc.). The base structures illustrated in Figure 2 use 8–10 electrical lines per device element, to offer maximal flexibility and device-monitoring capabilities. The fact that the microhotplate design is CMOS compatible allows one to readily integrate various types of circuits on-chip. In a 48-element array, to avoid having 384–480 lines, we have introduced on-board multiplexer circuitry (available through fabrication runs at some foundries) and a varied connectivity arrangement utilizing common bus lines in another case. The ability to adjust device features, such as the types of incorporated materials and wiring configurations, or to include on-chip electronic components is discussed further in Section V.

We conclude this subsection with a concise description of design and etching approaches we have employed. The mask set originally used to create the microhotplates at silicon foundries was designed using the MAGIC CAD tool* (4). (Other design tools, such as those produced by Tanner Research, Inc.*, Cadence Design Systems*, or other manufacturers, can also be used.) Regions of purposely exposed Si allow the devices to be micromachined in a manner that produces the thermally isolated, suspended-plate structure. (The necessary configuration of exposed substrate Si is defined by a modification in the design technology file called ''open'' (5)). We have had our microhotplate platforms

---

* Throughout the chapter, certain commercial instruments or suppliers are identified to adequately specify the experimental procedure. In no case does this identification imply endorsement by the National Institute of Standards and Technology.

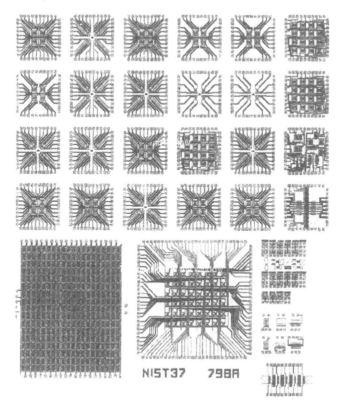

Figure 3 Computer image of a single 13-mm × 15-mm die (enlarged) showing the multiple device designs that are replicated across a wafer at a silicon foundry.

manufactured at three different foundries: the Metal Oxide Semiconductor Implementation Service (MOSIS)* (6), the Microelectronics Research Laboratory* (7), and the MIT—Lincoln Laboratories* (8). All offer certain advantages and certain limitations related to available types of materials, processing steps included, and delivery times and costs. Postprocessing etching steps (performed at NIST) involving various silicon etchants, including ethylene diamine pyrocatechol-water (EDP; anistropic etch) (9), XeF_2 (isotropic etch) (10), and tetramethyl ammonium hydroxide (TMAH; anistropic etch) (11), have been used to produce the suspended microhotplates without appreciably disturbing the surface electrode metals. Additional details connected with die layout, micromachining etchants, and packaging of the devices are given elsewhere (3).

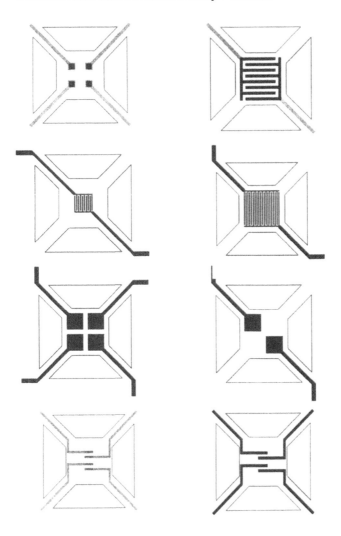

Figure 4 Examples of the surface contact variations included within microhotplate devices.

The cost of producing microarray platforms from wafer runs is somewhat difficult to assess for any given combinatorial application, for the desired fabrication materials and design complexity will vary. However, there is certainly the potential for producing research arrays with low unit cost, particularly if one can make production wafers with a small number of designs (and thus more total

a)

200 μm

b)

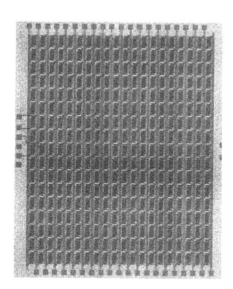

Figure 5 (a) Micrograph of a 16-element array chip with 65-μm × 65-μm elements. (b) Computer diagram of a 340-element array with 200-μm × 100-μm elements.

copies) and utilize standard foundry materials. Additional labor expenses will obviously exist for the steps of dicing, etching, packaging, wire bonding, and depositing experimental films on the microdevices.

III. DEPOSITING MICROSAMPLES ON MICROARRAY ELEMENTS

To study varied types of materials processing and performance using microscale array platforms (such as microhotplate arrays), methods must be developed for compatibly incorporating the materials of interest onto the miniature elements. Typically, films would be deposited after silicon has been etched away to produce the thermally isolated microhotplate structures. This is the case because etching chemicals used for micromachining are often too aggressive to properly maintain the integrity of predeposited microsample films. Then, however, one must deal with a three-dimensional, suspended device structure during film deposition. The etch pits are problematic, for example, when one attempts to use spin-on methods or, in some cases, if one uses lift-off processing. For combinatorial studies, methods must be developed to overcome these challenges in an efficient and reliable manner.

The processing methods we discuss here were developed for depositing functional materials for gas microsensors (either sensing, preconcentration, or filtering films). Varied sensing applications require that sensing films be incorporated with different compositions and microstructures for attaining acceptable performance when monitoring different gases or vapors. The methods employed are not general enough to directly address microsample deposition for the full spectrum of other technologies that could benefit from temperature-dependent microarray studies. However, the methods described do involve oxides, metals, and polymers produced and tailored by a relatively diverse set of techniques, and they therefore serve as a useful introduction to developing other specialized deposition procedures.

Among the methods that come to mind first for producing arrays of materials that can be processed and evaluated at different temperatures on individual microhotplates are (shadow) masking and micropipetting. The abilities to electronically address heating currents (temperatures) and potentials within a microarray, however, provide additional opportunities to locally deposit materials using thermally activated and electrochemical methods. The following discussion begins with these latter techniques.

For all deposition approaches, localized temperature control can be used during and/or after deposition. It should also be noted that depositions are done onto microsubstrate surface areas (see Figure 2a) consisting of metal contact material(s) and foundry SiO_2 [which may have received chemical mechanical polishing (CMP) treatment]. In cases where one is defining certain types of combinatorial libraries (those with materials of sufficient electrical conductivity), it may be advantageous to utilize the top-surface electrodes for continuously monitoring the deposition process (see upcoming Figure 7b).

A. Thermally Activated Chemical Vapor Deposition

Fabrication of most of our oxide-based, conductometric microsensors is accomplished by thermally activated CVD of the sensing oxide films. The approach is described schematically in Figure 6. By selecting precursor and reactant gases that form the desired oxide through a thermally activated process, localized ''self-lithographic'' deposition can be made to occur at a desired microhotplate that is heated to an appropriate temperature (12,13). By selecting appropriate precursors, reactants, and growth temperatures, varied types of oxide and metal films can be deposited on one or more microhotplate elements. Figure 7a illustrates the results of a self-lithographic CVD deposition of SnO_2 onto each element of a four-element array, in which the devices were sequentially heated to 450°C in tetramethyltin and oxygen. Due to the inherent thermal isolation, thermally activated CVD occurs only on the area of the addressed hotplates. The growth-monitoring capability of the microhotplates for overdeposited films is illustrated in Figure 7b, where the conductance traces follow the formation of SnO_2 films on four microhotplates. We have also used a similar thermally activated CVD procedure to sequentially deposit SnO_2 and ZnO (from diethylzinc and oxygen) and SnO_2 and Pt (from trimethyl[cyclopentadienyl] platinum and hydrogen) within the same array (2).

Process Steps

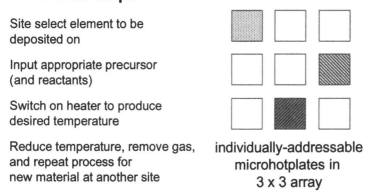

Site select element to be
deposited on

Input appropriate precursor
(and reactants)

Switch on heater to produce
desired temperature

Reduce temperature, remove gas,   individually-addressable
and repeat process for   microhotplates in
new material at another site   3 x 3 array

Figure 6   Schematic diagram of the thermally activated CVD processing steps that can be used to deposit different materials on the elements of a microarray.

a)

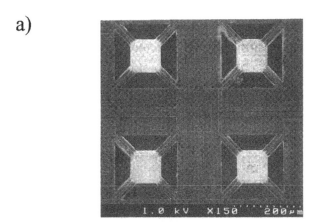

b)

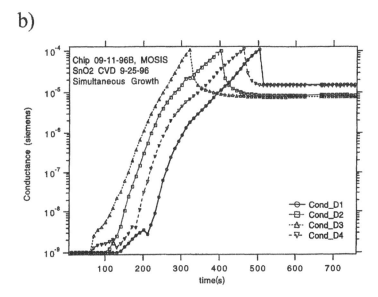

Figure 7 (a) Micrograph showing the localized deposition of SnO_2 films (lighter gráy) on a four-element microhotplate array accomplished by the self-lithographic, thermally activated CVD technique. (b) Conductance-versus-time plots that monitor the growth of CVD SnO_2 films in real time. (Time offsets between the conductance growth curves may arise from the differing nature of the random nucleation process in each case, as well as from possible small variations in absolute temperature between the elements during these particular experiments.)

B. Electrochemical Deposition

Electrochemical deposition is another technique that can make use of the ability to individually address microhotplates in an array to localize films. Using an interdigitated electrode structure, similar to the closely spaced "comb" electrodes shown in Figure 4, we have also deposited polypyrole from solution onto elements of a microhotplate array using addressable electrochemistry (14,15). The deposition solution consisted of anhydrous acetonitrile, 0.1 M LiClO$_4$, and 50 mM pyrrole. An electrode potential of $+0.1$ V (vs Pt) was applied for 5 minutes to obtain a polypyrrole film of about 10-μm thickness. The potential was switched to -0.5 V for 2 minutes after deposition to expel LiO$_4^-$ anions. A variety of other materials (metals, oxides, etc.) could also be deposited on the microelements via potential control.

Surface or subsurface electrodes could also be employed to create fields for controlling growth or aligning deposition of materials (including, for example, nanowires from a solution (16)).

C. Thermal Drying and Film Activation

Although the microhotplates are small, three-dimensional structures within an array, we have also succeeded in depositing (sensing) films onto them via other thermally based methods. Localized heating of the microhotplates can also be used indirectly to assist deposition by "fixing" solids deposited from suspensions or sol-gels through controlled-rate drying (14). Once the solution or sol-gel (dropped or spun onto the chip) has been heated to volatilize the carrier and to deposit a film on selected hotplate(s), the solution or sol-gel can be rinsed from other sections of the chip. Thus far, we have used such processing to deposit sol-gel SnO$_2$ and colloidal SnO$_2$ nanoparticle films (17). Related processing is now also being used to form porous SiO$_2$ over microhotplates from silsesquioxane/block copolymer mixtures in developing microscale preconcentrators (18).

D. Thermal Lithography

Thermal addressing of individual hotplates can also be used to define windows for deposition within "thermal resist" films (14). One lithographic technique with general applicability involves the use of an array element's microheater to volatilize a nitrocellulose coating that has been spun on an entire array. A (sensing) film can then be applied, for example, by spinning on a colloidal solution (17) or a sol-gel. Lift-off by dissolution of the nitrocellulose layer leaves behind only the film(s) on the selected hotplate(s).

Patterns of organosilane monolayers have also been successfully made on microhotplates. Organosilanes, such as (tridecafluoro-1,1,2,2-dihydrooctyl) trichlorosilane, readily react with surface hydroxyl groups that are present on the surface of the microhotplate arrays. By then heating specified microhotplates (nominally to 450°C for several seconds), the silane monolayer can be removed in a localized area (19). Secondary ion mass spectrometry (SIMS) has been used to establish the viability of this technique for creating ''clear'' deposition windows on selected microhotplates, and the approach has been used to deposit SnO_2- and Fe_2O_3-sensing films onto adjacent elements in a four-element microhotplate array.

E. Masking

Masking can also be effective for producing films on microhotplates by physical deposition methods, such as evaporation and sputter deposition. To produce microsamples, masks with very small apertures are necessary. Lithographic methods can be used to create metal shadow masks (available commercially) with apertures of ~ 50–100 μm. These masks must be carefully registered, in at least some cases, to the packaged and wire-bonded microhotplate arrays. Serial processing, with selected blocking of apertures, is required to attain a variety of film compositions within the array with such a method. We have utilized this approach to study the sensing effect of different surface-dispersed, low-coverage metals on SnO_2-based films (see also upcoming Section IV. C) on 36-element arrays (20). We have also deposited well-registered metal films through dedicated shadow masks onto 200-μm $\times$ 100-μm platforms of a 340-element array fabricated for studies of sensing mechanisms (21). While this procedure can work well for processes such as evaporation and sputter deposition, the small aperture size can be a problem for (liquid) spraying techniques, where particles and surface tension can clog the mask.

F. Micropipetting

Finally, for liquids that are not too viscous, one can also consider the use of micropipetting (without, or perhaps with, a mask) to place small volumes of materials on microhotplate elements. Precision alignment methods must be used for aligning a pipette dispenser to the microarray.

IV. EXAMPLES: PROCESSING/PROPERTY/ PERFORMANCE STUDIES ON GAS SENSOR MATERIALS

The challenge of selecting sensor materials for solid-state chemical microsensors that are properly matched to the analyte(s) encountered in a specific application and then optimizing parameters (e.g., microstructure, thickness, additive loading) for films of these materials presents an inherently combinatorial problem. Furthermore, if multiple analytes in a gaseous mixture are being monitored, different sensing films in an array format are often employed. The nature of the problem is expressed schematically in Figure 8. Varied compositions, microstructures, and operating temperatures can come into play in detecting and quantifying target species. In the examples that follow, we indicate several processing methods (see also Section III) to produce candidate materials, the use of characterization tools (external and on-chip) to define sample properties, and the use of on-chip electrical measurements to collect data for assessing performance (because we were interested in conductometric gas sensing). Thus, such studies allow correlations to be made for film processing, properties, and performance. The results all relate to pure or modified oxide materials investigated with 16-element or 48-element arrays (in which 36 elements were used). It is worth noting that the methodology can be applied to other types of sensor-related materials, such as catalytic metals in filters and high-area (porous) materials for preconcentrators (18). Aspects of the approach can also be used even more broadly in other technical areas.

A. Temperature-Dependent Microstructure of CVD TiO$_2$

The microhotplates within 16-element arrays have been used to study the temperature-dependent microstructures produced by CVD processing of titanium dioxide from titanium(IV) nitrate. This precursor is capable of depositing titanium oxide that is free of residual carbon without the use of an external oxidant (22). In this work, titanium(IV) nitrate was ground under inert atmosphere prior to transferring it to the precursor vessel, where it was maintained at 35°C during the growth experiments. High-purity argon was used as the carrier gas, with a flow rate of one sccm. Reactor pressure was typically 500–650 mtorr. Experiments were carried out under fixed-temperature conditions, with deposition times of 5 minutes for each sample. The 16-element arrays were used for deposition temperatures from ~100°C to ~400°C (23). Micrographs of the resulting films are shown in Figure 9. Films deposited above 170°C were of the same approximate thickness, while those deposited below 170°C were considerably thinner, consistent with reaction-limited growth. As the deposition temperature increases (175°C to 250°C), the large grains seem to subdivide along a particular crystallographic direction, resulting in a platelike microstructure (possibly caused by migration of adsorbed precursor more rapidly in some directions than in others). Increasing the deposition temperature further causes these plates to subdivide into smaller grains. Certain microstructures developed via these efficient, temperature-dependent processing surveys on multielement arrays have exhibited high analyte sensitivities (to chemical warfare agents) (24) and switching of response polarity (direction of conductance change), which can occur dependent on film composition,

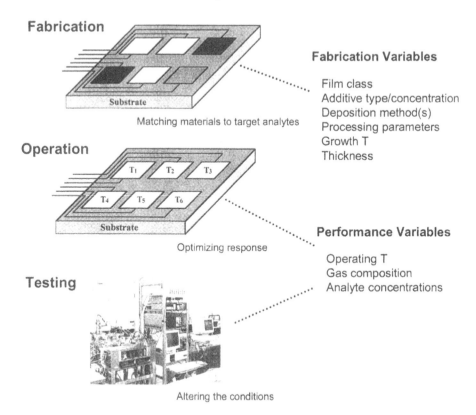

Figure 8 Illustration of the multiple factors involved in producing sensing films that are matched to monitoring applications and optimized for desired performance, showing the combinatorial nature of fabricating appropriate films.

microstructure, and temperature or analyte concentration (25). The discovered switching behavior can be very useful for distinctive analyte recognition.

We have also used temperature programmed deposition, in which varied time-dependent temperature profiles, such as those in Figure 10, were applied during deposition. Temporal variation of the growth temperature (possible over a range of frequencies) can alter the mode of growth, effectively switching the evolving film between different structures. The approach has been used to produce a number of new TiO_2 microstructures not observed in the fixed-temperature results of Figure 9 (23). This type of temperature-programmed deposition is easily implemented with the microheater elements.

In the case of CVD processing of oxides and metals, for example, microsamples can be formed not only with different hotplate temperatures, but also under

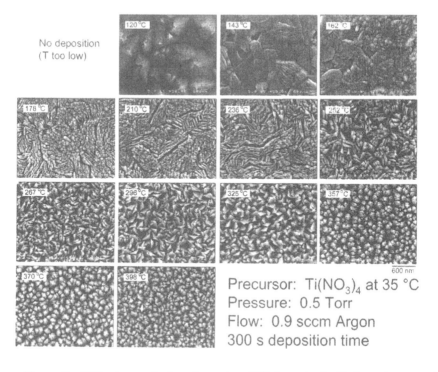

Figure 9 SEM micrographs from the centers of 13 elements of a 16-element array used in a temperature-dependent study of TiO$_2$ growth from titanium nitrate.

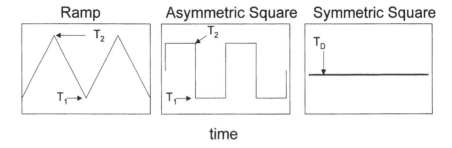

time

Figure 10 Schematic representation of some of the basic temperature excursions that can be run on individual microhotplate array elements in temperature-programmed deposition.

differing precursor fluxes, carrier/precursor concentration ratios, and pressures. Altering these variables can produce differing film stoichiometries, microstructures, defect densities, and degrees of interfacial dispersion and mixing (for metal/oxide systems). Microsamples with such varied properties will exhibit differing sensing characteristics. The differences are further enhanced if one deposits a different (catalytic) additive on the individual microsamples. Two further examples of sensor film processing now follow.

B.   Metal Particle Seeding in CVD Growth of SnO_2

Films fabricated by seeding the SnO_2 CVD growth (from tetramethyltin and oxygen) with small metal particles (Ni, Co, Fe, Cu, Ag) have been studied with 36-element arrays. Film reproducibility, as judged by SEM and sensing response, was investigated by making columns of six films in the array using the same recipe. The seed layers provide a larger number of distributed sites for nucleation, resulting in more rapid film growth with a finer grain microstructure (26). We have found that SnO_2 films prepared with seed layers of different composition result in different microstructures (27). Representative microstructures, shown in Figure 11, appear to correlate with the melting temperature of the seed layer metal. Low-melting-point materials like Ag were able to diffuse on the surface and produce larger but fewer nucleation sites, resulting in larger grains, while higher-melting-point metals like Fe and Ni created a larger number of small, stabilized nucleation sites, producing very fine ($\sim$30 nm) grains. While the fine microstructure generally enhanced the sensitivity to a wide range of gases, the different microstructures produced by using different seed layers also yielded sensor films that had different selectivities to the set of gases that was used for testing.

Temperature-dependent response data can be efficiently collected from the sensing materials deposited in microhotplate array formats. Dedicated multichannel electronic hardware, in fact, allowed simultaneous collection of sensor data from all 36 elements. Figure 12 shows a series of web or radar plots derived from samples in the 36-element seeding study on CVD-grown SnO_2 (samples like those shown in the micrographs in Figure 11). The data "snapshots" represent the relative sensitivities to the various test gases at 400°C for the differing microstructures [at analyte concentrations of 90 μmole/mole (ppm) in an air background]. Sensitivities are calculated from the conductance changes produced during exposure to the indicated analytes. The distance along each material axis represents the relative sensitivity of that type of film to the given analyte. The full database obtained from this study provides such information sets, at 50°C temperature increments, from 200°C to 450°C, and varied concentrations of each of the test gases from 10 μmole/mole (ppm) to 90 μmole/mole (ppm) (28). Data from replicate samples in these arrays have been used both for averaging and for obtaining

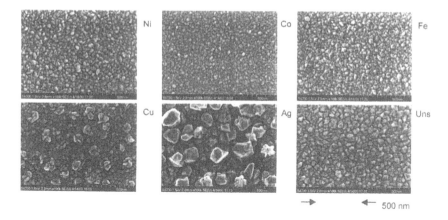

Figure 11 SEM micrographs of representative film microstructures for SnO₂ deposited using Ni, Co, Fe, Cu, and Ag seed particles as well as unseeded surfaces in a 36-element array study.

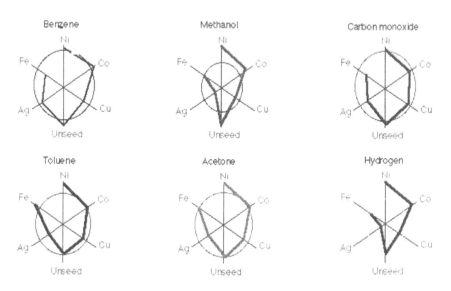

Figure 12 Radar plots representing a small fraction of the total gas sensing sensitivity data, obtained when the samples in the array of Figure 11 are operated at 400°C and exposed in separate measurements to 90-μmole/mole (ppm) in air, of benzene, methanol, carbon monoxide, toluene, acetone, and hydrogen.

statistical information on reproducibility. It is important to emphasize that multichannel operation and data collection ensures that the performances of the multiple sensors are compared under equivalent test conditions. Obviously the methodology utilized could be adapted to look at many other variables that affect sensing film properties (e.g., oxide thickness, deposition temperature, pressure, reactant ratios) in statistically designed experiments.

C. Surface-Dispersed Additives on CVD SnO$_2$

We have also examined the performance of various surface-dispersed catalytic additives (on equivalent CVD SnO$_2$ films) using 36-element arrays. The base SnO$_2$ was first deposited on all elements of the array by thermally addressed CVD using tetramethyltin and oxygen, as described earlier. Catalysts were deposited by evaporation to nominal thicknesses of 3 nm, and then the microhotplates were heated to affect the formation of noncontinuous layers of catalyst particles on the SnO$_2$ surfaces. The 36-sample library created for this study is shown schematically in Figure 13, where each type of (surface) additive metal is labeled. Note again that each sample type is produced six times, to examine process/materials reproducibility. In all of these multiple-element studies, the same array platform used to process the samples is employed to measure a database of sensing responses. In these studies, six different test analytes were introduced sequentially to the microsamples in an air background. Representative outputs are shown in Figure 14. These radar plots show relative sensitivities, calculated from conductance modulations from the air-signal baseline, for operation at three fixed temperatures (although data for many additional temperatures were also collected) (29). The sensitivities represent an average response observed from the multiple samples of each type.

Signal-processing studies that utilize chemometric (30) and neural networks (31) are able to extract information from these databases, permitting fabrication of sensing arrays with materials that are optimized with respect to certain response characteristics and for specific monitoring applications.

V. ADAPTATION OF MICROHOTPLATE ARRAY TECHNOLOGY FOR OTHER MATERIALS STUDIES

Aspects of the examples presented in the last section imply that microhotplate arrays can be valuable tools for materials studies in a multitude of technical areas, in addition to chemical sensing. In many of these other technical areas, modified versions of the illustrated microelement deposition methods can be used; however, to be of more general value, additional deposition techniques would need to be developed. Additional characterization techniques (beyond SEM and on-chip

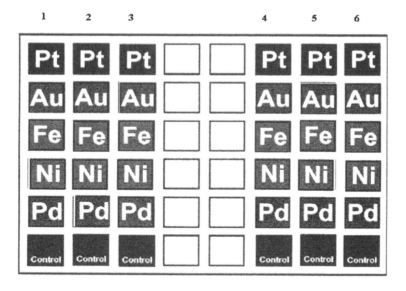

Figure 13   Library configuration of a 36-element array study of the sensing characteristics of CVD SnO$_2$ films that have either no surface additives or 3 nm of surface-dispersed (by heating) Pt, Au, Fe, Ni, or Pd. Each sample type is included six times to examine reproducibility of processing and performance.

conductance measurements) may also be required to adequately define sample properties and evaluate sample performance for other technical areas. For example, while the surface electrodes on the microhotplate elements allow for direct and rapid on-chip electrical characterization relevant to assessing gas-sensing performance, chemical diagnostics would be required to assess the yield from a catalytic microsample deposited on an array element.

The combinatorial thrust toward large sets of small samples, which are to be characterized by fast diagnostics, presents obvious issues relating to spatial resolution and signal-to-noise ratio when one probes small total volumes of experimental materials. Sample-fabrication issues, measurement challenges, and the assistance in these areas that can be derived from modifying the microarray platforms are now discussed, in that order. The relative merits of on-chip versus external characterization are also described briefly.

A.   Issues for Application-Specific Materials Processing

Functional issues often govern the manner in which a material must be fabricated for use in a given technology, and there are certainly deposition procedures that

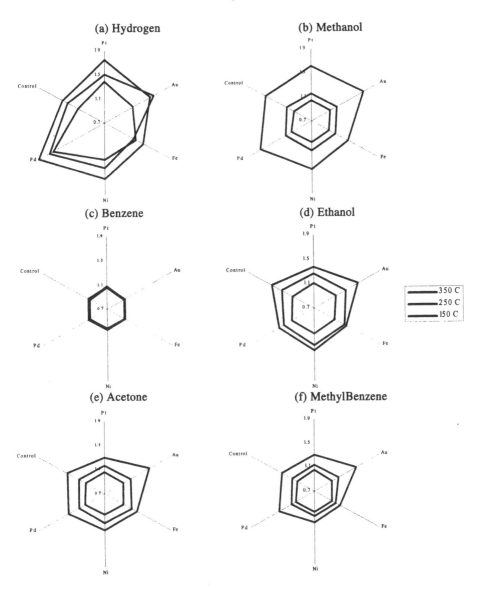

Figure 14   Radar plots representing a small fraction of the total gas-sensing sensitivity data, obtained when the samples in the library of Figure 13 are operated at 150°C, 250°C, and 350°C in (a) hydrogen, (b) methanol, (c) benzene, (d) ethanol, (e) acetone, and (f) methylbenzene test analytes in an air background.

will not be adaptable for microarray studies. However, the procedures for a considerable number of materials may be adapted, in some cases, with methodology similar to that discussed in the last two sections. It is possible that the microhotplate array approach could be employed to advantage in areas as diverse as biosystems, polymers, electronics, photovoltaics, and catalysis. However, a number of limitations can exist and should be considered. For example, the current generation of microhotplates can generally not be heated in excess of 750°C, which may exclude processing or evaluation of certain materials. In addition, the foundry-fabricated microhotplate elements (see Figure 2) can only offer certain foundry-compatible surface materials onto which microsamples are deposited. In some cases, to overcome this limitation, a more suitable coating can be deposited prior to depositing the test samples. One must also consider whether any properties of the microsubstrates the samples are to be placed on (whether precoated or not) will adversely affect later characterization. This issue could range from chemical mixing of the sample with the outer portion of the microhotplate to having an unacceptable emissivity in the case of optical characterization measurements.

B. Utilization of Compatible Analytical Tools

This subsection briefly considers examples of analytical techniques that might be employed in more generalized applications of microhotplate arrays as combinatorial tools. It is divided into external and on-chip methods.

1. External Measurements

A variety of analytical information would be required to define and evaluate samples for different application areas. Structural information could be acquired with tools including micro-x-ray diffraction, SEM, NSOM, and scanning probe methods. Chemical information directly on the films or on the local gas composition above individual elements could be collected from microhotplate-mounted samples by techniques including micro-Raman or micro-FTIR, small-area x-ray photoemission spectroscopy, scanning Auger spectroscopy, and energy-dispersive x-ray analysis. Physical characterization could involve micropyrometry (temperature), electrical characterization by external probes (resistance, capacitance, etc.), and various types of thickness measurements. The nominal 100-μm $\times$ 100-μm area of the individual samples must be considered when one selects characterization methods, because issues of spatial resolution, signal-to-noise ratio, and practical sampling times come into play. One must also deal with alignment of the analytical beams or probes onto the small array elements and the three-dimensional nature of the surface micromachined structures. Certain advantages can be realized for appropriate external analytical tools. For example, detector array–based optical methods can provide "snapshot" images of the total sample array. Optical and spectroscopic measurements can also have high information content (full spectral data). It is possible that the microhotplates can also serve as infrared sources during measurements/characterization.

Noncontact thermometry is one example of an external characterization method that can provide important baseline data on temperature uniformity across individual microhotplate elements. Noncontact thermometry could also play a critical role in monitoring temperature fluctuations that occur due to exothermic and endothermic reactions on the tiny elements when they are operated as microreactors. Highly resolved noncontact thermometric methods would also document thermal gradients. However, the external techniques can also require expensive instrumentation, especially if high-spatial-resolution options are necessary for the measurements. Compatibility with the array geometry may be nontrivial, and analytical complications may exist due to, for example, differences in optical emissivity across the array and array elements.

2. On-Chip Measurements

One type of on-chip electrical monitoring (resistance measurement) possible with element-addressable electrodes was illustrated in the last two sections. Such resistance (conductance) monitoring can be done in real time with multichannel acquisition for multiple array elements. This characterization and performance evaluation capability, which is of great value for the study of conductometric gas-sensing materials, would also be valuable for array-based temperature-dependent studies in a number of other areas, including the processing and optimization of photovoltaic materials. Again, this measurement can be made during or after deposition for sufficiently conductive materials.

Temperature can also be measured on-chip using calibrated data on the temperature coefficient of resistance from microhotplate components (such as the resistive heater and the heat distribution/thermometry plate). In fact, these types of measurements, used in the gas-sensing studies to set film-processing and sensor-operating temperatures, can also be used to monitor temperature variations relating to changes occurring at/on the microhotplate-mounted microsamples. Temperature variations arising from melting and other phase transitions as well as reaction-derived heating (which can occur for catalyst samples) can be monitored by microdevice temperature measurement (or the measurement of input power to maintain a given temperature) for microhotplates or by using microhotplate-related devices, such as microcalorimeters (32).

These types of on-chip measurements can generally be performed quickly (especially with multichannel acquisition hardware/software) for all array elements, and at low cost. There are technical issues to be dealt with, however, including the facts that gradient effects (in temperature or electrical properties) get averaged and that surface-bulk decoupling can occur (for example, if a microhotplate is pulse-heated at a high frequency, the surface of the sample may actually be cooler than the measurement from the microdevice indicates). Nevertheless, valuable information can be acquired from surveys that utilize on-chip measurements.

C. Modifications and Enhancements to the Microhotplate Array Platforms

While much of the sensor-related work described in this chapter has been done on arrays of 4–36 elements and with microhotplate elements of typical sizes of 50 μm × 50 μm or 100 μm × 100 μm, the layout for the elements and arrays can be readily modified. The ability to easily alter the microdevice structures was first indicated briefly in Section II. C. It is straightforward to produce arrays with larger numbers of elements for combinatorial work. The current design of the microhotplates can be employed for producing microsubstrates (suspended microhotplates) up to 200 μm × 200 μm. Larger substrate areas that would be desirable for certain methods of location-registered deposition can be accomplished through new architectures expressed in the computer-aided mask designs. As one increases the size of elements, it is eventually necessary to increase the size of the array chip.

In studies on the effects of operating proximal microheaters within an array adjacent to a gas microsensor, we have noticed signal changes of the order of 1–10% for 100-μm × 100-μm elements spaced ~350 μm apart (33). Such effects might also interfere with collection measurements of chemical yields from arrays of catalyst microsamples. One straightforward way to avoid such problems would be to increase the interelement spacing; one could also operate the elements in a properly sequenced order to collect temperature-dependent yields across the array using a schedule that does not heat neighboring elements at the same time. The first solution might require larger array chips, while the second would require more sophisticated operational electronics.

The ease with which computer-aided design can be used in modifying element and array design could also be applied to developing microarrays that are more compatible with various kinds of external probes. Certain other types of on-chip characterization configurations (besides conductance measurements, which were focused on in Section IV) could also be introduced to measure, for example, microsample capacitance or temperature changes (microcalorimetry).

Computer-aided design changes can be applied to incorporate different combinations of foundry materials, for example, replacing the typical polysilicon resistive heater component by a metal heater. In certain studies in our laboratories, where maximum temperature was more important than the uniformity of the temperature across each microhotplate surface, the metal heat-distribution plate was removed in the device mask design. This modification permitted the micro-hotplate maximum operating temperature to be increased from ~500°C to ~750°C, making possible the study of carbon nanotube growth. Figure 15 shows an example of nanotubes grown on microhotplates by CVD using evaporated and annealed metal catalyst particles as substrate-based growth sites (34). As micromachining processes are developed more broadly for microdevice materials such as SiC and Al_2O_3, the potential for arrays that can operate at higher tempera-tures (approaching or exceeding 1000°C) will be realized. The use of surface microheaters (surface-mounted resistive heaters), constructed with similar design methods from robust materials (and perhaps not requiring micromachining to produce thermally isolated configurations), presents another, related approach that might be most useful when one can adequately characterize sample areas even smaller than ~100 μm.

While this chapter has focused on studies based on discrete array elements, MEMS structures purposely designed to have thermal gradients can also be useful. Figure 16 shows a section of a microhotplate leg that included a considerable thermal gradient whenever the hotplate (without a heat-distribution plate) it was helping to support was heated to elevated temperatures. The CVD deposition of TiO_2 produced a variety of microstructures along this leg, as imaged by SEM. The structures in this gradient sample can be compared directly with microstruc-tures shown in Figure 9 (23).

The several examples provided here are but a few of the many that could be readily implemented to support specialized combinatorial experimentation. We close this subsection, however, with a brief description of one final, somewhat different enabling modification—that of on-chip electronics. Figure 17 shows a four-element microsensor chip that includes op-amp and multiplexer circuitry in addition to the microhotplate-based sensors (35,36). These electronics compo-nents can be linked directly to the MEMS structures from existing CMOS elec-tronics design libraries, because the microhotplates are (or can be) layed out using CMOS design rules. These types of on-chip electronics can foster the development of (nearly) "stand-alone" research microarrays for combinatorial materials re-search. Obviously, the value of components such as multiplexers increases as one goes to larger numbers of array elements.

Finally, it is noted that the types of microarrays described in the examples of Section IV can be included as parts of labs-on-a-chip, microanalytical systems and microreactor arrays with even greater capabilities for materials research. (Such integration is, in fact, consistent with the direction in which our gas micro-sensor efforts are moving.) To enable such microarray-based systems for combi-natorial research, one would consider integrating with the array platforms various components, such as microfluidics channels, valves, and preconcentrators appro-priate for a given study.

a)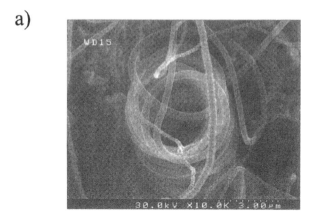

b)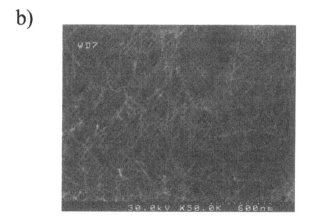

Figure 15   Example micrographs of (a) a helical carbon nanotube and (b) a nanotube mesh, grown on microhotplate elements under different conditions.

VI.   CONCLUSIONS

In this chapter the efficiency of studying temperature-dependent materials processing/property/performance relationships with MEMS-based microarrays has been described. Various types of microsamples ($\sim$100 μm $\times$ 100 μm in lateral dimension and of thicknesses between $\sim$10 nm and $\sim$1000 nm or more) have been deposited on microhotplate elements that were replicated to produce the arrays. Each array element is individually addressable through integrated electronic leads, so heating and temperature measurement can be performed on

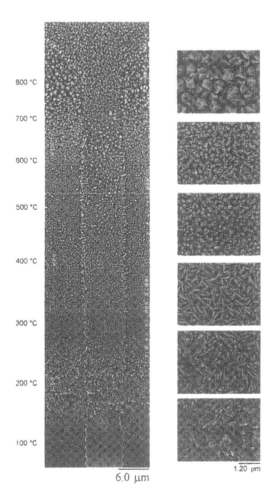

Figure 16 SEM micrograph (composite image) of CVD TiO$_2$ taken along one of the support legs of a microhotplate operated at $\sim$820°C, showing that the temperature gradient present produces a graded set of microstructures (some of which are shown at higher magnification at the right side).

selected elements, and localized microsamples can also be electrically character-ized. The microarray studies used in the illustrations of the approach were con-ducted to develop improved sensing films for solid-state gas microsensors, and they benefited greatly from the ability to use localized temperature control in depositing multiple samples and evaluating their performance. This application

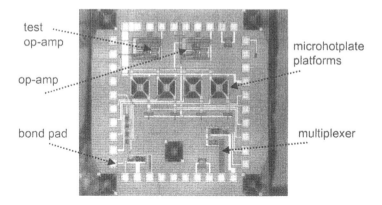

test ········
op-amp

op-amp ·····

bond pad
·········

microhotplate
platforms
····

multiplexer
········

Figure 17   Micrograph of an integrated sensing chip designed using CMOS rules and including four microhotplate devices as well as multiplexer and op-amp circuitry.

is an obvious one for MEMS microarrays, because the 16-element and 36-element arrays used in the example studies are essentially larger versions of the four-element arrays we are using to fabricate chemical microsensor prototypes. However, the potential value of the methodology for other technological areas, such as in developing catalysts, photovoltaics, or electronic materials, should also be clear from the presented work. (We, in fact, have ongoing microarray projects related to determining the flammability of polymer blends and appropriate conditions for porous SiO_2 growth.) As for the sensor materials case studies, other technical areas will have application-specific hurdles relating to design and fabrication of the most appropriate microarray platforms, localization of the library materials, characterization of microsample properties, and assessment of application-defined performance. In certain instances the rapid heating and cooling characteristics of the microhotplates will be especially important in providing novel thermal programs for depositing and testing films.

In closing, there appear to be two critical points that must be focused on to carry micromachined array methods forward, more generally, as an important tool in the area of combinatorial materials science. The first is a challenge to develop external probes that can provide, on an acceptably fast time scale, viable signals for characterizing and evaluating large numbers of samples with very small areas and total volumes. The second is a challenge of sorts, as well, for researchers to use the tremendous flexibility offered by evolving microfabrication technology (Si micromachining, as emphasized here, but also micromachining of other materials and a broad range of new techniques for constructing microdevices, nanodevices, and microanalytical systems) to tailor arrays with the specialized functionality necessary for a specific multielement materials study.

ACKNOWLEDGMENTS

The author wishes to acknowledge the significant efforts of a range of coworkers, including staff members, postdoctoral associates (National Research Council—NIST Postdoctoral Associateship Program), guest researchers, and other collaborators, who have contributed to the combinatorial-related efforts in the Chemical Microsensor Program at NIST. NIST staff members include: Jim Allen, Mike Carrier, Richard Cavicchi, Mike Gaitan, Ken Kreider, Jim Melvin, Chip Montgomery, and John Suehle; postdoctoral associates have included: Frank DiMeo, Chris Kendrick, Doug Meier, Nancy Savage, Chuck Taylor, Robin Walton, and Clay Wheeler; guest researchers and other collaborators have included: Yaqub Afridi (George Washington University), Zvi Boger (Israel IAEC), Baloo Panchapakesan (University of Maryland), and Don DeVoe (University of Maryland). [Affiliations represent the person's research institution at the time of their contribution.] Partial financial support of the Environmental Science Management Program of the U.S. Department of Energy, the Defense Threat Reduction Agency, and the National Aeronautical and Space Administration is also gratefully acknowledged.

REFERENCES

1. Suehle, J.; Cavicchi, R.E.; Gaitan, M.; Semancik, S. IEEE-Electron Device Lett. 1993, 14, 118–120.
2. Semancik, S.; Cavicchi, R.E. Accounts of Chemical Research. 1998; Vol. 31, 279–287. [See also S. Semancik, R.E. Cavicchi, M. Gaitan, J.S. Suehle, U.S. Patent #5,345,213 (1994); R.E. Cavicchi, S. Semancik, J.S. Suehle, M. Gaitan, U.S. Patent #5,356,756 (1994).].
3. Semancik, S.; Cavicchi, R.E.; Wheeler, M.C.; Tiffany, J.E.; Poirier, G.E.; Walton, R.M.; Suehle, J.S.; Panchapakesan, B.; DeVoe, D.L. Sensors and Actuators. 2001, B77, 579–591.
4. Ousterhout, J. K. IEEE Trans. Computer-Aided Design. 1984, CAD-3, 242–249.
5. Marshall, J. C.; Parmeswaren, M.; Zaghoul, M. E.; Gaitan, M. NIST Report 4845: Gaithersburg: MD; 1992.
6. Metal Oxide Semiconductor Implementation Service: Marina del Rey, CA.
7. Microelectronics Research Laboratory: Columbia, MD.
8. MIT-Lincoln Laboratories: Lexington, MA.
9. Bean, K. E. IEEE Trans. Electron Devices. 1978, ED-25, 1185–1192.
10. Chang, F.I.; Yeh, R.; Lin, G.; Chu, P.B.; Hoffman, E.; Kruglick, E.J.J.; Pister, K.S.J. Proceedings of the SPIE 1995 Symposium on Micromachining and Microfabrication: Austin: TX; October 23–24, 1995; Vol. 2641, 117–128.
11. Paranjape, M.; Pandy, A.; Brida, S.; Landsberger, L.; Kahrizi, M.; Zen, M. J. Vac. Sci. Technol. 2000, A18, 738–742.

12. Semancik, S.; Cavicchi, R.E.; Kreider, K.G.; Suehle, J.S.; Chaparala, P. Sensors Actuators. 1996, B34, 209–212.

13. Cavicchi, R.E.; Suehle, J.S.; Kreider, K.G.; Shomaker, B.; Gaitan, M.; Chaparala, P. Appl. Phys. Lett. 1995, 66, 286–288.

14. Walton, R. M.; Kendrick, C.; Cavicchi, R. E.; Semancik, S.; Panchapakesan, B.; DeVoe, D. L. Proc. of the 10th International Conf. on Solid-State Sensors and Actuators: Sedai: Japan; June 1999; Vol. 1, 676–679.

15. Kendrick, C. personal communication.

16. Smith, P. A.; Nordquist, C. D.; Jackson, T. N.; Mayer, T. S.; Martin, B. R.; Mbindyo, J.; Mallouk, T. E. Appl. Phys. Lett. 2000, 77, 1399–1401.

17. Cavicchi, R. E.; Walton, R. M.; Aquino-Class, M.; Allen, J. D.; Panchapakesan, B. Sensors Actuators. 2001, B77, 145–154.

18. Meier, D. C.; Fasolka, M. J.; Semancik, S. Polymer–silsesquioxane composites as a route to high-surface-area materials for MEMS sensor applications (in preparation).

19. Savage, N.O.; Roberson, S.; Gillen, G.; Tarlov, M.J.; Semancik, S. Thermolithographic patterning of sol-gel metal oxides on microhotplate sensing arrays using organosilanes (Analytical Chemistry, in press).

20. Tiffany, J.; Cavicchi, R.E.; Semancik, S. Proc. of the SPIE Advanced Environmental and Chemical Sensing Technology Symposium. 2000, 4205, 240–247.

21. Wheeler, M. C.; Cavicchi, R. E.; Poirier, G. E.; Semancik, S. A microarray technique for measuring adsorption/desorption kinetics, (in preparation).

22. Gilmer, D. C.; Colombo, D. G.; Taylor, C. J.; Roberts, J.; Haugstad, G.; Campbell, S. A.; Kim, H. S.; Wllk, G. D.; Gribelyuk, M.; Gladfelter, W. L. Chem. Vapor Deposition. 1998, 4, 9–11.

23. Taylor, C. J.; Semancik, S. Chem. Mat. 2002, 14, 1671–1677.

24. Meier, D. C.; Taylor, C. J.; Cavicchi, R. E.; Semancik, S. Sensing chemical warfare agents with MEMS conductometric microsensors, (in preparation).

25. Taylor, C.J.; Meier, D.C.; Cavicchi, R.E.; Semancik, S. Response reversals for TiO_2 gas sensing films, (in preparation).

26. Panchapakesan, B.; Devoe, D.; Cavicchi, R.E.; Semancik, S. Mat. Res. Soc. Symp. Proc. 2000, 574, 213–218.

27. Panchapakesan, B.; Devoe, D.L.; Widmaier, M.; Cavicchi, R.E.; Semancik, S. Nanotechnology. 2001, 12, 336–349.

28. Panchapakesan, B. Ph.D. dissertation; University of Maryland, 2001.

29. Cavicchi, R. E.; Montgomery, C.J.; Semancik, S. Microarray studies of the gas sensing performance for catalytic additives on SnO_2, (in preparation)—and S. Semancik, Department of Energy Final Report (EMSP65421, January 2002.

30. Dable, B.; Booksh, K.; Cavicchi, R.; Semancik, S. Calibration of microhotplate conductometric sensors by nonlinear multivariate regression methods, (in preparation).

31. Boger, Z.; Cavicchi, R.E.; Semancik, S. Analysis of conductometric microsensor responses in a 36-element array by artificial neural networks modeling'', Proc. of the 2002 International Symposium on Olfaction and Electronic Noses: Rome; 2002, In press.

32. Cavicchi, R. E.; Afridi, M.; Berning, D.; Hefner, A.; Suehle, J.; Gaitan, M.; Semancik, S.; Montgomery, C. Micro-differential scanning calorimeter for combustible gas sensing, (in preparation).

33. Wheeler, M. C.; Cavicchi, R. E.; Walton, R. M.; Semancik, S. Sensors Actuators. 2001, B77, 167–176.

34. Taylor, C. J. personal communication.

35. Afridi, M. Y.; Suehle, J.S.; Zaghloul, M. E.; Berning, D. W.; Hefner, A. R.; Semancik, S.; Cavicchi, R.E. Proc. of the IEEE Symposium on Circuits and Systems. May, 2002, II-732–II-735.

36. Afridi, M. Y.; Suehle, J.S.; Zaghloul, M. E.; Berning, D. W.; Hefner, A. R.; Cavicchi, R.E.; Semancik, S.; Montgomery, C. B.; Taylor, C. J. A monolithic CMOS microhotplate-based gas sensor system, (in preparation).

10

X-Ray Techniques for Characterization of Combinatorial Materials Libraries

William Chang
Advanced Research and Applications Corporation, Sunnyvale, California, U.S.A.

I. INTRODUCTION

The usefulness of combinatorial materials synthesis for discovery of new functional materials relies ultimately on the effectiveness of rapid characterization of combinatorial libraries. Materials properties of interest, such as magnetic, electronic, and photonic properties, are all directly or indirectly governed by structures and elemental compositions of synthesized materials. X-ray analytical techniques are capable of providing accurate structural and compositional information that allows us not only to understand the newly observed materials properties, but also to generalize and to predict properties for other new potential materials. Because of the nature of the combinatorial libraries, analytical techniques must also meet the challenge of rapid characterization of a large number of samples with very small sample volumes (tens to hundreds of microns).

In this chapter, two x-ray analytical techniques for the characterization of combinatorial samples will be discussed: (a) x-ray diffraction for structural identification and (b) x-ray fluorescence for elemental quantification. Several examples of combinatorial thin-film materials research using x-ray techniques are provided to demonstrate their utility and related instrumentation issues. In terms of instruments, the emphasis is given to laboratory x-ray systems that can be specialized for generating x-ray microbeams as well as high characterization throughput.

II. X-RAY ANALYTICAL TECHNIQUES

A. X-Ray Diffraction

X-ray diffraction (XRD) is an indispensable tool for solid-state and metallurgical materials research. X-ray photons probe materials by scattering off electrons bound to crystal structures, and x-ray coherent scattering occurs when Bragg's law, $2d_{hkl} \sin \theta = \lambda$ is satisfied by the diffraction process. In general, the x-ray wavelength is comparable to the unit-cell dimensions of crystalline materials (typically 1–10 Å), and x-ray diffraction is ideal for identifying unit-cell dimensions and atomic basis positions within the unit cells. X-ray diffraction is sensitive to small perturbations in crystal structures, such as lattice distortion caused by interstitial or substitutional crystal point defects, and by internal lattice strain due to mismatch of unit-cell dimensions of thin epitaxial or nonepitaxial films. Other applications of x-ray diffraction include the state of order–disorder of solid-solution materials.

Bragg's law, $2d_{hkl} \sin \theta = \lambda$, states that x-ray photons with a given wavelength λ can diffract at an angle, θ, from a crystalline material that consists of a set of atomic planes (h, k, l) whose interplane distance is d_{hkl}. hkl is an all-integer notation known as *Miller indices*, which identify a particular atomic plane in a crystal. The value d_{hkl} is not only hkl dependent, but also a function of the type of crystal lattice, such as cubic, tetragonal, orthorhombic, or hexagonal, common to solid-state and metallurgical materials. For a specific lattice, d_{hkl} can be expressed in hkl and the lattice parameters. For example, for a cubic lattice of lattice constant a_o, $d_{hkl} = a_o/(h^2 + k^2 + l^2)^{1/2}$, whereas for an orthorhombic lattice of a_o, b_o, and c_o, $d_{hkl} = [(h/a_o)^2 + (k/b_o)^2 + (l/c_o)^2]^{-1/2}$. Various atomic arrangements and different positions within a lattice determine a particular structure. For example, within a cubic lattice, Fe atoms are arranged in body-centered cubic (bcc) structure, Ni atoms are in face centered cubic (fcc) structure, and $BaTiO_3$, a perovskite structure, with Ti being at the body center, O at the face center, and Ba at the corner of the lattice.

Given a mosaic crystal, the intensity of diffracted x-rays at a specific diffraction plane, hkl is proportional to the structure factor $|F_{hkl}|^2$, where F_{hkl} is expressed as the following:

$$F_{hkl} = \sum_{k=1}^{m} \sum_{j=1}^{n} f_{jk} \exp[-2\pi i(hx_{jk} + ky_{jk} + lz_{jk})] \qquad (1)$$

where f_{jk} is the atomic scattering factor of the jth atom of type k and the coordinates, x_{jk}, y_{jk}, and z_{jk}, represent the position of the jth atom with respect to the origin. For crystal structure with a monoatomic basis, the value of F_{hkl} normally indicates whether a diffraction, from the hkl plane is allowed or forbidden, depending upon whether the exponential factor is zero or nonzero, and also upon the product of the atomic scattering factor and the exponential factor in the polyatomic basis case. For example, F_{100} for Cu, an fcc structure, is zero, whereas F_{100} for Cu_3Al, also an fcc structure, with Au atoms located at the corners and Cu atoms at the centers of the faces, is equal to $f_{Au}-f_{Cu}$.

Neglecting the effects of the temperature factor, the reflecting power of diffracted x-rays from a mosaic crystal can be expressed as

$$P_{hkl} = \lambda^3 / V^2 \mid F_{hkl} \mid^2 r_e^2 (1 + \cos^2 2\theta)/(2 \sin 2\theta)/2\mu \tag{2}$$

where $(1 + \cos^2 2\theta)/(2 \sin 2\theta)$ is called the *Lorentz-polarization factor* for an unpolarized x-ray incident radiation, $r_e = e^2/mc^2$ is the classical radius of an electron, μ is the linear absorption coefficient, λ is the x-ray wavelength, V is the volume of the unit cell, and θ is the Bragg angle. Based on the mosaic crystal model, (1) the crystal is assumed to consist of many tiny crystal blocks misoriented or displaced to each other. There is no coherence among crystal blocks diffracting x-rays, and there is no interaction among incoming and traveling x-rays inside the crystal. The reflecting power of diffracted x-rays outside the crystal is therefore the sum of the amplitude of each individual block. The mosaic crystal is sometimes called an *ideal imperfect crystal*, and Eq. (2) is the result of a kinematic theory describing the diffraction process of this type of crystal model. In practice, mosaic crystals can be used to model real crystals containing defects such as dislocations, stacking faults, and some type of imperfection.

Another characteristic of a mosaic crystal is the range of angles, mosaic spread, that satisfies the Bragg diffraction condition. Since each block is not perfectly aligned or is slightly misaligned with other crystal blocks, a Gaussian distribution is normally used to describe their misorientations, centered at the exact Bragg angle. The mosaic spread is usually an indicator of the degree of imperfection of the crystal. In contrast to the mosaic spread, the intrinsic, or the Darwin, width is the angular width that characterizes perfect crystals such as silicon wafers. In reference to our mosaic crystals, the intrinsic width characterizes the angular range at the Bragg angle for each individual block. In this case, a dynamic theory is needed to describe the process of the interference between the incident and diffracted x-rays inside crystals. There are a number of excellent general references on x-ray diffraction in crystals (2–5) and in materials characterization (6–8).

It can be seen from Eq. (2) that P_{hkl} is proportional to $|F_{hkl}|^2$, which implies that diffracted x-rays are sensitive not only to the lattice type and the set of *hkl* planes having appreciable Bragg diffraction peaks, but also to the crystal structures. In practice, crystal structure or phase identification can be achieved by directly comparing an x-ray diffraction pattern, the d_{hkl} spacing calculated from diffraction peak positions and peak intensity values, to a database of x-ray diffraction patterns, Powder Diffraction Files (PDF) (9). The PDF consist of a collection of single-phase x-ray powder diffraction patterns published in tables listed with *d*-spacing, relative intensities, chemical formula, and PDF number. X-ray diffraction is particularly useful to characterize multiphase materials, since each phase can be identified with its fingerprint, namely, peak positions and integrated intensities. Once a phase is determined, the lattice constant of that structure can be readily calculated using the formula discussed earlier. Another indispensible application of x-ray diffraction is the ability to perform a precision lattice-constant measurement, which is especially important for a solid solution or mixed crystal materials, because the intensity and peak-shift measurements will provide a basis for elucidating the nature of order–disorder or crystal imperfection.

Experimentally, diffractometer scanning is commonly employed to measure diffraction patterns. An x-ray tube, a detector, and the sample are mounted on a focal circle based on the Bragg–Brentano configuration. The primary x-ray beam is incident on the sample surface, with an incident angle, θ, and the diffracted x-rays are detected by a detector, also located at an angle θ from the sample surface. Thus, an entire diffraction pattern can be recorded by incrementally rotating the sample surface by an angle θ and rotating the detector by 2θ. For mosaic crystal samples in a thin-film form, this method is effective, since the primary *hkl* planes are parallel to the crystal surface: Films are well oriented and parallel to their substrates. A more versatile version, based on the Bragg–Brentano geometry, is shown in Figure 1. In addition to the theta-two theta scan about the center of the focal circle, the sample can also rotated about three additional axes, ω, ϕ, and χ, centered at the tangent point of the sample surface and the focal circle. Considered to be complementary to the theta scan, the ω scan is usually applied to find the film orientation if misorientation is suspected. As shown in Figure 1, the ω scan is implemented by fixing the x-ray source and the detector position according to the expected Bragg diffraction *hkl* condition and rotating only the sample about the ω axis, until the particular *hkl* diffraction peak is found. The difference between the expected and the current film angular position, $\Delta\omega$, indicates the actual degree of film misorientation. Similar to the ω scan, the χ scan is for the orientation measurement along the χ axis, which becomes more important in stress–strain measurements. The ω scan is also used to perform routine rocking-curve measurements, which provide the qualitative value of the mosaic spread, namely, the full width at half maximum (FWHM) of the rocking curve. The ϕ scan is usually used to verify whether the sample possesses a rotational symmetry and how well the sample is oriented along the ϕ axis, through the measurement of the FWHM of the angular spread.

B. X-Ray Fluorescence

X-ray fluorescence (XRF) analysis is a widely used spectroscopic method for qualitative or quantitative measurement of the elemental composition of materials.

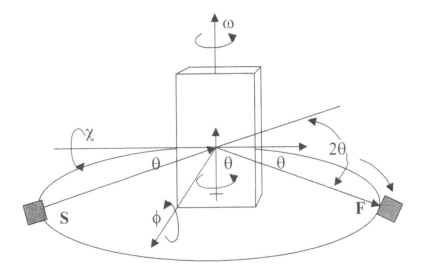

Figure 1 Schematic of the general diffraction geometry. Source S and detector F are located in the diffraction plane.

Some well-cited references (10–12) provide a broad and in-depth discussion of this field. The basic principle of the method is to detect both the energy and the number of x-ray fluorescence photons simultaneously for determination of the element and its concentration, respectively. The physical process of the x-ray fluorescence generation can be briefly outlined as follows. An x-ray beam used as an excitation source is incident on a sample. Then the x-ray photons are attenuated and absorbed as they penetrate the sample. The energy loss is transferred to eject electrons from atoms and create vacancies, or holes, which subsequently are filled by down-transitions of outer-shell electrons. The process of electron down-transition causes an emission of characteristic x-rays with an energy equal to the difference between the initial and final energy levels associated with the down-transition. The potential energy required to produce a specific energy of x-ray photons corresponds to the ionization potential of specific electrons. For example, an excitation x-ray beam with an ionization energy of 8.980 keV or higher is required to eject K-shell electrons of a Cu atom and to produce Cu Kα (K ← L) and Cu Kβ (K ← M), characteristic x-ray photons with energies of 8.048 keV and 8.905 keV, respectively.

The theoretical expressions relating elemental concentrations to measured corresponding characteristic x-ray fluorescence intensities have been analytically derived (13). The elemental concentration in a sample matrix can be calculated directly from measured specific characteristic x-ray intensities. One of the widely used techniques is the fundamental parameter method, or the FP method (14–16), which relies on a set of fundamental parameters, such as x-ray tube spectral distribution, mass absorption coefficients, fluorescence yields, and absorption edge jump ratios. Given for a monochromatic x-ray incident beam and a flat, thick, and homogeneous sample, the basic expression for the fluorescence intensity of an element j in the sample is written as follows:

$$I_j = I_o C_j Q_j \mu_{1m} / (\mu_{1m} + f\mu_{3m}) \times \Omega / 4\pi \qquad (3)$$

where C_j is the concentration of element j, in weight percent, and Q_j is defined as the fluorescence factor, which is a product of three fundamental parameters (fluorescence yield, absorption jump ratio, probability of emission of a particular K line with respect to all lines within the K series for element j), and μ_{1m} and μ_{3m} are the mass absorption coefficients $\mu_m = \mu/\rho$, tabulated from the literature. The factor $\Omega/4\pi$ is the solid angle for the detector. For multielement samples, μ can be calculated by using $\mu_m = \Sigma\mu_j C_j$ ($j = 1, 2, \cdots m$). f is equal to sin ϕ/sin ψ, where ϕ and ψ are the incident and take-off angles measured from the sample surface, respectively. μ_{1m} accounts for the absorption process, when the incident x-ray photons are penetrating the sample, whereas μ_{3m} accounts for the excited x-ray fluorescence traveling outward from the sample.

Commonly, the relative intensity, I_{jr}, defined as the intensity I_j of unknown concentration with respect to intensity $I_{j=100\%}$ of pure element j, is used for calculation, since the factors Q_j and $\Omega/4\pi$ can be cancelled out:

$$I_{jr} = C_j \mu_{1m} / (\mu_{1m} + f\mu_{3m}) \qquad (4)$$

Thus, for a given energy and geometry, the relative intensity can be calculated by providing an initial guess at C_j and comparing the first I_{jr} value directly to the measured intensity, which requires spectra processing, such as background subtraction, and peak-overlap correction. An iterative process is usually needed to find C_j, which yields a match between the measured intensity and the calculated relative intensity. For multielement samples, Eq. (4) can be applied to each individual element of interest, one at time, while holding the concentration value of the elements constant.

There are two common spectroscopic methods for detecting and analyzing emitted x-ray fluorescence photons. They are energy-dispersive spectroscopy (EDS) (17) and wavelength-dispersive spectroscopy (WDS) (18). With a solid-state detector, EDS can be used to acquire, for a given time, an entire spectrum, including all energies associated with elements present in the sample and the number of x-ray photons associated with each specific energy. Equipped with an analyzing crystal, WDS, however, is normally employed to record one spectrum at a time for one wavelength (energy) identified by the analyzing crystal through an angular scanning motion. In combinatorial materials research, the stoichiometry of the material is a critical parameter that directly affects data analysis. For example, the dielectric constant of the ferroelectric material barium-strontium titanate is strongly dependent on the atomic ratios of Ba/Sr and (Ba + Sr)/Ti. A 10% change from Ba/Sr = 1.0 will cause a 30% decrease in dielectric constants (19). There is a separation of ~320 eV between the Ba Lβ (4.828 keV) and Ti Kα (4.511 keV) x-ray fluorescence peaks, which implies that EDS, normally with an energy resolution of 250 eV at 6 keV, is not adquate to resolve these two peaks. Supposing a bent crystal of silicon used as the analyzing crystal for WDS, the energy resolution is determined mainly by the width of the reflection curve for the silicon (111) reflection. In this case, the angular width for the Ti Kα line is 2.8×10^{-4} radians, which is equivalent to an energy width of 2.6 eV, which clearly eliminates the peak-overlap problem.

III. APPLICATIONS TO COMBINATORIAL THIN-FILM LIBRARIES AND COMPOSITION SPREADS

Using a synchrotron x-ray microbeam at the National Synchrotron Light Source, Eric Issacs et al. (20) applied several x-ray analytical techniques simultaneously to characterize phosphors containing materials such as rare earth–activated Gd(La, Sr)AlO$_3$. A 128-member library, in a chip format, was synthesized by using a number of sputtering targets: La$_2$O$_3$, GdF$_3$, Al$_2$O$_3$, EuF$_3$, and SrCO$_3$. Within each chip, the elemental concentration of Gd, Sr, Al, and Eu were all varied. X-ray diffraction was used to probe the structural properties of samples within each library. Results from several discrete sample showed that (Gd, Sr)AlO$_3$ was the predominant phase, with a cubic perovskite structure. Other phases, such as Al$_2$O$_3$ and SrAl$_2$O$_4$, were also present in some samples. In addition, LaAlO$_3$, also a cubic perovskite structure, was identified inside a La-free sample, and it is believed to be a contamination from an adjacent sample containing La. With an x-ray microprobe, they have demonstrated that chemical and structural information of the synthesized film could be mapped based on elemental compositions. The results also demonstrated that the potential of the microprobe for characterization of a large number of library samples present on a relative small-scale area. In this case, a probe beam of 20 μm was used for 1000 samples in a 20-mm-long strip. As the authors indicated, the enhancement of the x-ray probe and the automation of the characterization procedures could eventually lead to completing the same work in less than an hour.

The optics used in this work was a pair of mirrors positioned perpendicularly to each other in a so-called K-B configuration (21). The mirror collects the x-ray beam from a set of monochromators, focuses the beam down in one dimension after reflection from the first mirror, and then focuses down in the other direction after passing the second mirror. Normally, the mirror is figured to an elliptical surface, which is suited to point-to-point focusing, especially for the synchrotron x-ray source with its source brightness and low beam divergence.

Combinatorial intermetallic films were also studied by the synchrotron x-ray microbeam at Advanced Photon Sources. Y. K. Yoo et al. (22) have demonstrated the phase structure change in $Ni_{1-x}Fe_x$ thin films. Different from other library samples, this library chip was prepared as a single continuous compositional spread sample, which required a linear scan of 200 points (beam size = 50 μm) along the direction of the compositional spread. A diffractometer scan was generated for each point, and the (111) reflection for Ni and the (110) reflection for Fe were used for lattice constants calculations. Variation in lattice-constant obtained from the diffractometer scans was clearly shown to be a function of Fe and Ni composition shown in Figure 2. The Fe (bcc) and Ni (fcc) phase transition was observed at $x \sim 0.12$, and the gradual phase transition was in agreement with the conventional bulk property of this binary alloy. The relative saturation magnetization, M_s, was measured by a scanning Hall probe instrument, and the so-called invar behavior near the phase-transition region corresponding to a dip in M_s was observed. However, a quite different compositional dependence was observed from the relative Kerr rotation measurements. A fairly steep drop near the phase-transition point, $x = 0.14$, implied that the occurrence of an anomaly in the Kerr rotation during the structural phase transition.

The x-ray microbeam in this experiment was devised by using a pair of slits to achieve a 50-μm beam spot. The theta-two theta scan was used to measure the peak positions, while the ω scan was used to measure the FWHM of the rocking curve as ~ 0.1 degree. The ϕ scan was also performed to measure the mosaic spread for the in-plane orientation.

Laboratory x-ray systems have also been used to analyze combinatorial materials libraries. A scanning microdiffractometer called general area detector diffraction system (GADDS) has been manufactured by Bruker-AXS. In this system, the x-ray microbeam is generated by first applying a graded multilayer mirror or monocapillary optics (discussed later) for collecting the divergent x-ray beam generated from the x-ray tube and then aperturing it down to a diameter as small as 10 microns. For data collection, this system uses a photon-counting multiwire area detector without intrinsic background. A GADDS was used to characterize in situ deposited $(Ba,Sr)TiO_3$ composition spreads. Based on the theta-two theta measurement; continuous change in the lattice constant of the spread film was observed, and the change was consistent with the compositional spread measured by the Rutherford backscattering technique (23).

As discussed in Chapter 8, K. Omote et al. (24) used a laboratory concurrent x-ray diffraction system, a modified diffractometer, to investigate combinatorial epitaxial thin films $[(SrTiO_3)_{10}/(BaTiO_3)_n]_{30}$ ($n = 5, 10, 15$) grown on a $SrTiO_3$ substrate by using laser molecular beam epitaxy. For three different compositions

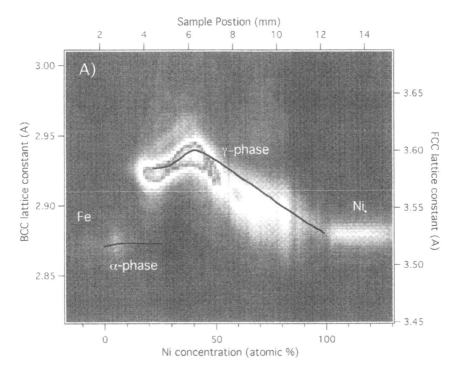

Figure 2 Lattice constant of the $Ni_{1-x}Fe_x$ thin film as a function of compositional parameter x. (From Ref. 22.)

of $(BaTiO_3)_n$ obtained from a separate x-ray fluorescence measurement, the authors demonstrated the corresponding structural change revealed through the change of the angular peak positions of the diffracted x-rays.

The major emphasis of the work was to improve the x-ray diffraction throughput for the combinatorial materials characterization. The concurrent x-ray diffraction approach was implemented by using a convergent x-ray incident beam to eliminate the necessity of angular rotation required by the conventional diffractometer configuration. For analyzing a prominent diffraction peak, the angular scan range needed is normally less than one degree. In this case, a two-degree convergent angle of diffracted x-rays from the bent crystal geometry was achieved. Thus, the conventional theta-two theta diffractometer scan was replaced by a combination of a convergent beam and a two-dimensional detector with the sample at a fixed position. Although the spatial resolution for the analysis is limited to about 100 μm, the approach is certainly attractive because of its improvement on the throughput.

Recently, a prototype instrument specially designed for combinatorial libraries characterization was jointly developed at ARACOR and Intematix (25). A thin-film sample of $Ba_ySr_{1-y}Ti_{1-x}V_xO$, grown on a LaO substrate, with x varying from 0 to 0.3 over the entire film length of 15 mm, was characterized.

In contrast to Eric Issacs' work, which was done on a synchrotron beam line, this work, was carried out with a laboratory x-ray source to implement simultaneously x-ray diffraction and fluorescence techniques. Since the total variation of the concentrations of titanium and vanadium was only ~30%, we chose to have a 2.5% interval between library samples, each of which varied approximately 0.1% within a 50-square-micron region. The film was linearly scanned along its length at 13 discrete points. At each point, a theta-two theta scan was made, and the entire x-ray fluorescence spectrum was collected. The (200) reflection was selected for probing the change of the lattice constant as a function of the elemental composition variation.

As seen in Figure 3a, the diffraction results clearly reveal the continuing (200) peak shift as vanadium concentration increases. A qualitative interpretation was made based on each peak position and its corresponding fluorescence spectrum. The lattice for the film is tetragonal, and the tetragonal d-spacing formula is $1/d^2 = (h^2 + k^2)/a^2 + l^2/c^2$, where h, k, and l are the Miller indices for a particular crystal plane and a and c are the lattice constants. Since the (200) reflection was used, the expression became simply $a = 2d$. If we differentiate Bragg's law, $2d \sin \theta = \lambda$, then the ratio $\Delta d/d$ yields $\Delta a/a = \Delta\theta \cot \theta_B$, where θ_B is the Bragg Angle and $\Delta\theta$ is the amount of angular shift with respect to the value where vanadium concentration is zero. The results, shown in Figure 3b, illustrate the increasing trend of the ratio $\Delta a/a$ as the concentration of vanadium increased, which is consistent with the fact that the ionic radius of vanadium, 1.35Å, is smaller than that of titanium, 1.46Å. It is also noticed that at the vanadium concentration of 0.24, the ratio of $\Delta a/a$ decreased as the vanadium concentration continued increasing. This occurrence could be attributed to the fact that at this vanadium composition state, the lattice starts to elongate along the c axis, and contract along the a axis. Since only the (200) reflection, which was associated with lattice constant a, was measured, there was no data on the ratio $\Delta c/c$. It is evident that for noncubic lattice films, the choice of the reflection peaks and, in turn, the d-spacing needs to be related to various lattice constants.

Another example of characterization is the phase mapping of a ternary thin film Fe-Co-Ni combinatorial library. The sample was made on an equilateral triangle shape, and the elemental concentration varies from zero to 100 percent for each two elements along each triangle edge. X-ray diffraction patterns (θ-2θ), collected at each grid point on the sample shown in Figure 4a, were characterized by fitting the peaks with a Gaussian curve. The 2θ-scan range of the diffraction pattern is 4 degrees, which is sufficient to cover the angular range of the characteristic diffraction peaks of Fe, Co, Ni, and some well-known compound

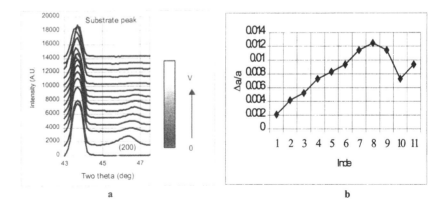

Figure 3 (a) Diffractometer scans of a thin-film composition spread Ba_ySr_{1-y} $Ti_{1-x}V_xO$, at different positions of the spread. (b) Plot of lattice-constant change versus film composition.

phases. Some indexed peaks were identified and assigned their crystal structure phases based on known references. Some unidentified diffraction peaks could be attributed to crystal structures in transition states between different phases, or to some possible new structure phases, which would require more thorough characterization. In Figure 4a, only bcc, fcc, and Co were categorized phases. The corresponding detailed crystal structure phases are mapped as shown in Figure 4b. The mapping revealed presence of seven known crystal structure phases, four two-phases mixed regions, and four regions with unknown structure phases. They are all listed in Table 1.

We found that the structure phases along the three edges are qualitatively consistent with their corresponding binary phase diagrams for bulk materials. The nature of those unidentified structure phases regions 5, 6, 9, and 11 could be revealed by extending the angular scan range of examine other possible diffraction peaks.

The prototype instrument used for this consists of a conventional diffraction tube with a Cu anode, a microbeam optic, a glass monocapillary, and two detectors: an energy-dispersive PIN detector for fluorescence and a scintillation detector for diffraction. The key element of this system is the capillary optic, which can effectively generate a low beam divergence and a controllable beam size ranging from 100 microns down to submicrons. In addition, the instrument is automated to facilitate the acquisition and processing of both diffraction and fluorescence data at each programmed scanning position.

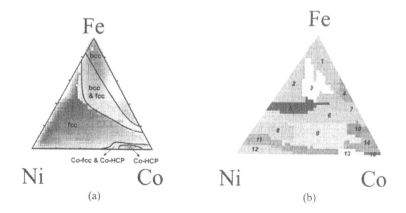

Figure 4 Ternary phase mapping with boc, fcc, and Co phases (a) and corresponding identification of detailed crystal structure phases (b). Refer to Table 1.

Table 1 Mapped Crystal Structure Phases for Fe-Co-Ni Sample

Color code	Element/Compound	Reflection	Crystal structure	2θ (°)
12	Ni	(111)	FCC	44.35
1	Fe	(110)	BCC	44.55
14	Co (2)	(111)	FCC	44.10
13	Co (1)	(002)	HCP	46.20
15	Co(1)/Co(2)	(002)/(111)	HCP/FCC	46.20/44.10
8	$FeNi_3$	(111)	$CuAl_3$	44.10
7	FeCo	(110)	CsCl	44.85
4	Fe/FeCo	(110)/(110)	BCC/CsCl	44.55/44.85
2	$Fe_{0.8}Ni_{0.2}$	(111)	C (Pm3m)	43.48
3	$Fe/Fe_{0.8}Ni_{0.2}$	(110)/(111)	BCC/C (Pm3m)	44.55/43.48
10	FeCo/Co(2)	(110)/(111)	CsCl/FCC	44.85/44.10
6	$Fe_{0.26-0.35}Co_{0-0.6}Ni_{0.08-0.7}$	—	—	43.85–43.90
9	$Fe_{0.05-0.26}Co_{0.25-0.8}Ni_{0.50-0.90}$	—	—	44.00–44.05
5	$Fe_{0.35-0.42}Co_{0-0.4}Ni_{0.25-0.6}$	—	—	43.80
11	$Fe_{0.05-0.15}Co_{0-0.7}Ni_{0.3-0.9}$	—	—	44.20–44.25

IV. INSTRUMENTATION

A laboratory based x-ray instrument specifically for characterization of combinatorial materials requires optimized performance of screening throughput. Basically, there are two approaches for throughput improvement. One is the sequential screening shown in Figure 5a, similar to several cases discussed earlier, where, at each point on the sample, the x-ray diffraction pattern is recorded by scanning the sample over an angular range of θ, and of 2θ for the detector. During the angular scan, the x-ray fluorescence spectrum can also be taken to measure the compositions of each element at point X. This data acquisition cycle will be repeated for a number of different positions X on the sample. This is accomplished by a linear scan along the X direction. To improve the throughput, it is necessary to decrease the x-ray diffraction scan time at each point, which requires either a more powerful x-ray tube or advanced x-ray optics to bring down the detector counting time.

The other option is parallel screening, shown in Figure 5b, which is similar to the approach taken by K. Omote et al. (24). In that case, the x-ray diffraction scan is replaced by a two-dimensional detector and an x-ray optic, which produces a line with a large convergence angle, Ω, although the angular range is still limited to approximately two degrees. In fact, this line beam coupled with the two-dimensional detector can eliminate the need for the linear scan, by aligning the line beam with the direction of composition spread, the X direction, provided the angular spread along the X direction is small, to avoid line broadening. If the line beam is 10 mm long and 200 μm wide, then the linear dispersion will be 10 mm/0.2 mm = 50 for one x-ray exposure, as compared to 50 exposures for a linear dispersion of 1 in the sequential case. The angular resolution of the fan-shaped beam is considered to be controllable, normally less than 1 mrad, and determined by the reflection curve width, namely, the broadened intrinsic width (Darwin width) caused by the crystal bending and focusing geometry effects. If an angular scan step of 0.1 degree is required in the sequential case, then a convergence angle of 5 degrees in this case will provide a 50-fold improvement in angular scan time.

Although the throughput improvement is apparently significant in the parallel case, it is difficult to produce a line beam width comparable to that of a point beam, because of the broadening of the curved crystal reflection curve width and the finite size of the x-ray source. A beam size of 100 μm or more will limit the ability to characterize high-composition-gradient samples. Furthermore, the XRD/XRF parallel detection will be compromised, because the emission of x-ray fluorescence at each point on the sample is isotropic and omnidirectional, as shown in Figure 5b, and limits the capability of the elemental position-sensitive detection. Therefore, both approaches could be complementary to one another in the entire combinatorial material screening process. In the following sections, general aspects of the instrumentation for both cases are given.

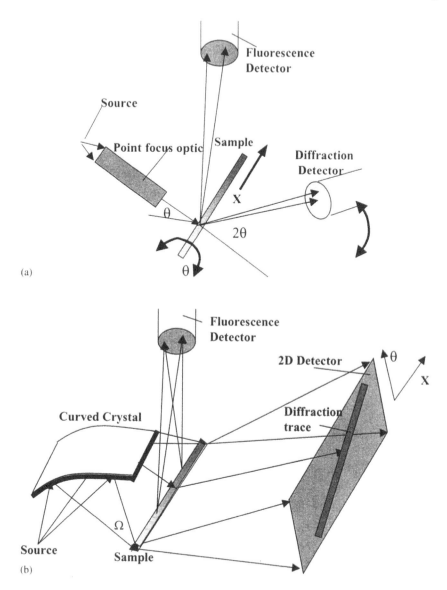

Figure 5 Laboratory instrument configurations: (a) sequential screening, (b) parallel screening.

Two key requirements in the x-ray characterization of combinatorial libraries are high throughput and high spatial resolution, in the range of 10–50 μm. For a laboratory x-ray instrument, it is particularly challenging to meet the requirement for high spatial resolution while maintaining a reasonable throughput performance. A brief discussion on x-ray tubes and x-ray detectors will be given. Issues related to combinatorial materials characterization will be addressed in terms of x-ray diffraction and x-ray fluorescence. In addition, the emphasis will be placed on x-ray optics, considered to be a key element in improving instrument performance, such as photon flux for an x-ray microbeam.

A. X-Ray Sources

For laboratory-based x-ray analytical instruments, the x-ray tube used as an x-ray source is of the electron-impact type, which is most common and most cost effective for x-ray diffraction and fluorescence work. A conventional x-ray tube consists of a filament generating electrons through thermionic emission and an anode producing x-rays under the bombardment of accelerated electrons. This type of x-ray tube is normally sealed, and no vacuum pumping is needed. There are two kinds of x-ray emission spectra from the tube, continuous and characteristic radiations (26–27). The continuous spectrum depends only on the operating voltage, and its total intensity is nearly proportional to the atomic number of the anode. For example, the tungsten anode generates a higher total continuous x-ray intensity than that of the copper anode, given the same operating voltage and current. The characteristic emission spectrum is generated when electrons are down-transitioned to lower atomic energy levels under the bombardment of an accelerated electron beam. The characteristic radiation is conventionally characterized by a series of lines, K, L, and M, corresponding to transitions between atomic levels, and is produced when the applied operating voltage is higher than the ionization potential of the corresponding levels. For example, the copper Kα emission line energy is 8.05 keV, and the ionization energy is 8.98 keV, whereas it is 59.32 keV and 69.53 keV, respectively, for the tungston Kα. Hence, operating voltages of a minimum of 9 kV and 70 kV are needed to excite Cu Kα and W Kα lines, respectively.

Most x-ray diffraction work applied to the characterization of combinatorial libraries is based on Bragg diffraction, which utilizes an incident monochromatic x-ray beam. Copper is one of the most common materials for the anode because of its intense Kα line. To achieve a high angular resolution, an x-ray monochromator is usually required to reduce the x-ray line width and to decrease the continuous radiation, which is always present in an x-ray tube spectrum. Monochromatic beam is normally not that critical in x-ray fluorescence excitation, since the production efficiency of the x-ray fluorescence is of primary interest here, although the monochromatic excitation will result in a reduced background in a spectrum.

Therefore, a high Z anode, with its continuous spectrum for excitation, is preferable to provide a high intensity of x-ray excitation.

In terms of anode selection for performance enhancement, considerations need to be given to both XRD and XRF. For XRD, as discussed earlier, a more intense characteristic x-ray line is favored, such as a Cu Kα line. However, the Cu Kα line is not capable of generating Kα fluorescent lines of elements with higher atomic number than Cu. The Kα fluorescent lines are usually preferred over L lines for analysis, because they are more intense and narrower in energy width, which reduces uncertainty during spectrum processing. Thus, high-Z anodes, such as W or Rh anodes, are common for XRF work, due to their relatively high energy for exciting Kα fluorescent lines for a large range of elements, and high spectrum intensity for efficient fluorescence excitation. For example, W anodes can be used to generate a wide range of elements, such as the transition elements, Fe, Ni, and Cu using W L lines, and Ba, In, and Sn using the spectrum near the W K line. However, the L lines of W are less intense and of a wider energy line width when compared with the Cu Kα line. This implies that the incident x-ray intensity for diffraction work will be compromised in this case. Although the x-ray tube anode is not a critical factor in the selection process, it does depend upon the emphasis on either XRD or XRF analysis and on the specific type of combinatorial libraries.

The spot size of an x-ray tube is considered an important factor in source selection, because it will influence the coupling efficiency of the optics and detector components and, subsequently, instrument performance. In general, for producing a microbeam of x-rays at a given distance, a microfocused source is preferred (28), due to the fact that it will reduce source broadening and will increase the optics focusing or transport efficiency. However, power loading for these microfocused tubes is limited to approximately 1–2 μm/watt. For example, to generate a 10-μm x-ray beam, an x-ray source size of 10 μm or smaller would be desirable, which implies that the maximum power load of the x-ray tube is about 10 watts. However, in some cases, x-ray optics is governed by condensing, not imaging from the source to the sample, such as is the case in capillary optics, for which it would be beneficial to have a larger source, ~100–400-μm range with a significantly higher power loading, 2–3-kW diffraction tubes, or ~15-kW rotating anode tubes. In most cases, utilizing condensing optics to generate a 10-μm beam from a 300-μm source with a higher power loading would be more effective in overall performance than employing a microfocused source. Some tradeoffs will be discussed in Section IV. C, on x-ray optics. Some of the conventional x-ray tubes that pertain to our applications are listed in Table 2. It should be pointed out that for the microfocused type, a power of 60 watts is the maximum value at which the spot size of 10 μm becomes unrealistic; a range of rather, 30–60 μm is preferred. It can be seen that the difference in power loading between the microfocused and the rotating anode is on the order of 250, while the spot size is only approximately 10. To obtain a comparable performance using a micro-focused source, the x-ray optics have to be 250 times more efficient, which is not easy to achieve. The same trend is true for the difference between diffraction tubes and microfocused tubes.

B. X-Ray Detection

A short survey is given for common x-ray detectors and detection configurations for x-ray diffraction and fluorescence techniques relevant to combinatorial materials research. Detectors such as the gas-filled proportional counter, the scintillation counter, the solid-state detector, and the two-dimensional detector (29–30) are included for discussion.

The principle of x-ray detection for these detectors is based on the photoelectric absorption of x-ray photons and the generation of photoelectrons. In the gas-filled counter, the photoelectrons lose their energy in ionizing gas, producing ion–electron pairs. In an applied electric field, these ion–electron pairs undergo further collision and ionization. A significant increase in the number of ion–electron pairs causes an avalanching process, which is the basis of internal electron multiplication in gas-filled proportional counters. In the scintillation counter, the photoelectrons transfer their energy to valence-band electrons in the scintillator crystal, into higher energy levels, resulting in an emission of light photons when these electrons return to their ground states. The number of light photons produced is proportional to the energy of the x-ray photon, which provides the basis for the intensity of the detector output pulse. In the Si(Li) semiconductor detector, photoelectrons expend their energy to excite valence-band electrons into the conduction band, producing electron–hole pairs. The lithium drifted silicon detector, sometimes called a PIN diode, consists of a single-crystal silicon fabricated as a PIN junction; i.e., an intrinsic layer is sandwiched between p-type and n-type layers (31). A compensated region is formed by diffusing (drifting) lithium atoms (a fast diffuser) into the p-type region to accommodate the dopants already present in the layer. The detector proportionality is based on the number of electron–hole pairs, which is proportional to absorbed x-ray energy.

Table 2 Some X-Ray Tube Characteristics

Tube type	Beam size ($\mu m^2 s$)	Mode	Power (W)	Anode
Microfocused	10^2	Spot	60	Cu-W
Diffraction	400×800	Line/spot	1000–4000	Cr-Mo
Rotating anode	100^2	Spot	15,000	Cu-Mo

Scintillation counters are usually used in conventional x-ray diffractometers to measure individual reflections from the sample in a one-dimensional space by scanning through an angular range. A gas-flow proportional counter is commonly applied in x-ray fluorescence detection in conjunction with a wavelength-dispersive device, as mentioned in Section I.2. By rotating the analyzing crystal about the center of the Rowland circle, based on Bragg's law, the crystal selects out a particular wavelength of the x-ray fluorescence emitted from the sample at a corresponding angle. The fluorescence is diffracted from the crystal and then focused into the proportional counter, after which the pulse height distribution is analyzed and the concentration of this element can be qualitatively determined. Wavelength-dispersive spectroscopy is basically a sequential process, since each x-ray wavelength is dispersed through each corresponding discrete angle through the crystal's rotation. A Si(Li) detector is widely also used in the detection of x-ray fluorescence, however, in an energy-dispersive mode. X-ray fluorescence excited from the sample will be collected at once into the Si(Li) detector. The number of the electron–hole pairs is directly proportional to the energy of individual x-ray photons. Thus, the detector output spectrum contains all the elemental fluorescence peaks present in the sample, except, due to significant air absorption, for elements with atomic numbers lower than Al. For detecting low-Z elements, a vacuum environment is usually required, similar to the practice in scanning electron microscopes. Energy-dispersive spectroscopy is normally considered a parallel-detection process, although the energy resolution is poorer.

Detectors discussed so far are primarily not position sensitive; nevertheless, their detection principles have been the basis for two-dimensional detectors. Analogous to photosensitive grains in photographic films, a two-dimensional detector contains a large active area, or a window, consisting of an array of pixels sensitive to x-ray photons. The x-ray photons detected at each pixel then either further ionize the gas atoms, creating electron–ion pairs, as in multiwire proportional chambers (32), or further create a metastable color center inside a photostimulable phosphor containing barium halide doped with europium, as in image plates (33–34). Since each individual pixel inside the active window works as an independent detector, multiple x-ray reflections can be recorded in parallel, as in photographic film, but processed digitally.

An image plate consists of a flexible plastic sheet, one side of which is coated with an active layer of photostimulable phosphor. The absorption of x-ray photons by the image plate creates color centers, causing the phosphors to be in excited states. The output data are obtained by using a red laser to read out the image plate pixel by pixel. Deactivation of phosphors in each pixel induces the emission of blue-light, which is proportional to the amount of x-ray photons absorbed. The blue light photons are subsequently measured and recorded in a digital format. The image plate can be reused after a complete exposure to white light.

A multiwire chamber has the form of a gas chamber containing a series of parallel wires sandwiched between two electric field plates (29). Each wire, of diameter on the order of 20–30 μm, acts as an independent gas proportional counter. The occurrence of electron avalanche can also be detected along the wire by using charge division of a resistive wire method. A fraction of millimeter accuracy in registration on the wire can be achieved. The spatial resolution of the two-dimensional multiwire chamber is commonly not larger than 300 μm.

C. X-Ray Optics

Optics, including curved crystals, multilayers, and capillaries, are included in this survey, and the emphasis of the discussion will be given to cases associated with producing point and line-shaped x-ray microbeams that are relevant to combinatorial library characterization with a laboratory-based x-ray instrument. A qualitative comparison is given for those optical elements in terms of performance in a given configuration.

1. Curved Crystals

Single crystals, such as silicon, germanium, and quartz crystals, can be used to collimate, focus, and monochromatize x-rays, based on Bragg's law, $2d \sin \theta_B = \lambda$. Crystals can be conformed to various geometric forms to realize different collimation or focusing geometry. Many curved crystal geometries, including Johann and Johansson, von Hamos, log-spiral, and Berreman, are applied in the broad field of x-ray research (35–40). Only Johann and Johansson focusing geometry, shown in Figure 6, will be discussed here. Analogous to the Rowland circle in light optics, a Johann crystal (35) is curved to a radius of $2R$, where R is the radius of the Rowland circle. The crystal lattice planes in this case are curved and parallel to the curved surface. An x-ray beam emanated from the source, S, located on the circle, will be diffracted from the curved crystal and focused onto the focal point, F, on the circle, as shown in Figure 6a. Under the focusing condition, the amount of x-rays that can be diffracted from the crystal is determined by Bragg's law and the reflection width of the curved crystal. The size of the crystal, which can be efficiently used for diffraction, depends on geometrical parameters, such as the radius of curvature of the crystal, the distance from the x-ray source, and the reflection curve width (41–42). The convergence angle of the diffracted x-ray beam from the crystal can be calculated by

$$\Omega \approx 4[\delta\theta \tan\theta_B]^{1/2} \tag{5}$$

where $\delta\theta$ is the reflection curve width, either the mosaic spread or the intrinsic width of the curved crystal, depending upon the nature of the crystal used, and θ_B is the Bragg angle. $\delta\theta$ usually ranges from 50 μrad to 5 mrad, determined by

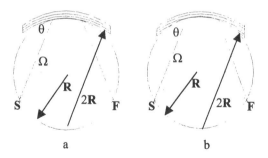

Figure 6 Schematic of the bent crystal focusing geometries: (a) Johann and (b) Johansson.

the x-ray energy, the radius of curvature, and the crystal materials (43). The reflection curve of a curved crystal can be calculated by various methods; some are rigorous (44–45), and others are simplified (46–48). The calculated results, the curve width, the peak height, and the integrated intensity of the reflection curve can be used for subsequent calculations. For a 20-cm-radius curved silicon crystal with the (111) reflection, the value of $\delta\theta$ is about 0.5 mrad if a Cu anode is chosen as the x-ray source. According to Eq. (5), the convergence angle is estimated to be 2.6 degrees. Theoretically, Johann focusing geometry is not an exact focusing scheme. All the x-rays diffracted from the curved crystal are focused not into a single point but onto a broadened spot. The broadening at focus is due to the fact that not all the diffracted rays from the curved crystal obey Bragg's law and satisfy the Rowland circle focusing condition simultaneously, because it is only at the central region of the crystal surface that these two conditions are fulfilled.

Johannson geometry (36) (see Figure 6b), on the other hand, was invented to compensate for the shortcoming of Johann geometry. The crystal is first curved to the $2R$ radius, where R is the Rowland circle radius, and then the crystal surface is modified to a radius of curvature of R. It can be seen that every point on the $2R$ curved lattice plane intersects the R curved surface. Thus Johansson geometry is claimed to be exact focusing geometry. This feature has been applied primarily to improve the x-ray collection solid angle, since the curved crystal area satisfying the Bragg diffraction condition is large, at least in principle, depending only upon crystal size. Consequently, the convergence angle becomes larger, given the same diffraction conditions, as in the case of Johann, and it can be expressed as

$$\Omega \approx 64\delta\theta \sin \theta_B^{\ 2} \times (R^2 / L^2) \qquad (6)$$

where L is the width of the crystal, which is assumed to be within the effective diffraction region. The focal width at the sample for both Johann and Johansson geometries can be estimated by using the following expression:

$$\delta f = [(q/p\,\delta s)^2 + (2R \sin \theta_B \delta\theta)^2]^{1/2} \tag{7}$$

where δs is the x-ray source size and p and q are the distance between source and optic and optic and sample, respectively. Clearly, only the source and reflection curve width broadening are taken into account in Eq. (7), since the geometry broadening is usually less significant. In terms of measuring the throughput of the optical element, the x-ray collection solid angle is usually used as a benchmark for qualitative comparison and estimation purposes (49). The collection solid angle can be expressed, in general, as the following:

$$\delta\omega \approx \Omega L/q \tag{8a}$$

and for the symmetric geometry $(p = q)$ case, $\delta\omega$ can be reduced, for Johann, to

$$\delta\omega \approx 4[\delta\theta/\sin 2\theta_B]^{1/2}(L/R) \tag{8b}$$

and, for Johansson geometry, to

$$\delta\omega \approx 32\delta\theta \sin \theta_B (R/L) \tag{8c}$$

The diffracted intensity can then be estimated to be

$$I_d \approx I_s \delta\omega P \tag{9}$$

where I_s is the x-ray source intensity, in photons/s/W/sterad, and P is the crystal peak reflectivity.

2. Multilayers

Multilayers are essentially an extension of natural crystals (50), and they are fabricated to artificially create periodic lattice spacing that can coherently diffract x-rays. As an x-ray optical element, the multilayer provides a bridge in lattice spacing between natural crystals and gratings. Synthetic multilayers are fabricated by repeatedly depositing paired materials of controlled thicknesses onto a prefigured substrate to achieve the collimation and focusing effect. The characteristics of the reflection curves, such as the width and the peak height, are related to the number of layer pairs and to the paired materials used (51–52). The capability to control thickness and uniformity with high precision, down to the angstrom level, has resulted in the creation of graded multilayer mirrors. The graded multilayers shown in Figure 7 can be implemented by gradually changing the thickness of each pair in depth or spatially on a curved or a flat substrate (53–54).

Given a fixed wavelength, the d-spacing can be tailored to compensate for various reflection angles, providing a large collection solid angle for the x-ray source. As a focusing optic, a graded multilayer is based on an elliptical focusing configuration, and yields an increase in d-spacing on the surface as the distance from the source increases to compensate for the decreasing angular acceptance, indicated as Λ_1 and Λ_2 in Figure 7. For multilayers, the d-spacing, Λ, is defined as $d_1 + d_2$, where d_1 and d_2 are the thicknesses of two different layers, and can also be expressed based on Bragg's law:

$$\Lambda = \lambda/[2\sin\theta_0(1-2\delta/\sin^2\theta_0)^{1/2}] \qquad (10)$$

where δ is the real part of the index refraction and θ_0 is the grazing angle in vacuum. Thus, for a given elliptical curve, θ_0 can be expressed as

$$\theta_0 = \cot^{-1}(bx/ay) \qquad (11)$$

where a and b are the major and minor axes of the ellipse, respectively. Substituting θ_0 into Λ, the required d-spacing grading can be directly related to points (x,y) on the elliptical curve. The convergence angle, Ω in this case, can be written as

$$\Omega = \theta_1 - \theta_2 = \arcsin[(\lambda^2 + 8\delta\,\Lambda_1^2)/4\,\Lambda_1^2]^{1/2}$$
$$- \arcsin[(\lambda^2 + 8\delta\,\Lambda_2^2)/4\,\Lambda_2^2]^{1/2} \qquad (12)$$

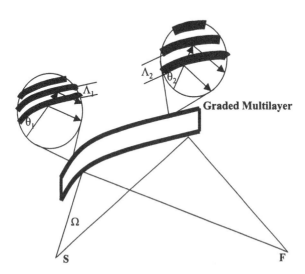

Figure 7 Schematic of a multilayer mirror with graded layer thickness.

where Λ_1 and Λ_2 are d-spacing values at the boundary of the elliptical arc. As mentioned earlier, the minimum value of Λ is approximately 25Å, which practically determines the largest angle at the arc boundary. There is no limit on maximum Λ. However, normally the length of the optic limits the smallest angle at the other end. Thus, the value of Ω for a Cu Kα radiation is in the range of approximately 1 degree on less. For a given paired material combination, the width is inversely proportional to the number of layer pairs. The reflectivity of the multilayer is influenced largely by the smoothness of the surface. Similar to the curved crystal case, the collection solid angle can be expressed as

$$\delta\omega \approx \Omega L / p \tag{13}$$

where p is measured from the center of the multilayer. A crude comparison between multilayers and crystals would indicate that curved crystals offer a large convergence angle, while multilayers provide larger collection solid angles, which in turn produces more x-ray photons at the sample.

3. Capillaries

A monocapillary is a figured glass tubing, as shown in Figure 8. Based on the x-ray external reflection principle (55–56), an x-ray beam can be guided through the figured bore of the glass tubing. The inner glass wall serves as the reflecting surface for incident x-rays, with a grazing angle equal to or less than the critical angle. The critical angle is defined as

$$\theta_{crit} = (2\delta)^{1/2} \tag{14}$$

where δ is the index refraction of the glass at a given x-ray wavelength. Thus, there is a cutoff wavelength below which the reflectivity of the x-rays will be diminished. Any incident x-rays with a longer wavelength than the critical wavelength can be reflected at the glass surface, and the output beam then yields a large energy band pass of the incident x-ray spectra.

Monocapillaries can be grouped into two categories: condensing and imaging capillaries (57). For imaging capillaries, analogous to an optical mirror reflection, incident x-rays reflect only once from the inner glass surface, usually elliptical or parabolic in shape; while the incident x-rays reflect multiple times over

Figure 8   Schematic of a monocapillary as a condensing optic.

the inner glass surface for condensing capillaries. Hence, the imaging capillary produces a convergent x-ray beam at the focus, whereas the condensing capillary produces a divergent beam, as shown in Figure 8. Note that the convergent beam is actually in the form of annular cone, which is different from the fan-shaped cone generated by crystals and multilayers.

Here we discuss primarily the properties of condensing monocapillaries, since they are more commonly applied in laboratory-based instruments (58–61). For linearly tapered capillaries, typical entrance diameters range from 50 to 350 microns, with an exit diameter between 1 and 50 microns, over approximately a 10-cm length of glass. The size and the divergence of the exit beam can be controlled by shaping the inner glass wall to linearly tapered, elliptical, or other figured surfaces. The exit diameter of the condensing capillary is the key for generating micron and submicron beams. Although monocapillary optics is not suitable for producing convergent beam incidence, it is capable of generating micron and submicron-size x-ray beams. The effective collection solid angle is normally expressed as the following:

$$\delta\omega_e = \delta\omega T(\delta s, p, L, \beta, id) \tag{15}$$

where $T(\delta s, p, L, \beta, id)$ is a transmission factor, a function of the x-ray source size, δs, the source distance, p, the capillary length, L, the capillary taper angle β, and the exit inner diameter, id. $\delta\omega$ can be calculated as $\pi(L\beta + id)^2/4p^2$. The divergence of the exit x-ray beam, $\delta\theta$, is determined largely by the critical angle and the taper angle. The larger the taper angle is, the more bounce there is in the x-rays traveling through the capillary. Thereby, the x-rays exiting from the capillary yield a beam with a larger divergence, but not larger than twice the critical angle. The critical angle also influences the divergence, because a large critical angle will allow more x-rays to bounce through a capillary with a large angle, which clearly contributes to the enlargement of the divergence of the beam.

Another class of capillary optics is polycapillary optics (62), which consists of a bundle of thin-walled glass capillaries configured into various shapes to be used to collimate and focus x-ray beams. Compared with other optics, polycapillaries are efficient for guiding or transporting x-ray beams with a large divergence, because of their estimated collection solid angle, which is approximately an order or more of magnitude larger than the other optics discussed in this section. Similar to graded multilayers, polycapillary optics has also been applied to generate a parallel x-ray beam for diffraction analyses, but with a large angular spread, ∼ 6 mrad, if Cu Kα radiation is used. The large angular spread could also be a limiting factor for applying the large convergent beam generated by polycapillary optics to parallel screening of combinatorial libraries.

Zone plates (63–64) and Bragg–Fresnel lenses (65–66) need to be mentioned, since they are capable of generating micron or submicron x-ray beams.

As a diffraction grating, a zone plate consists of a set of concentric transparent rings alternating with opaque rings. Based on the Fresnel diffraction principle, a zone plate can focus a quasi-parallel x-ray incident beam to a point determined by the focal length of the zone plate. The Bragg–Fresnel lens, with a set of elliptical concentric rings fabricated onto a single crystal, can produce not only a focused beam, based on the Fresnel diffraction, but also a monochromatic beam, based on the Bragg diffraction. Here, the concern of angular spread in the capillary optics for x-ray diffraction is no longer pertinent, since the angular spread of the focused beam is determined by the crystal reflection curve width.

Finally, a figure of merit, FOM, is suggested here to provide a guide to evaluate and select suitable optics for a specific application:

$$\text{FOM}_L = \delta\omega_e \Omega L / (\delta\theta \ \delta f) \text{ and } FOM_P = \delta\omega_c / (\delta\theta \ \delta f^2) \tag{16}$$

where FOM_L and FOM_P are the figures of merit for line focus and point focus cases, respectively. $\delta\omega$, Ω, and δf have been previously defined and can be calculated by using Eqs. (8), (12), and (7), respectively. The reflection curve width, $\delta\theta$, for both crystal and multilayer cases, can be calculated by using the x-ray crystal optics program (67). ω is introduced into the figure of merit owing to its importance for improving the throughput, although the convergence angle is considered less important in the point focus case. A test case is selected to illustrate the evaluation process for various optical components, and the results are listed in Table 3. A Cu anode x-ray tube with a spot size of 400 μm is selected; the source-to-sample distance is 160 mm. Depending on the nature of the optical component, the distance of an optical component to the x-ray tube and the sample is not fixed. The detailed calculations are shown in the Appendix to this chapter.

It is more meaningful to compare the values of $\delta\omega_e$ when Ω is negative, as in the case when a slit, a pinhole, or a capillary is used to produce divergent x-rays. It can be seen that the significant difference in FOM_L between the multilayer and the curved crystals is the convergence angle. The convergence angle might be capable of reaching ~ 1 degree for multilayer optics, but there are more options to increase the convergence angle in the crystal case. It should be pointed out that the magnification of the Johann geometry used for the curved crystal is 1 ($p = q$), and the focus size is merely the image of the source, ~ 400 μm. It is possible to decrease the focus size by optimizing the magnification, achieving the narrowest focus size and the largest convergence angle. The source size of the x-ray tube is another parameter that can effect a decrease in the focus size. In fact, conventional diffraction tubes (68) can be operated in a line mode 40 μm wide and 8 mm long, which could decrease the focus size by a factor of 10 in this case. In terms of increasing the convergence angle, many parameters, such as the crystal type, the bending radius, and the crystal reflection, can be used to optimize the crystal optics. In general, the limitation of the line focus is the ability

Table 3 Performance Comparison of Various X-Ray Optics

	Focus size, δf (μm)	Angular spread, $\delta\theta$ (mrad)	Conv. angle, Ω (degrees)	Effective solid angle, $\delta\omega_c$ (msrad)	FOM_L (msrad)
Line focus					
Slit	400	2.5	-1	0.3	-12
Curved crystal, Johann	402	0.5	3.4	0.52	1526
Multilayer mirror	96	0.87	0.7	0.27	559
					FOM_P (mrad/mm^2)
Point focus					
Pin-hole aperture	10	2.5	-1	3.1×10^{-6}	0.01
Monocapillary	10	6.0	-1	2.1×10^{-3}	3.50
K-B multilayer mirrors	106	0.87	0.7	2.8×10^{-3}	32.20
Doubly curved crystal, Johann	412	0.6	3.4	3.3×10^{-4}	5.50

to produce a narrow line width compared with that of other point focus optics. It is difficult to produce a line focus with a width of less than 100 μm; if an aperture is inserted at the focus, the integrity of the convergent x-rays will be compromised. In the point focus case, a large convergent x-ray beam is not considered effective, because it is difficult to control the uniformity of the convergence in two directions. Moreover, it is not straightforward to interpret the diffraction pattern when the convergent x-rays are not well defined.

Several points deserve to be addressed. First, for a 10-μm beam, monocapillary appears to be a robust optic to use, even though the FOM is almost an order of magnitude lower than that of a K-B multilayer. The effective solid angle, however, is comparable in these two cases. The obtainable focal spot size is 100 μm for the K-B multilayer, before inserting a 10 μm aperture, which apparently is not readily available. If a 10 μm spot size could be relaxed, then K-B multilayer mirrors would be more desirable. Second, compared with K-B multilayer mirrors, a significant drawback of the capillary option is the large angular spread, which will degrade the angular resolution of the incident x-rays and affect the measurements of diffraction peak positions.

V. CONCLUDING REMARKS

X-ray analytical techniques, diffraction and fluorescence analysis for combinatorial materials characterization, have been discussed. Emphasis has been given to

Table 4 Line Focus Parameters

	w/L (mm)	f_c	p/q (mm)	Ref./θ_B (deg)	Ellipse a/b (mm)	θ_1/θ_2 (deg)	$\delta\theta/P$ (mrad)/%
Slit	20/0.4	1.0	$\sim$160/$\sim$0	—	—	—	—
Crystal	9.1/10	0.11	87.1/87.1	Si(220)/ 23.66	$a = b$ $= 217$	—	0.5/0.7
Multilayer	40/15	0.28	130/30	—	80/1	0.773/1.479	0.87/0.8

developing special instruments in this field. The material surveyed here, intended to serve as guidance for further development, is far from complete in its depth and breadth. Design goals should be set by the type and the range of materials libraries to be characterized and by the projected throughput improvement and precision requirement. Trade offs among all the components must be taken into consideration in the process of developing an x-ray tool for specific combinatorial applications.

Appendix

The calculation parameters are listed in Tables 4 and 5 for the line focus and the point focus cases, respectively. Equation (7), $\delta f = [(q/p\ \delta s)^2 + (2R \sin\theta_B \delta\theta)^2]^{1/2}$, intended for the line focus case, is also used for the point focus case with certain approximations. δs is the source size of 0.4 mm, and $2R$ is the radius of curvature, 217 mm, for curved crystals. In terms of slit, $\delta f = w$, the width of the slit opening. Toroidal geometry is a point-focusing configuration, an extension of Johann geometry to include an additional sagittal curvature with a radius of $2R \sin\theta_B^2$ and a meridian curvature of $2R$. In both the K-B multilayer and the toroidally curved crystal cases, the focal spot size is approximated to be the same in both directions. For the sake of comparison between the multilayer and curved crystal and the pinhole and monocapillary, a common 10-μm size is imposed onto the case of K-B multilayer and toroidally curved crystal.

Table 5 Point Focus Parameters

	w/L (mm)	β (mrad)/d (mm)	p/q (mm)	f_c	g_c	T (%)	$\delta\theta/P$ (mrad)/%
Pinhole aperture	$\sim$0	$\sim$0/0.01	$\sim$160/$\sim$0	—	1.0	100	—
Monocapillary	130	2.23/0.01	30/$\sim$0	—	1.0	2.7	—
K-B multilayer	40/15	—	$\sim$130/30	0.28	0.011	—	$0.87/0.8^2$
Toroidal curved crystal	9.1/10	—	87.1/87.1	0.11	6.2×10^{-4}	—	0.5/0.7

The convergence angles for both line focus and point focus cases are calculated by using Eq (5) and (12) for curved crystals and multilayers, respectively. A Ni/B_4C combination is used for the graded multilayer optic with a 40-Å period of thickness ratio $\Gamma = 0.5$ and 100 layers. Λ_1 and Λ_2 are computed from Eqs. (10) and (11) to be 57.14 Å and 29.85Å, respectively, which yield the 2θ values listed in Table 4. Values of $\delta\theta$ are obtained by using some of the online utilities (69–70).

The effective solid angle is expressed as $\delta\omega_e = \delta\omega f_c P$, where P is the peak reflectivity of the reflection curve for both crystals and multilayers, whereas f_c, the finite source size correction factor, is expressed as $2R \sin \theta_B \, \delta\theta/\delta s$ for curved crystals and $p \, \delta\theta/\delta s$ for multilayers. Note that the collection solid angles, $\delta\omega$, for curved crystals and multilayers are approximated using Eqs.(8) and (13).

It should be pointed out that in the toroidally curved crystal case, the value of $\delta\omega$ is likely to be underestimated, because the reflection curve width is usually increased when the crystal is under 2D bending. In addition, the length of the line at focus will also be shrunk to a size smaller than δf, 402 μm. The assumption is also made for the K-B multilayer case, that the second multilayer mirror accepts the entire x-ray beam reflected off the first multilayer, and the reflectivity for both mirrors are identical. Another factor, g_c, is used for scaling the focal area to 10 μm^2 to maintain consistency of the same focal size for the point focus case. Thus, the expressions for $\delta\omega_e$ are $\delta\omega f_c P^2 g_c$ and $\delta\omega f_c P g_c$ for the K-B multilayer and the toroidally curved crystal, respectively.

In terms of the monocapillary case, the effective solid angle is calculated using Eq. (15). The transmission, T, is obtained by using the 2D ray-tracing program developed at ARACOR. The results of the ray-tracing program also include the divergence of the exit x-ray beam, which is defined here as the angular spread, $\delta\theta$. In general, the divergence is less than or equal to twice the critical angle, θ_{crit}, which is calculated using Eq. (14).

REFERENCES

1. Darwin, C. G. Roetgen-ray reflection. Phil. Mag. 1914, 27(2), 315–333.
2. James, R. W. The Optical Principles of the Diffraction of X-Rays; G. Bell and Sons: London, 1962.
3. Zachariasen, W.H. Theory of X-Ray Diffraction in Crystals. 1967.
4. Warren, B. E. X-Ray Diffraction; Addison-Wesley: Reading, MA, 1969.
5. Batterman, B.W.; Cole, H. Dynamical diffraction of x-rays by perfect crystals. Rev. Mod. Phys. 1964, 36, 681–717.
6. Guinier, A. X-Ray Diffraction in Crystals, Imperfect Crystals, and Amorphous Bodies; Dover: New York, 1994.
7. Cullity, B. D. Elements of X-Ray Diffraction; Addison-Wesley: Reading, MA, 1967.

8. Klug, H.P.; Alexander, L.E. X-Ray Diffraction Procedures. 1954.
9. Powder Diffraction Files. International Center for Diffraction Data: 12 Campus Boulevard, Newtown Square, PA, 19073-3273.
10. Bertin, E.P. Introduction to X-Ray Spectrometric Analysis; Plenum Press: New York, 1978.
11. Jenkins, R.; de Vries, J.L. Practical X-Ray Spectrometry, 2nd ed.; Springer Verlag: New York, 1969.
12. Tertian, R.; Claisse, F. Principles of Quantitative X-Ray Fluorescence Analysis; Heyden & Son: London, 1982.
13. Shiraiwa, T.; Fujino, N. Theoretical calculation of fluorescent x-ray intensityies in fluorescent x-ray spectrochemical analysis. Jpn. J. Appl. Phys. 1966, 5, 886–899.
14. Criss, J.W.; Birks, L.S. Calculation methods for fluorescent x-ray spectrometry. Anal. Chem. 1968, 40, 1080–1086.
15. Sparks, C. J., Jr. Quantitative x-ray fluorescent analysis using fundamental parameters. Advances x-ray. Anal. 1976, 19, 19–51.
16. Mantler, M. Advances in fundamental-parameter methods for quantitative XRFA. Advances x-ray. Anal. 1986, 30, 97–104.
17. Jaklevic, J.M.; Giauque, R.D. Energy Dispersive X-ray fluorescence analysis using x-ray tube excitation. In: Handbook of X-Ray Spectrometry; Van Grieken, R.E., Markowicz, A.A., eds.; Marcel Dekker: New York, 1993, 151–179.
18. Helsen, J.A.; Juczumow, A. Wavelength-dispersive x-ray fluorescence. In: Handbook of X-Ray Spectrometry; Van Grieken, R.E., Markowicz, A.A., eds.; Marcel Dekker: New York, 1993, 75–145.
19. Remmel, T.; Werho, D. Development of an XRF metrology method for composition and thickness of barium strontium titanate thin films. Advances x-ray Anal. 2000, 42, 99–108.
20. Isaacs, E.D.; Marcus, M.; Aeppli, G.; Xiang, X.-D.; Sun, X.-D.; Schultz, P.; Kao, H.-K.; Cargill, G. S., III; Haushalter, R. Synchrotron x-ray microbeam diagnostics of combinatorial synthesis. Appl. Phys. Lett. 1998, 73, 1820–1822.
21. Kirkpatrik, P.; Baez, A. V. Formation of optical images by x-rays. J. Opt. Soc. Am. 1948, 38, 766–774.
22. Yoo, Y.K.; Ohnishi, T.; Wang, G.; Duewer, F.; Xiang, X.-D.; Chu, Y.S.; Mancini, D.C.; Li, Y.-Q.; O'Handley, R.C. Continuous mapping of structure–property relations in $Fe_{1-x}Ni_x$ metallic alloys fabricated by combinatorial synthesis. Intermetallics. 2001, 9, 541–545.
23. Chang, K.S.; Aronova, M.; Famodu, O.; Takeuchi, I.; Lofland, S.E.; Hattrick-Simpers, J.; Chang, H. Multimode quantitative scanning microwave microscopy of in situ grown epitaxial $Ba_{1-x}Sr_xTiO_3$ composition spreads. Appl. Phys. Lett. 2001, 79, 4411–4413.
24. Omote, K.; Kikuchi, T.; Harada, J.; Kawasaki, M.; Ohtomo, A.; Ohtani, M.; Ohnishi, T.; Komiyama, D.; Koinuma, H. A convergent beam parallel detection x-ray diffraction system for characterizing combinatorial epitaxial thin films. Proc. SPIE. 2000, 3941, 84–91.
25. Unpublished.

26. Compton, A.H.; Allison, S.K. X-Rays in Theory and Experiment, 2nd ed; D. Van Nostrand: New York, 1935, 69–93.

27. Norman, A. Dyson X-Rays in Atomic and Nuclear Physics; Longman: London, 1973, 7–132.

28. XTG UltraBrite Microfocus Source. Oxford Instruments: X-ray Technology Group, 275 Technology Drive, Scotts Valley, CA 95066.

29. Buckley, C. J. X-ray detectors. In X-Ray Science and Technology; Michette, A. G., Buckley, C. J., eds.; IOP: London, 1993, 225–253.

30. Rehak, P.; Smith, G.C.; Warren, J.B.; Yu, B. First results from the micro pin array detector (MIPA). IEEE Trans. Nucl. Sci. 2000, 47, 1426–1429.

31. Woldseth, R. X-Ray Energy Spectrometry; Kevex Corporation: Burlingame, CA, 1973, 2.1–2.15.

32. Charpak, G.; Bouclier, R.; Bressani, T.; Favier, J.; Zupancic, C. The use of multiwire proportional counters to select and localize charged particles. Nucl. Inst. Meth. 1968, 62, 262.

33. Amemiya, Y.; Miyahara, J. Nature. 1993, 336, 89.

34. Dauter, Z.; Wilson, K. S. Imaging plates in synchrotron and conventional X-ray crystallographic data collection. Acta. Phys. Pol. A. 1994, 86, 477–486.

35. Johann, H. H. Die Erzeugung lichtstarker Roentgenspektren mit Hilfe von Konkavkristallen. Z. Physik. 1931, 69, 185–206.

36. Cermak, J. A contribution to the theory of x-ray monochromators with approximate focusing of Johann's type. J. Phys. E. 1970, 3, 615–620.

37. Johansson, T. Ueber ein neuartiges genau fokussierendes Roentgenspektrometer. Z. Physik. 1931, 82, 507–528.

38. v. Hamos, L. X-ray image method of chemical analysis. J. Mineral Soc. Am. 1938, 23, 215–226.

39. de Wolff, P.M. Focusing Monochromators and Transmission Techniques. Norelco Rep. 1968, 15, 44–49.

40. Berreman, D. W. Curved crystal x-ray monochromator efficiency. Phys. Rev. 1979, B19, 560–567.

41. Wittry, D. B.; Sun, Songquan. X-ray optics of doubly curved diffractors. J. Appl. Phys. 1990, 67, 1633–1638.

42. Chang, W. Z.; Foerster, E. X-ray diffractive optics of curved crystals: focusing properties on a diffraction-limited basis. J. Opt. Soc. Am. 1997, A14, 1647–1653.

43. Caciuffo, R.; Melone, S.; Rustichelli, F.; Boeuf, A. Monochromators for x-ray synchrotron radiation. Phys. Rep. 1987, 152, 1–71.

44. Takagi, S. A dynamical theory of diffraction for a distorted crystal. J. Phys. Soc. Jpn. 1239–1253, 26, 1969.

45. Taupin, D. Théorie dynamique de la diffraction des rayons X par les cristaux déformés. Bull. Soc. Fanç. Minér. Crist. 1964, 87, 511–469.

46. Penning, P.; Polder, D. Anomalous transmission of x-rays in elastically deformed crystals. Philips Res. Rep. 1961, 16, 419–440.

47. Chukhovskii, F. N.; Chang, W. Z.; Foerster, E. X-ray focusing optics. II Properties of doubly bent crystals with an extended x-ray source. J. Appl. Phys. 1995, 77, 1849–1854.

48. Sánchez del Río, M.; Ferrero, C.; Mocella, V. Computer simulations of bent perfect crystal diffraction profiles. SPIE Proc. 1997, 3151, 312–323.

49. Wittry, D. B.; Chang, W. Z.; Li, R. Y. X-ray optics of diffractors curved to a logarithmic spiral. J. Appl. Phys. 1993, 74, 3534–3540.

50. Gilfrich, J. V.; Nagel, D. J.; Barbee, T. W., Jr. Layered synthetic microstructures as dispersing devices in x-ray spectrometers. Appl. Spectr. 1982, 36, 58–61.

51. Michette, A. G. Reflective optics for x-rays. In: X-Ray Science and Technology; Michette, A. G., Buckley, C. J., eds.; Institute of Physics: London, 1993, 280–296.

52. Spiller, E. Multilayers. In: Handbook of Optics; Bass, M., ed.; McGraw-Hill: New York, 2001, 24.1–24.13.

53. Schuster, M.; Goebel, H. Parallel-beam coupling into channel-cut monochromators using taylored multilayers. J. Phys. D. 1995, 28, A270.

54. Verman, B.; Jiang, L.; Kim, B.; Smith, R.; Grupido, N. Confocal graded d-spacing multilayer beam conditioning optics. Adv. X-Ray Anal. 2000, 42, 321–332.

55. Stern, E. A.; Kalman, Z.; Lewis, A.; Lieberman, K. Simple method for focusing x-rays using tapered capillaries. Appl. Opt. 1988, 27, 5135.

56. Hirsch, P. B. X-ray microbeam techniques. In: X-Ray Diffraction by Polycrystalline Materials; Peiser, H. S., Rooksby, H. P., Wilson, A. J. C., eds.; IOP: London, 1955, 288–297.

57. Bilderback, D.H.; Franco, E.D. Single capillaries. In: Handbook of Optics; Bass, M., ed.; McGraw-Hill: New York, 2001, 29.1–29.8.

58. Yamamoto, N.; Hosokawa, Y. Development of an innovative 5-μm focused x-ray beam energy dispersive spectrometer and its applications. Jpn. J. Appl. Phys. 1988, 27, L2203.

59. Carpenter, D. A. An improved laboratory x-ray source for microfluorescence analysis. X-ray Spectro. 1989, 18, 253.

60. Engstroem, P.; Larsson, S.; Rindby, A.; Stocklassa, B. A 200-μm x-ray microbeam spectrometer. Nucl. Instru. Meth. 1989, B36, 222.

61. Chang, W. Z.; Kerner, J.; Franco, E. Analytical micro x-ray fluorescence spectrometer. Adv. X-Ray Anal. 2001, 44, 325–328.

62. MacDonald, C.A.; Gibson, W.M. Polycapillary and multichannel plate x-ray optics. In: Handbook of Optics; Bass, M., ed.; McGraw-Hill: New York, 2001, 30.1–30.17.

63. Kirz, J. Phase zone plates for x-rays and the extreme UV. J. Opt. Soc. Am. 1974, 64, 301–309.

64. Lai, B.; Yun, W. B.; Legnini, D.; Xiao, Y. H.; Chzas, J. Hard x-ray microimaging techniques based on phase zone plates. SPIE. 1992, 1741, 180–185.

65. Aristov, V. V. Bragg–Fresnel optics: principles and prospects of applications. In; Sayre, D., Howells, M., Kirz, J., Rarback, H., eds. X-Ray Microscopy II; Springer Verlag: Berlin, 1987, 108–117.

66. Chang, W. Z.; Uschmann, I.; Foerster, E. Application of an InSb flat crystal with elliptically shaped modulated structures to x-ray point focusing. Appl. Phys. Lett. 1996, 69, 872–874.

67. Uschmann, I.; Foerster, E.; Gaebel, K.; Hoelzer, G. X-ray reflection properties of elastically bent perfect crystals in Bragg geometry. J. Appl. Cryst. 1993, 26, 405–412.

68. Fine-focus ceramic x-ray tubes, manufactured by Bruker AXS, Inc.; 6300 Enterprise Lane: Madison, WI, 53719–1173.

69. http://www-cxro.lbl.gov/optical_constants/multi2.html Center for X-ray Optics; Lawrence Berkeley National Laboratory: Berkeley, CA.

70. http://www.esrf.fr/computing/scientific/xop European Synchrotron Radiation Facility: Grenoble, France.

11

High-Throughput Screening of Electrical Impedance of Functional Materials by Evanescent Microwave Probe

Gang Wang and Xiao-Dong Xiang
Intematix Corporation, Moraga, California, U.S.A

1. INTRODUCTION

Electrical impedance is one of the most important and frequently referred to properties of a material. The quantity is a complex number with strong frequency dependence. One of the major challenges of combinatorial materials science is reliable, rapid, nondestructive, and quantitative screening of electrical properties of materials chips. No existing tool with these capabilities was available at the time of our initial effort in combinatorial materials science. Most inspection tools in the semiconductor industry employ light as a probing signal. However, a material's index of refraction (i.e., the electrical impedance) at light frequencies (10^{16}/s) has almost no direct correlation with the electrical impedance at working frequencies (10^9/s) of microelectronics. For electronic applications, lower-range microwave frequencies (i.e., $1-10$ GHz) are most relevant and best suited to characterize electrical impedance. However, the wavelengths of propagating microwaves in this frequency range are very long (order of centimeters), which limit the spatial resolution of the measurements. Rapid development in both the microelectronics industry and materials science has also raised the demand for nondestructive and spatially resolved quantitative characterization of electrical impedance. During our endeavor in combinatorial materials sciences, a scanning evanescent microwave probe with capability of satisfying these demands has been invented and developed. In this chapter we will discuss its development.

The evanescent-wave microscope can be classified as a type of a scanning probe microscope (SPM). The first scanning probe microscope was probably the evanescent photon microscope envisioned by E. H. Synge in 1928 (1). Then, Fraint and Soohoo, working independently, had demonstrated this idea at microwave frequencies in 1959 and 1962, respectively (2,3) (although the work by Ash and Nicholls 10 years later (4) is often credited as the first work in literature). In recent years, due to the impact of the invention of the scanning tunneling microscope (STM) (5), which can be considered an evanescent de Broglie wave microscope, many different SPMs have been proposed and developed (6). However, until recently, performing quantitative microscopy of different materials properties, such as the complex dielectric constant and conductivity, has been almost impossible. The difficulty arises from two major barriers. First, in all SPMs, the microscopy signal is a convolution of topography and physical properties. Separating them requires measuring at least two independent signals simultaneously. The development of the scanning near-field optical microscope (SNOM) (7,8) has provided this capability by implementing shear-force detection in addition to optical signal detection. Second, a detailed field configuration in the tip-sample region has to be solved, which subsequently gives rise to solutions that relate the signals explicitly to tip-sample distance and physical properties. These relations can then also be used for feedback control of tip-sample distance to obtain simultaneously quantitative topography and physical property images. Although numerical methods based on a finite element analysis have been used to solve the field distribution around an SNOM tip (9), this approach is generally not practical in routine applications.

In order to access the electrical impedance relevant to a broad range of applications, evanescent (or near-field) microwave microscopy was suggested. In previous efforts (2–4), aperture or tapered waveguide probes had been used. Operating below the cutoff frequency, these probes suffer severely from waveguide decay (tapered waveguide probes were widely used in SNOM with typical attenuation of 10^{-3}–10^{-6}). In these probes, a linear improvement in resolution will cause an exponential reduction in sensitivity, as realized by Soohoo (3). As a result, it is very hard to reconcile the conflict between resolution and sensitivity. Fee et al. realized the conflict and suggested using a transmission line probe with a reduced cross section and a sharp tip (10). However, because the resolution is also determined by the cross section, further improvement to submicron resolution (if practical) still causes significant transmission line decay. In addition, unshielded far-field components around the tip in the transmission line probe significantly limit the resolution and ability for quantitative analysis.

In 1995, we developed a novel scanning evanescent microwave probe (SEMP), with a shielding structure designed so that the propagating far-field components are shielded within the cavity, whereas the nonpropagating evanescent waves are generated at the tip. We also developed theoretical models to obtain near-field analytic solutions, which allowed quantitative microscopy of microwave electrical impedance of materials with submicron resolution (11–13).

The technology was a breakthrough in nondestructive and quantitative electrical impedance measurements of a wide range of materials, from insulating dielectrics (14), to semiconductors, to highly conducting metals (15). The evanescent microwave probe (EMP) sends evanescent microwaves and detects their interaction with the sample through a conducting tip. This interaction depends on complex electrical impedance (including both the real part and the imaginary part) of the sample, which causes changes in the resonant frequency and the quality factor of the resonator. The EMP can simultaneously measure the real and imaginary parts of a material's electrical impedance and surface topography through the measurements of the shifts in resonance frequency and the quality factor of the sensing resonant probe (and their derivatives with respect to various modulations).

II. FUNDAMENTAL PHYSICS

We first give a brief discussion of the fundamental physics of evanescent-wave microscopy. The evanescent waves in this context refer to electromagnetic waves with wavevectors of imaginary number not originating from dissipation. In fact, the evanescent electromagnetic waves are the photon equivalent of quantum mechanical electron waves in the classically forbidden region (within a barrier). In the far-field description of electromagnetic waves, an orthogonal eigenfunction set of Hilbert space is chosen as the plane waves whose wavevectors are any real number satisfying the Helmholtz equation (as a consequence, these plane waves are propagating waves). Any propagating wave (for example, a propagating spherical wave) can be expanded as the superposition of these plane waves. The magnitudes of the wavevectors are determined solely by the frequency and speed of light according to the Helmholtz equation, i.e., $k = 2\pi f\sqrt{\varepsilon\mu} = 2\pi/\lambda = \sqrt{k_x^2 + k_y^2 + k_z^2}$. For propagating waves, k_x, k_y, and k_z are real numbers and thus must be smaller than k (in free space, $k = k_0$). These waves only have resolving power on the order of λ. These plane waves cannot construct, for example, a spherical wave whose wave front has a radius less than the wavelength λ. Therefore, a true complete set of Hilbert space should include plane waves whose wavevectors are any complex numbers satisfying the Maxwell equation to construct such a spherical wave. Since imaginary wavevectors are allowed, the components (k_x, k_y, k_z) can then be any value and still satisfy the Maxwell equation. Here the "plane waves" whose lateral components $k_r = \sqrt{k_x^2 + k_y^2}$ are larger than k will have higher lateral resolving power (on the order of $1/k_r$). However, since they must have imaginary components k_z to satisfy the Helmholtz equation, these waves are "evanescent" and cannot propagate much more than a wavelength λ.

A metal sphere or tip fed by a wave source with a radius of R_0 ($<< \lambda$) will generate evanescent waves (to form a spherical wave on the metal surface satisfying the boundary conditions) whose wavevectors range up to $k_r \sim 1/R_0$ and resolving power up to $\sim R_0$. Interaction between the tip and a sample (with a high effective dielectric constant) may further increase the resolving power (e.g., resolution $\sim \sqrt{gR_0}$ for conducting material, where g is the tip–sample distance) as a result of decreasing the effective tip radius from the polarizing effect. Since these waves decay over a distance R_0 in free space, the sample has to be brought to within R_0 of the tip to obtain strong interaction. Note, these waves are not necessarily evanescent in conducting materials, since wavevector $k_c = 2\pi f \sqrt{\epsilon\mu(1 + i\sigma/\omega\epsilon)}$ is many orders of magnitude larger than that in free space. Here σ is the conductivity of the sample.

III. HISTORIC ACCOUNTS

Here, we give a historical perspective of various techniques and efforts in this field. Frait and Soohoo first independently demonstrated similar evanescent microwave microscopes using the microwave cavity with a small aperture in 1959 and 1962, respectively (2,3). Soohoo used it to study the local property of magnetic materials based on the ferromagnetic resonant absorption of microwaves. Soohoo is also the first one to realize the conflict between spatial resolution and sensitivity in the aperture approach. Bryant and Gunn were probably the first (1965) to use a tapered coaxial transmission line probe to study the local conductivity of materials (resolution of 1 mm) (16). Ash and Nicholls published a paper in 1972 emphasizing the aperture approach (4). They are probably the first to demonstrate the superresolution on dielectrics (but they are also often mistakenly credited as the first to demonstrate an evanescent microwave microscope in the literature). Note, many studies later often demonstrated the resolution on metals, which is misleading since the wavelengths of microwaves in metal are many orders of magnitude (at least four orders) smaller than that in air and smaller than the resolution those studies demonstrated. Massey, in 1984, also discussed the microscopy with scanned evanescent waves from an aperture and tested the theory at 450 MHz (17). Those early pioneers were not aware of the theoretic proposal by Synge in 1928 (1), and none of them were probably aware of each other's work. Fee, Chu, and Hänsch published a paper in 1988 (10) pointing out explicitly the limitation of the aperture-type probe in near-field microscopy and suggested using a coaxial transmission line with a very small cross section to obtain high resolution with much less loss. They also tested such a probe on metal and pointed out that protruding and sharpening the center conductor could improve the resolution. Wang et al., in 1987 and 1990, demonstrated an evanescent microwave microscope based on a scanned, tapered, open (electric dipole) and closed (magnetic dipole) end of a microstrip resonator, respectively (18). Tabib-Azar et al., in 1993, discussed a similar approach (19). We published a unique design of an evanescent wave microscope in 1996 (11), which we will discuss in detail later in this review. Several other groups also have been actively implementing transmission-line-type probes in evanescent microwave micros-copy studies, including Keilmann et al. (20) and Vlahacos et al. (21). Recently, there has been renewed interest in designing new cavity- or waveguide-based structures. Golosovsky and Davidov, in 1996, demonstrated an open, narrow slit structure in a waveguide (22) (similar to what Gutmann et al. demostrated in 1987 (18a), which has much improved energy transmission over the aperture structure. Bae et al. discussed a similar approach in 1997 (23). Grober et al. proposed yet another bow-tie antenna structure as the scanning probe (24).

IV. EXPERIMENTAL TECHNIQUE

The core of our EMP is the design for obtaining relatively pure evanescent micro-waves near the tip and at the same time maintaining very high Q of the microwave sensor (i.e., resonator). An example of an EMP sensor is illustrated in Figure 1. The probe is based on a high-quality-factor (Q) microwave coaxial resonator with a sharpened metal tip mounted on the center conductor. Since the tip is an integral part of a very sensitive detector (microwave resonator with Q of a few thousands), the sensitivity of the instrument can be extremely high. The tip extends beyond an aperture formed on a thin metal shielding endwall of the resonator. The tip and the shielding structure are designed so that the propagating far-field compo-nents are shielded within the cavity, whereas the nonpropagating evanescent waves are generated at the tip. The design created a near-perfect situation, where the tip functions as a monopole evanescent-wave antenna, especially when the tip is very close to the sample. This feature is crucial for both high resolution and quantitative analysis. If both evanescent and propagating waves (leaked from the resonator) have to be considered and calculated, the local quantitative micros-copy will be impossible, since the total signal will depend on sample features far away from the tip. In contrast to the conventional far-field antenna probes, this probe does not emit significant energy (and therefore has very high Q to boost the sensitivity). Only when the tip is in close range of the sample will the evanescent waves on the tip interact with the materials. The interaction gives rise to resonant-frequency and Q changes of the cavity and, consequently, in the microscopy of the electrical impedance.

The resonant frequency (f_r) and Q can be measured by sweeping the fre-quency of a digital synthesizer, measuring, and fitting the entire response (11).

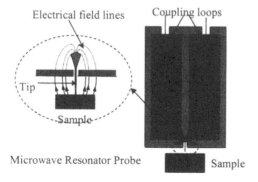

Figure 1   Structure of an EMP resonator.

As an alternative technique (12), a phase-sensitive detection can be used. Figure 2 shows the simplified data-acquisition scheme of the EMP2001, developed by Ariel Technologies Inc. The microwave generated by a direct digital synthesizer (DDS) (which has frequency stability of better than 10^{-9}) or a phase-lock-loop–based synthesizer is coupled to the coaxial resonator probe. The transmitted signal is sent to an I/Q mixer to yield the in-phase (i) and quadrature (q) signals (90° phase-shifted from in-phase signal) by phase-coherent detection. The phase shifter is adjusted so that the in-phase output is zero and the quadrature output is a maximum at the resonant frequency. The digital signal processor 1 (DSP1) then calculates f_r and Q using the output signals of the I/Q mixer. DSP2 uses f_r and Q information to control motion components for tip–sample positioning and other imaging functions. This method allows data rates around 100 kHz to 1 MHz (limited by the bandwidth of the resonator) and frequency sensitivity below 1 kHz / ($\Delta f_r/f_r \approx 10^{-6}$–$10^{-7}$). Alternatively, a voltage-controlled oscillator (VCO) can be used to replace DDS if a lower frequency sensitivity can be accepted. The in-phase output is then used to feedback control the output frequency of VCO to follow the f_r of the resonator. The error signal of the phase-lock loop is proportional to the f_r shift. The change in Q is determined by measuring the amplitude change of the quadrature output signal of the mixer. Both schemes are significantly faster than the frequency-sweeping technique. The images are obtained by scanning the sample under the resonator-tip assembly while recording the changes in f_r and Q.

Two different approaches have been used to scan samples. First, for insulating materials, physical contact between tip and sample surface can be used to obtain constant (zero) tip–sample distance. A very soft spring-loaded cantilever sample stage with a force estimated (under a typical operating condition) to be less than 20 μN was placed under the sample to provide soft contact of the sample

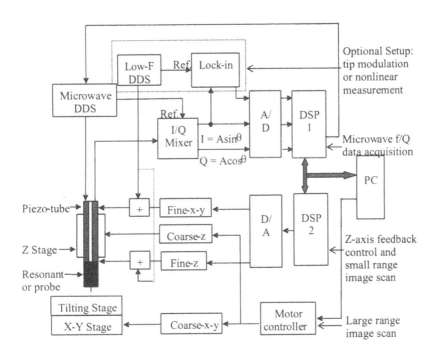

Figure 2 EMP2001 schematics.

with the tip. The tip with radius < 1 μm can be kept from damaging during the scan with this contact force. The microscope in this configuration is therefore not sensitive to topographic features on the sample surfaces. For conducting materials, an insulating layer can be coated on the tip or a sample surface for a soft contact scan.

Recently, independent tip–sample distance feedback control techniques were used to obtain noncontact scan images. First, since the EMP measures f_r and Q simultaneously, one of the quantities can be used to control the tip–sample distance. For highly conducting materials, f_r can be used to keep a constant tip–sample distance, since f_r is a function of tip–sample distance only, not conductivity (25). For a semiconducting sample, f_r is a function of conductivity; however, constant-f_r mode can still be employed in many cases if a careful analysis method is used. Figure 3a and b) demonstrate simultaneous imagining of sample surface topography and conductivity of a selectively doped (with line patterns) silicon and a periodically poled nonlinear dielectric crystal using a distance feedback-control technique, respectively (26,27).

In general cases, a tip–sample distance control that is independent of the sample electrical property is desired. Atomic force–based tip–sample distance control is therefore probably best suited for this purpose. We have implemented an atomic force–based tip–sample distance control. A vibration modulation is introduced by a small piezoelectric element mounted on the resonator wall. As the tip vibrates at its mechanical resonant frequency, a modulation in the microwave envelope is introduced and detected at the output of the mixer. This signal is then detected by a lock-in amplifier and fed into the third signal channel of data acquisition to obtain tip–sample distance control and imaging. Figure 4 demonstrates the range and sensitivity of the atomic force between tip and sample. Note that our detection scheme of atomic force is very different from conventional

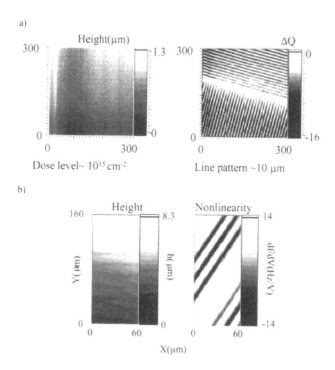

Figure 3   Imaging of a sample surface with a distance feedback-control technique: (a) simultaneous topography and conductivity profiles of doped silicon; (b) simultaneous topography and ferroelectric domain imaging of poled LiNbO$_3$.

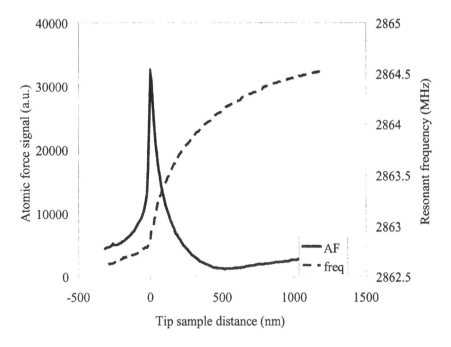

Figure 4   Atomic force signal measured by modulation of the EMP tip.

methods, ranging from laser interference and deflection, capacitive, piezo-resistive to piezoelectric sensing. Microwave detection of atomic force–induced mechanical resonance changes are very sensitive and convenient to implement in our system.

The measured EMP signals, namely, changes in resonant frequency and Q, have specific functions dependent on the tip–sample distance with given physical parameters, as we discuss in detail later in our theoretical analysis. Therefore, the measured signal as a function of the tip–sample distance can be used to determine precisely the physical parameters. We implemented a tip–sample approaching mode using a nanopositioning device with an accurate encoder. In this mode, EMP records the resonant frequency (f_r) and quality factor (Q) while the tip approaches the sample with an accurate step-size measurment (with accuracy of about a few angstroms). Based on our quasi-static theroetical model, discussed later, all necessary parameters can be obtained by fitting the approaching curve. Since there are hundreds to thousands of data points in a curve, the fitted results are more reliable than the one-point measurement result in the simple contact mode, where the dielectric constant and loss tangent are calculated by only one set of f_r and Q measured at the contant point. The fidelity of approach-curve data acquisition relies on accurate measurement of the position of the tip at each approaching step. However, the ordinary PZT actuator has hysteretic and nonlinear behavior, making it very hard to know the true position of the tip with high precision without an accurate encoder. We adopted a nanopositioning device, manufactured by PI, with a capacitance gauge to obtain the tip–sample approaching curve. The device has a closed-loop resolution of 0.05 nm and linearity of 0.03%. Figure 5 shows experimental data of the tip–sample approaching curve fitted to the theoretical result, illustrating the power of this method. Precise control of tip–sample approaching also provides a way to avoid damaging the fine tip, since we take the measurement when the tip is in atomic force contact range.

V. QUANTITATIVE MICROSCOPY

A quantitative microscopy should be based on the theoretical relations that predict the specific function relationship between the tip radius (R_0), the tip–sample distance (g), and physical parameters, such as real and imaginary dielectric constants. Some overall geometric constants can be determined by proper calibration.

A. Equivalent Circuit Analysis of Scanning Evanescent Microwave Probe

The resonator detector (here, specifically for an ideal quarter-wave resonator) can be analyzed using an equivalent lumped series resonant circuit, as shown in Figure 6 with effective capacitance C, inductance L, and resistance R:

$$C = \frac{16\varepsilon_r \varepsilon_0}{\pi \ln(b/d)} l$$

$$L = \frac{\mu_r \mu_0}{4\pi} l \ln(b/a)$$

$$R = \frac{1}{4\pi\sigma\delta}(\frac{1}{b}+\frac{1}{a})l \qquad (1)$$

where $l \approx \lambda/4$ is the effective cavity length, $\delta = (2/\mu_r\mu_0\omega\sigma)^{1/2}$ is the skin depth of the cavity metal, σ is the conductivity of cavity material, b and a are the radii of center and outer conductors, respectively, ε_0 and μ_0 are the permittivity and permeability of free space, respectively, ε_r and μ_r are the relative permittivity and permeability of the cavity material, respectively, and ω is the microwave frequency. In Figure 6, M_1, M_2, L_1, and L_2 are the mutual and self-inductance of the input- and output-coupling loop, and $C_{\text{tip–sample}}$ is the capacitance introduced by the tip–sample interaction.

a) Approaching curve fitting on LAO

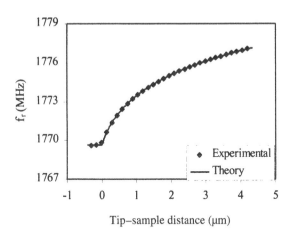

Tip–sample distance (μm)

b) Approaching curve fitting on MgO

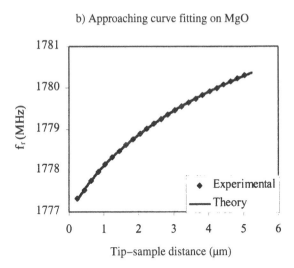

Tip–sample distance (μm)

Figure 5 Approach curve fitting on different samples.

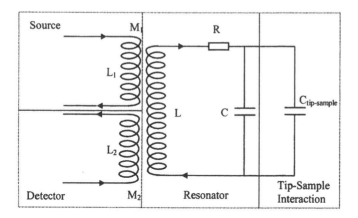

Figure 6   Equivalent lumped series resonant circuit for SEMM

If we neglect the effect of $C_{\text{tip-sample}}$, the unloaded (Q_u) and loaded (Q_l) quality factors of the resonant system are given by:

$$Q_u = \frac{1}{\omega_r CR}$$

$$Q_l = \frac{1}{\omega_r CR} \frac{1}{1+\beta_1+\beta_2} \tag{2}$$

where

$$\omega_r = 2\pi f_r = 1/\sqrt{LC} \tag{3}$$

f_r is the resonant frequency, and β_1, β_2 are the input and output-coupling coefficient, respectively. Generally, $L_1 \ll M_1$ and $L_2 \ll M_2$, so β_1 and β_2 can be expressed approximately as

$$\beta_i = \frac{R_i}{R_c} \quad i = 1, 2 \tag{4}$$

where R_1 and R_2 is the input and output resistance, respectively, and R_c is the characteristic resistance of the resonator.

$$R_i = \frac{(\omega M_i)^2}{R} \quad i = 1, 2$$

$$R_c = \sqrt{\frac{L}{C}} \tag{5}$$

β_l and β_2 can also be obtained by measuring the S parameters of the resonator.

To allow quantitative calculation of the cavity response to a sample with a certain dielectric constant, a detailed knowledge of the electric and magnetic fields in the probe region is necessary. The most general approach is to apply an exact finite element calculation of the electric and magnetic fields for a time-varying three-dimensional region. This is difficult and time consuming, particularly for the tip–sample geometry shown in Figure 7. Since the tip is sharply curved, a sharply varying mesh size has to be implemented. For routine applications, analytic solutions are most desirable for determining the cavity response as a function of tip geometry, tip–sample distance, and physical parameters. Since the spatial extent of the region of the sample–tip interaction is much less than the wavelength of the microwave radiation used to probe the sample ($\lambda \sim 28$ cm at 1 GHz, ~ 14 cm at 2 GHz), the quasi-static approximation can be used; i.e., the wave nature of the electric and magnetic fields can be ignored. This allows the relatively easy solution of the electric field inside the dielectric sample. We outline the calculation of the relation between the complex dielectric constant and SEMP signals as follows.

For the first-order effect, the loss can be modeled by assuming a complex value for $C_{tip\text{-}sample}$:

$$C_{tip-sample} = C_r + iC_i \tag{6}$$

where C_r and C_i are, respectively, the real and imaginary parts of the tip–sample capacitance. To relate the measured signal to physical parameters, perturbation theory[28] for microwave resonators is used to calculate the fr and Q shifts caused by the sample:

$$\frac{\Delta f}{f_0} = -\frac{\int_v (\Delta\varepsilon \vec{E}_1 \bullet \vec{E}_0 + \Delta\mu \vec{H}_1 \bullet \vec{H}_0) dv}{\int_v (\varepsilon_0 E_0^2 + \mu_0 H_0^2) dv} \tag{7}$$

$$\Delta\left(\frac{1}{Q}\right) = \frac{\int_v (\Delta\varepsilon'' \vec{E}_1 \bullet \vec{E}_0 + \Delta\mu'' \vec{H}_1 \bullet \vec{H}_0) dv}{\int_v (\varepsilon_0 E_0^2 + \mu_0 H_0^2) dv} \tag{8}$$

where $\Delta f = f_r - f_0$,

$$\Delta\left(\frac{1}{Q}\right) = \frac{1}{Q} - \frac{1}{Q_0}$$

and f_0 and Q_0 are the resonant frequency and quality factor of the resonator when there is no sample near the tip, respectively. $\vec{E}_0$, $\vec{F}_0$ and $\vec{E}_1$, $\vec{F}_1$ refer to the electric and magnetic field before and after the perturbation, respectively. $\varepsilon = \varepsilon' + j\varepsilon''$

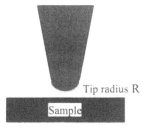

Tip radius R

Sample

Figure 7 Tip–sample geometry of an EMP.

and $\mu = \mu' + j\mu''$ are complex dielectric constants and magnetic permeability of the samples(13). This is equivalent to, $C_{\text{tip-sample}} << C$ (cavity capacitance):

$$\frac{\Delta f}{f_0} = -\frac{C_r}{2C} \tag{9}$$

$$\Delta\left(\frac{1}{Q}\right) = -\left(\frac{1}{Q_0} + \frac{2C_i}{C_r}\right)\frac{\Delta f}{f_0} \tag{10}$$

B. MODELING OF THE CAVITY RESPONSE FOR BULK DIELECTRIC MATERIALS

The complex dielectric constant measured can be determined by an image charge approach. By modeling the redistribution of charge when the sample is brought into the proximity of the sample, the complex impedance of the sample for a given tip–sample geometry can be determined with the measured cavity response (which is described later). Since the tip geometry will vary appreciably among different tips, we require a model with an adjustable parameter to describe the tip. Since the region close to the tip predominately determines the sample response, we can model the tip as a metal sphere of radius R_0. Figure 8 illustrates the infinite series of image charges used to determine the tip–sample impedance for thick samples. For dielectric samples, the dielectric constant is largely real or the loss tangent (tan δ) is small:

$$\varepsilon = \varepsilon_0(\varepsilon_r + i\varepsilon_i)$$

$$\tan\delta = \frac{\varepsilon_i}{\varepsilon_r} < 0.1 \tag{11}$$

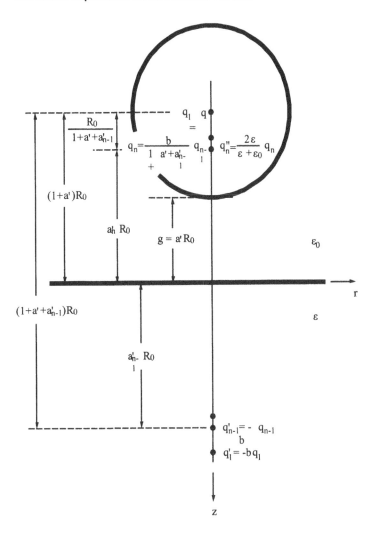

Figure 8 Image charge distribution for a thick sample in contact with the tip; the q_n series represents the charge redistribution on the tip, the q'_n series represents the effect of polarization of the dielectric sample, and q''_n is the effective value of ρ_n inside the sample.

where ε_r and ε_i are the real and imaginary parts of the dielectric constant of the sample, respectively. Therefore, the real portion of the tip–sample capacitance, C_r, can be calculated directly, and the imaginary portion of the tip–sample capacitance, C_i, can be calculated by simple perturbation theory. Using the method of image charge, we find that the tip–sample capacitance is given by

$$C_r = 4\pi\varepsilon_0 R_0 \sum_{n=1}^{\infty} \frac{bt_n}{a_1 + a_n} \tag{12}$$

where t_n and a_n have the following iterative relationships:

$$a_n = 1 + a' - \frac{1}{1 + a' + a_{n-1}} \tag{13}$$

$$t_n = \frac{bt_{n-1}}{1 + a' + a_{n-1}} \tag{14}$$

with $a_1 = 1 + a'$, $t_1 = 1$,

$$b = \frac{\varepsilon - \varepsilon_0}{\varepsilon + \varepsilon_0} \quad \text{and} \quad a' = \frac{g}{R_0}$$

where ε is the dielectric constant of the sample, ε_0 is the permittivity of free space, g is the tip–sample separation, and R is the tip radius.
    This simplifies to

$$C_r = -4\pi\varepsilon_0 R_0 \left[\frac{ln(1-b)}{b} + 1 \right] \tag{15}$$

as the tip–sample gap approaches zero.
    Since the dielectric constant of dielectric materials is primarily real, the loss tangent (tan δ) of dielectric materials can be given by

$$\tan \delta \propto \frac{C_i}{C_r} \tag{16}$$

Finally, we combine Eqs. (10)–(14) with Eqs. (7) and (8) and obtain the relationship between f_r, Q, ε, and tan δ:

$$\frac{\Delta f}{f_0} = \begin{cases} -A\sum_{n=1}^{\infty} \dfrac{bt_n}{a_1 + a_n} & g > 0 \\ A\left[\dfrac{ln(1-b)}{b} + 1 \right] & g = 0 \end{cases} \tag{17}$$

$$\Delta\left(\frac{1}{Q}\right) = -(B_Q + B_Q' \tan \delta)\frac{\Delta f}{f_0}$$

(18)

where $A = 2\pi\varepsilon_0 R_0/C$ is a constant determined by the geometry of the tip-resonator assembly, and

$$\Delta\left(\frac{1}{Q}\right) = \frac{1}{Q} - \frac{1}{Q_0}$$

(18a)

Q and Q_0 are the quality factor of the resonator, with or without sample present near the tip, respectively.

Figure 5 in shows measured approaching curves on two different single crystals fitted to theoretical results. The agreements between theory and experimental data are excellent (with a fitting error of less than 10^{-6}). Using this method, we can routinely obtain a precise value of ε. Figure 9 shows a comparison between the experimentally determined dielectric constant of various samples and theoretical results (solid line) plotted as the change in resonant frequency vs. dielectric constant.

In principle, we can obtain a precise measure of loss tangent following the same method. However, more complex procedures are required. Since the extra current required to support the charge redistribution induces additional resonator loss (mainly from the tip), even with lossless sample, a coefficient B_Q is introduced in Eq. (18). B_Q is a constant in the zeroth-order approximation and equal to $1/Q_0$ in an idealized case, i.e., an isolated sphere tip. Experimentally B_Q as a constant is a good approximation if the tip radius is very small (Figure 10).

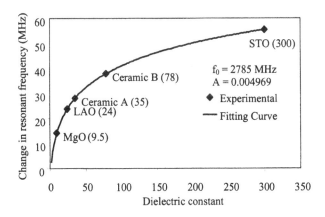

Figure 9 Experimental data and theoretical fitting for the change in resonant frequency as a function of dielectric constant

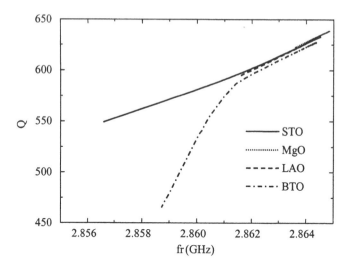

Figure 10 F_r–Q curve of different samples measured with a tip radius smaller than 5 μm.

However, in reality B_Q is a frequency-dependent parameter that varies with the details of the tip and resonator geometry. B_Q needs to be subtracted in order to obtain an accurate measurement of the loss tangent. This has been a major problem in the experimental determination of the loss tangent to a level below 10^{-2}.

Figure 10 shows the experimentally measured Q as function of f_r of different single-crystal samples. Since the loss tangent of MgO, STO, and LAO are all lower than 10^{-3}, which is the detection limit of the current EMP system, these curves are very close to each other and close to a straight line, indicating a near constant B_Q parameter. In this case, we have used a very small tip for the measurement. However, if we use a tip with a larger tip radius to increase the sensitivity of the EMP in order to measure the tangent loss as low as 10^{-3}, the higher-order effects become significant (i.e., B_Q is a complex function of frequency, which has been proven in experiments). The simplest way to eliminate this higher-order background effect (or frequency-dependent B_Q) is to measure the f_r–Q curve experimentally on a known low-loss-tangent sample. We used a calibration sample with loss tangent lower than our EMP sensitivity value to obtain the f_r–Q curve. In addition, to cover a larger f_r range for the requirement of variety of samples, the dielectric constant of the calibration standard sample should be as large as possible (e.g, we selected STO). Considering all of these issues, we have the following formula:

$$\frac{1}{Q} - \frac{1}{Q_{STO}} = \Delta\left(\frac{1}{Q}\right) - \Delta\left(\frac{1}{Q}\right)_{STO} = -B_Q' \tan\delta\frac{\Delta f}{f_0} \tag{19}$$

where Q is the measured quality factor when there's contact with the sample, and Q_{STO} is the quality factor with the tip-approaching sample STO at resonant frequency. Q_{STO} at any specific resonant frequency can easily be calculated by interpulating the f_r–Q curve. Using this method, the complex effect of B_Q coefficient can be totally substracted from the data. The experimental data for several single crystal samples after subtraction of B_Q are shown in Figure 11.

In Eqs. (18) and (19), B_Q' is another constant that needs to be calibrated by a sample with known loss tangent. For BTO—a well-known material with large loss tangent—the curve in Figure 11 shows an obvious difference from other samples. We use this difference to calibrate the parameter B_Q'. The Q changes for MgO and LAO are very small relative to STO because their low loss tangent values ($<10^{-3}$) are similar to STO, while the Q change for BTO is quite different due to its high loss tangent. Given the loss tangent of BTO, the B_Q' parameter can be calibrated through the Q change at the resonant frequency, where the tip–sample gap is zero (the highest value on the Q curve of BTO in

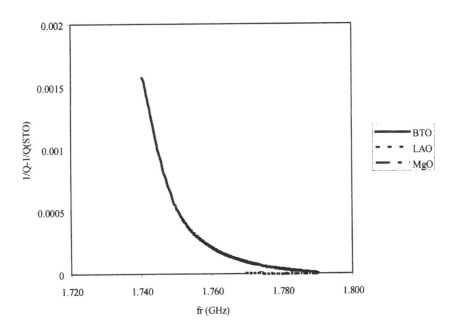

Figure 11 $1/Q$ change as a function of the resonant frequency of different samples. The background Q measured by approaching STO is subtracted from the data.

Figure 11). The result yields $B'_Q = 0.14$, assuming BTO tan $\delta = 0.4$, which has been proven with other techniques. Using this systematic method, we can precisely determine the tangent loss from the measurement of Q. The sensitivity is limited by the systematic error in determining the complex background B_Q parameter. To this end, we have determined this error and our sensitivity to be under 10^{-3}.

In the foregoing discussion, we assume B'_Q is constant. A detailed analysis shows that B'_Q is also a complex function of the sample dielectric constant (ε) and tip–sample gap (g). For example, we plot in Figure 12 the theoretical relationship between B'_Q and ε for the case of contact mode ($g = 0$).

Sometimes, it is useful to have an explicit formula for the electric field solved by the image change method, as we show in the following. The perturbed electric field inside the sample is

$$\vec{E}_1(\varepsilon,d) = \frac{q}{2\pi(\varepsilon+\varepsilon_0)} \sum_{n=1}^{\infty} t_n \frac{r\vec{e}_r+(z+a_n)\vec{e}_z}{\left[r^2+(z+a_n)^2\right]^{3/2}} \tag{20}$$

where $q = 4\pi\varepsilon_0 R_0 V_0$, V_0 is the tip voltage, and $\vec{e}_r$ and $\vec{e}_z$ are the unit vectors along the directions of the cylindrical coordinates r and z, respectively.

C. Modeling for Dielectric Film on Highly Conductive Substrates

Let's consider a dielectric thin film sandwiched between the tip and the conductive substrate. In the extreme case, when the dielectric constant of thin film is 1, the model becomes the simple case of a tip on the metal with an air gap equal to the

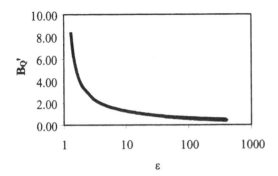

Figure 12 B'_Q as a function of ε in contact mode.

film thickness (t). The tip–sample capacitance ($C_{tip-sample}$) can be calculated by the formula in the bulk case with $\varepsilon = +\infty$ (see the section about conductive material for a more detailed analysis). When the dielectric constant of the thin film increases, $C_{tip-sample}$ will increase accordingly. On the other hand, $C_{tip-sample}$ will increase in a similar way if we keep the dielectric constant unchanged and decrease the tip–sample gap. This is very similar to a parallel plate capacitor with capacitance $C \propto \varepsilon/t$ (t is the distance between the two plates and ε is the dielectric constant of the filling material between the plates). Also, when $t << R_0$ (tip radius), the electric field lines will be concentrated right under the end of the tip, which is very similar to the field distribution of a parallel plate capacitor, assuming the tip end is approximately a flat surface. This implies that if we replace the film thickness by a reduced thickness defined as $t' = t/\varepsilon$, we can treat the film as an air gap of t' between tip and substrate. As a consequence, the analytical results from the theoretical model based on image charge theory can be used. Therefore, we have the following equations for this situation:

$$\frac{\Delta f}{f_0} = -A \sum_{n=1}^{\infty} \frac{b_n}{a_1 + a_n} \tag{21a}$$

where

$$a_n = 1 + a' - \frac{1}{1 + a' + a_{n-1}} \tag{21b}$$

$$b_n = \frac{b_{n-1}}{1 + a' + a_{n-1}} \tag{21c}$$

and $a_1 = 1 + a'$, $b_1 = 1$, $a' = t'/R_0'$ and $t' = t/\varepsilon$ is the reduced film thickness. This formula is basically the formula for bulk conducting materials with replacement of t by effective thickness t'. The curve in Figure 13 shows the resonant frequency as a function of the reduced film thickness, together with experimental data, indicating very good agreement.

Although Figure 13 indicates that Eq. (21a) can be used as an approximate formula to calculate dielectric films on the conducting substrate, the accuracy of the first-order approximation (or the error) needs to be further considered. We analyzed this error from the difference between more a realistic finite element method calculation and a result given by Eq. (21a). Figure 14(a) shows that the overall agreement between them is very good, and Figure 14b gives the small difference between them as a function of film thickness and dielectric constant. The difference is approximately proportional to the product of ε and t/R_0 when both of them are small. Considering this term as a correction to the analytical formula, eq. (21a) becomes:

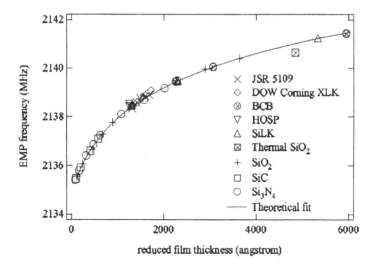

Figure 13 EMP data plots vs. reduced thickness t', showing all data points collapsed onto a single curve. The solid line is the fitting result using Eq. (21); the markers are the experimental data on the nine types of low-K films listed in the figure. All films are grown on the same highly doped silicon substrate (dopent level $> 10^{19}$ cm^{-3}).

$$\frac{\Delta f}{f_0} = -\left(A\sum_{n=1}^{\infty} \frac{t_n}{a_1 + a_n} + B \cdot t \cdot \varepsilon \right) \qquad (22)$$

B is another parameter that needs to be calibrated for a given system. For example, from Figure 14b we can derive B in our system to be 1.16×10^{-4}. This approach provides us a much-simplified method to measure dielectric thin films on highly conductive substrates or substrates with a high dielectric constant. One needs only one standard film material with known dielectric constant and various thicknesses to fit the parameters in Eq. (22).

D. Modeling for Thin Films on a Dielectric Substrate

For the situation of a dielectric film on a dielectric substrate, the boundary condition for the image charge problem becomes more complex. Figure 15 illustrates the steps to calculate $C_{tip-sample}$. Strictly speaking, the image charge approach will not be applicable to thin films due to the divergence of the image charges, as shown in Figure 15. However, if we can model the contribution of the substrate to the reaction on the tip properly, the image charge approach is still a good approximation. We expect that all films can be considered as bulk samples if the

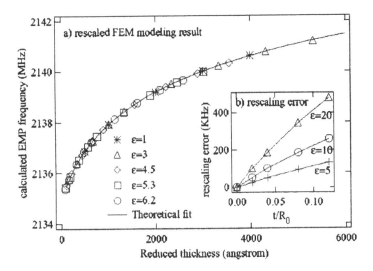

Figure 14 (a) FEM modeling result plotted by reduced thickness t'. The solid line is the fitting result from Eq. (21)(b). The rescaling error as a function of ε and t/R_0. It shows that the error is proportional to the product of ε and t/R_0.

tip is sharp enough, since the penetration depth of the field is only about R. The contribution from the substrate will decrease with increases in film thickness and dielectric constant.

We model this contribution by replacing the effect of the reaction from the complicated image charges with an effective charge with the following formula:

$$b_{eff} = b_{20} + \left(b_{10} - b_{20}\right)\exp\left(-D\frac{a}{1-b_{20}}\right) \tag{23}$$

where

$$b_{20} = \frac{\varepsilon_2 - \varepsilon_1}{\varepsilon_2 + \varepsilon_1}, \quad b_{10} = \frac{\varepsilon_1 - \varepsilon_0}{\varepsilon_1 + \varepsilon_0} \tag{23a}$$

and ε_2 and ε_1 are the dielectric constants of the film and substrate, respectively. $a = d/R_0$, and d is the thickness of the film. This format reproduces the thin- and thick-film limits for the signal. The coefficient D can be calibrated from a standard sample. One of our published papers reports that $D = 0.18$, which was obtained by calibrating against interdigitated electrode measurements at the same frequency on $SrTiO_3$ thin film. (Note that this value could be changed when using different resonators.) Alternatively, the coeffcient D can be obtained through a

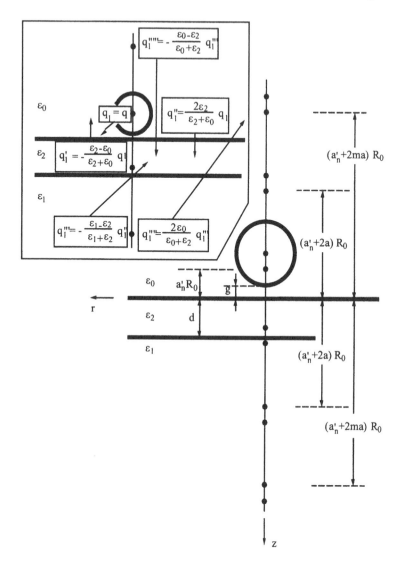

Figure 15   Image charge distribution for a thick sample in contact with the tip.

theoretical simulation. It is conceivable that the coeffcient D takes different values in different ranges of dielectric constant of the thin films.

Following a process similar to the previous derivation, we have:

$$C_r = 4\pi\varepsilon_0 R_0 \sum_{n=1}^{\infty} \sum_{m=0}^{\infty} b_{\mathit{eff}}^{n-1} b_{21}^m b_{10}^m \left(\frac{b_{20}}{n+1+2mna} - \frac{b_{21}}{n+1+2(m+1)na} \right) \tag{24}$$

$$C_i = 4\pi\varepsilon_0 R_0 \sum_{n=1}^{\infty} \sum_{m=0}^{\infty} b_{\mathit{eff}}^{n-1} b_{21}^m b_{10}^m \left(\begin{array}{c} \tan\delta_2 \left(\dfrac{1}{n+1+2mna} - \dfrac{1}{(n+1+2(m+1)na)} \right) \\ + \dfrac{2\varepsilon_1\varepsilon_2 \tan\delta_1}{(\varepsilon_2+\varepsilon_1)(\varepsilon_2+\varepsilon_0)} \dfrac{1}{(n+1+2(m+1)na)} \end{array} \right) \tag{25}$$

where

$$b_{21} = \frac{\varepsilon_2 - \varepsilon_1}{\varepsilon_2 + \varepsilon_1}$$

and $\tan\delta_2$ and $\tan\delta_1$ are the loss tangents of the film and substrate, respectively. Alternatively, we have:

$$\frac{\Delta f}{f_0} = -A \sum_{n=1}^{\infty} \sum_{m=0}^{\infty} b_{\mathit{eff}}^{n-1} b_{21}^m b_{10}^m \left(\frac{b_{20}}{n+1+2mna} - \frac{b_{21}}{n+1+2(m+1)na} \right) \tag{26}$$

$$\Delta\left(\frac{1}{Q}\right) = A \sum_{n=1}^{\infty} \sum_{m=0}^{\infty} b_{\mathit{eff}}^{n-1} b_{21}^m b_{10}^m \left(\begin{array}{c} \tan\delta_2 \left(\dfrac{1}{n+1+2mna} - \dfrac{1}{(n+1+2(m+1)na)} \right) \\ + \dfrac{2\varepsilon_1\varepsilon_2 \tan\delta_1}{(\varepsilon_2+\varepsilon_1)(\varepsilon_2+\varepsilon_0)} \dfrac{b_{21}}{(n+1+2(m+1)na)} \end{array} \right) - B\left(\frac{\Delta f}{f_0} \right) \tag{27}$$

Figure 16 shows the EMP-measurement result of a series of BST films.

E. Modeling for Nonlinear Dielectric Materials

The component of the electric displacement D perpendicular to the sample surface is given by

$$D_3 = P_3 + \varepsilon_{33}(E_l + E_m) + \frac{1}{2}\varepsilon_{333}(E_l + E_m)^2 + \frac{1}{6}\varepsilon_{3333}(E_l + E_m)^3 + \cdots \tag{28}$$

where P_3 is the spontaneous polarization, ε_{ij}, ε_{ijk}, ε_{ijkl}, $\cdots$ are the second-order (linear) and higher-order (nonlinear) dielectric constants, respectively. Since the field distribution is known for a fixed tip–sample separation, we can estimate the nonlinear dielectric constant from the change in resonance frequency with

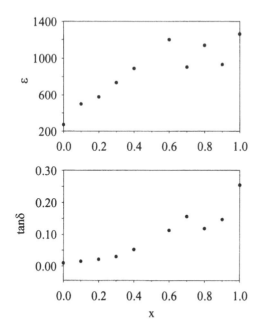

Figure 16 EMP-measured ε and $\tan \delta$ as a function of x for a thin-film composition of $(Ba_xSr_{1-x}TiO_3)$ with a thickness of 300 nm on a MgO substrate deposited by laser ablation.

applied voltage. For tip–sample separations much less than R, the signal comes mainly from a small region under the tip where the electric fields (both the microwave electric field E_m and the low-frequency-bias electric field E_l) are largely perpendicular to the sample surface. Therefore, only the electric field perpendicular to the surface needs to be considered.

From Eq. (28), the effective dielectric constant with respect to E_m can be expressed as a function of E_l:

$$\varepsilon_{33}(E_l) = \frac{\partial D_3}{\partial E_m} = \varepsilon_{33} + \varepsilon_{333}(E_l + E_m) + \tfrac{1}{2}\varepsilon_{3333}(E_l + E_m)^2 + \dots \tag{29}$$

and the corresponding dielectric constant change caused by E_l is

$$\Delta\varepsilon = \varepsilon_{333}E_l + \tfrac{1}{2}\varepsilon_{3333}E_l^2 + \dots \tag{30}$$

The change in f_r for a given applied electric field E_l is related to the change in the energy stored in the cavity. Since the electric field for a given dielectric constant is known and the change in the dielectric constant is small, this can be calculated by integrating over the sample:

$$\frac{df_r}{f_r} = -\frac{\int_{V_s} \Delta\varepsilon E_m^2 dV}{\int_{V_t} (\varepsilon E_0^2 + \mu H_0^2)dV} = -\frac{\int_{V_s} \left(\varepsilon_{333} E_l + \frac{1}{2}\varepsilon_{3333} E_l^2 + ...\right) E_m^2 dV}{\int_{V_t} (\varepsilon E_0^2 + \mu H_0^2)dV} + ...$$ (31)

where V_s is the volume of the sample containing electric field, H_l is the microwave magnetic field, and V_t is the total volume containing electric and magnetic fields, possibly with dielectric filling of dielectric constant ε. E_m is given by Eq. (20). The application of a bias field requires a second electrode, located at the bottom of the substrate. If the distance from the bottom electrode to the tip is much larger than the tip–sample distance and the tip radius, Eq. (20) should also hold for E_l. The upper portion can be calculated by integrating the resulting expression. The lower portion of the integral can be calibrated by measuring the dependence of f_r versus the tip–sample separation for a bulk sample of known dielectric constant. If the tip–sample separation is zero, the formula can be approximated as

$$C_r(V) = C_r(V = 0) + 4\pi\varepsilon_0 R \frac{1}{32} \frac{A}{\varepsilon_{33}} \frac{V}{R} \frac{\varepsilon_{33} + \varepsilon_0}{2\varepsilon_0} \varepsilon_{333}$$ (32)

where V is the low-frequency voltage applied to the tip. This calculation can be generalized in a straightforward fashion to consider the effects of other nonlinear coefficients and thin films (30).

Figure 17 shows images of topography and ε_{333} for a periodically poled single-crystal LiNbO$_3$ wafer. The crystal is a 1-cm × 1-cm single-crystal substrate, poled by periodic variation of dopant concentration. The poling direction is perpendicular to the plane of the substrate. The dielectric-constant image is essentially featureless, with the exception of small variations in dielectric constant due to the variation in dopant concentration or piezoelectric strain effect. The nonlinear image is constructed by measuring the first harmonic of the variation in the output of the phase detector using a lock-in amplifier. Since ε_{ijk} reverses when the polarization switches, the output of the lock-in switches sign when the domain direction switches. The value (-2.4×10^{-19} F/V) is within 20% of bulk measurements. The nonlinear image clearly shows the alternating domains.

In principle, changes in the polarization direction do not induce changes in the linear dielectric constant (for most crystal symmetries). If the variations in the dielectric constant are sufficiently small, the frequency shift may be used to regulate the tip–sample separation while ε_{333} is simultaneously imaged (Figure 3b). Since ε_{ijk} is a third-rank tensor, it reverses sign when the polarization switches, providing an image of the domain structure. For the case of the periodically poled LiNbO$_3$ single crystal imaged in this work, small variations in the

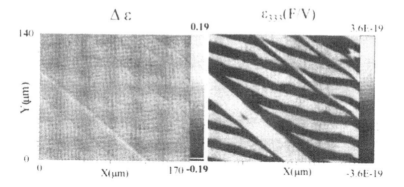

Figure 17   Simultaneous imaging of the linear and nonlinear dielectric constants of LiNbO₃.

linear dielectric constant occur, due presumably to piezoelectric-induced strain at the boundary or other process-related defects. However, these changes are in the range of 10^{-2} in the dielectric constant and will induce negligible variations in the tip–sample separation on the order of nanometers.

VI.   CONDUCTIVE MATERIALS

A.   Low Conductivity

Previously derived Eqs. (12)–(14) could also be used for conductive materials by simply replacing the dielectric constant with the effective dielectric constant of a conductive material at a given frequency f:

$$\varepsilon = \varepsilon_0 \left(\varepsilon_r + i \frac{\sigma}{2\pi\varepsilon_0 f} \right) \tag{33}$$

where ε_r is the real part of the permittivity (relative value) and σ is the conductivity. The quasi-static approximation should be applicable when the wavelength inside the material is much larger than the tip–sample geometry. For $R_0 \sim 1$ μm and $\lambda \approx 14$ cm,

$$\varepsilon_{max} \approx \varepsilon_0 \left(\frac{\lambda}{R_0} \right)^2 \approx 2 \times 10^{10} \varepsilon_0$$

or

$$\sigma_{max} \approx 2\pi\varepsilon_0 f \varepsilon_{max} \approx 2 \times 10^9 \frac{1}{\Omega \cdot m} \tag{34}$$

For $\sigma \ll \sigma_{max}$, the quasi-static approximation remains valid. $C_{tip-sample}$ can be calculated by the method of image charges. Each image charge will be out of phase with the driving voltage. By calculating the charge (and phase shift) accumulated on the tip, driven by a voltage V at frequency f, one can calculate a complex capacitance. For moderate tip–sample separations, it is primarily a capacitance with a smaller real component.

By writing $b = b_r + ib_i = |b|e^{i\phi}$, we can separate real and complex capacitances:

$$C_r = 4\pi\varepsilon_0 R_0 \sum_{n=1}^{\infty} \frac{|b|^n \cos(n\phi) g_n}{a_1 + a_n} \tag{35}$$

$$C_i = 4\pi\varepsilon_0 R_0 \sum_{n=1}^{\infty} \frac{|b|^n \sin(n\phi) g_n}{a_1 + a_n} \tag{36}$$

where $g_1 = 1$ and g_n is given by

$$g_n = \frac{g_{n-1}}{1 + a' + a_{n-1}} \tag{37}$$

Figure 18 illustrates f_r and $\Delta(1/Q)$ as a function of conductivity. The curve peaks approximately where the imaginary and real components of ε become equal. It is clear from the plot that f_r and $\Delta(1/Q)$ both change as conductivity varies in the low-conductivity range if the tip–sample distance is fixed. This relation can be plotted in a different way so that the resonant frequency is fixed while the tip–sample distance is allowed to vary. Figure 19 shows $\Delta(1/Q)$ as a function of the conductivity of semiconducting samples measured at a fixed resonant frequency. This method can provide very reproducible measurement results, since the resonant frequency can be kept constant (or normalized) for a long period of time whereas the tip–sample distance is very difficult to be kept constant.

B. High Conductivity

For $\sigma \geq \sigma_{max}$, both the electric and the magnetic fields should be considered. For the electric field, the image charge method is still effective. Eq. (12) can be used here to calculate the real portion of $C_{tip-sample}$ by using $b = 1$. In this limit, the formula can also be reduced to a sum of hyperbolic sines:

$$C_r = 4\pi\varepsilon_0 R_0 \sinh \alpha \sum_{n=2}^{\infty} \frac{1}{\sinh n\alpha} \tag{38}$$

where $\alpha = \cosh^{-1}(1 + a')$.

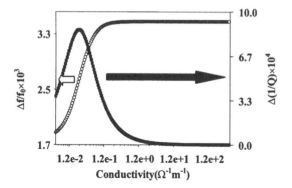

Figure 18 Complex capacitance vs. sample conductivity.

To calculate the contribution from the magnetic field, knowledge of the magnetic field distribution around the tip-sample is necessary. Assuming the tip length protuding out of the resonator is much longer than the tip–sample distance, the magnetic and electric fields at the surface of the conductor are given by taking the semi-infinite wire approximation:

$$\vec{E}_s(r) = \frac{R_0}{2\pi\varepsilon_0} \sum_{n=1}^{\infty} \frac{a'_n q_n}{[r^2 + (a'_n R_0)^2]^{3/2}} \vec{e}_z \tag{39}$$

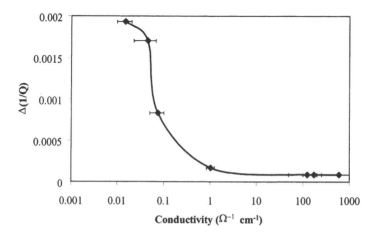

Figure 19 Conductivity of wafers vs. EMP-measured Q at a fixed frequency.

$$\vec{H}_s(r) = -i\frac{\omega}{2\pi r}\sum_{n=1}^{\infty} q_n \frac{[r^2 + (a_n' R_0)^2]^{\frac{1}{2}} - a_n' R_0}{[r^2 + (a_n' R_0)^2]^{\frac{1}{2}}} \vec{e}_\phi \tag{40}$$

The electrical and magnetic contributions can be calculated by substituting the field expressions (39) and (40) into Eqs. (7) and (8). The detailed calibration and calculation in this direction remain to be carried out in the future.

Since the tip–sample capacitance is independent of the conductivity for good metals, we can use $C_{tip\text{-}sample}$ as a distance measure and control. Figure 20 illustrates the tip–sample distance approach curve for a conducting thick silver layer. From the curves, a frequency $f_{reference}$ can be chosen to correspond to some tip–sample separation. We can then regulate the tip–sample separation to maintain the resonance frequency at $f_{references}$. This can be accomplished digitally through the use of the digital signal processor in the system. We have also used an analog method using a phase-locked loop. Figure 21 illustrates the measurement results of various samples of different conductivity measured at a fixed frequency (and therefore a fixed tip–sample distance).

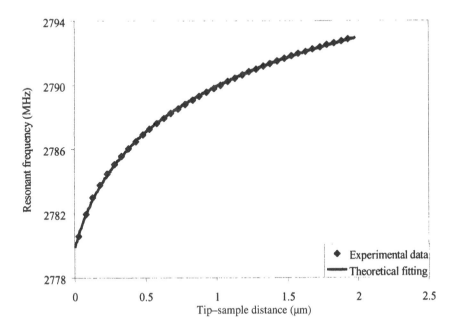

Figure 20 Approaching curve fitting on silver; fitting result; $f_0 = 2709.803$, $A = 0.00369$, tip radius $R_0 = 60$ μm.

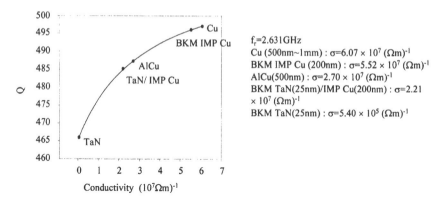

Figure 21   EMP-measured Q vs. conductivity of metal films on silicon.

VII. CONCLUSION

In this chapter, we have discussed experimental and theoretical methods using EMP to obtain spatially resolved and quantitative electrical impedance measurements. The technique has been routinely applied to combinatorial high-throughput screening of material chips (31–33). The experimental details and data on different material chips are discussed in Chapters 4 and 8 in this book and were be repeated here. Future development effort is being made to expand the capability to electron spin resonance and ferromagnetic resonance measurements. Broad applications of this technique in the semiconductor and biotech industries are also being actively pursued and will be discussed elsewhere (34).

REFERENCES

1. Synge, E.H. Philos. Mag. 1928, 6, 356.
2. Frait, Z.; Czechoslov, Z. J. Phys. 1959, 9, 403.
3. Soohoo, R.F. J. Appl. Phys. 1962, 33, 1276.
4. Ash, E. A.; Nicholls, G. Nature. 1972, 237, 510.
5. Binnig, G.; Rohrer, H.; Gerber, C.; Weiber, E. Phys. Rev. Lett. 1982, 49, 57.
6a. Binning, G.; Quate, C.F.; Gerber, C. Phys. Rev. Lett. 1986, 56, 930.
6b. Wiesendanger, R. Scanning Probe Microscopy and Spectroscopy; Cambridge University Press: Cambridge: UK, 1994.
7. Phol, D.W.; Denk, W.; Lanz, M. Appl. Phys. Lett. 1984, 44, 651.
8. Lewis, A.; Isaacson, M.; Murray, A.; Harootunian, A. Biophys. J. 1983, 41, 405a.
9. Grard, C.; Dereux, A. Rep. Prog. Phys. 1997, 59, 657.

10. Fee, M.; Chu, S.; Hänsch, T.W. Optics Commun. 1988, 63, 219.

11. Wei, T.; Xiang, X.-D. Appl. Phys. Lett. 1996, 68, 3506.

12. Gao, C.; Wei, T.; Duewer, F.; Lu, Y.; Xiang, X.-D. Appl. Phys. Lett. 1997, 71, 1872.

13. Gao, C.; Xiang, X.-D. Rev. Sci. Instrum. 1998, 69, 3846.

14. Lu, Y.; Wei, T.; Duewer, F.; Lu, Y.; Ming, N.-B.; Schultz, P.G.; Xiang, X.-D. Science. 1997, 276, 2004.

15. Takeuchi, I.; We, T.; Duewer, F.; Yoo, Y. K.; Xiang, X.-D.; Talyansky, V.; Pa, S. P.; Chen, G. J.; Venkatesan, T. Appl. Phys. Lett. 1997, 71, 2026.

16. Bryant, C. A.; Gunn, J. B. Rev. Sci. Instrum. 1965, 36, 1614.

17. Massey, G. A. Applied Optics. 1984, 23, 658.

18a. Gutmann, R. J.; Borrego, J. M.; Chakrabarti, P.; Wang, M.-S. IEEE MITT-S Digest. 1987, 281.

18b. Wang, M. S.; Borrego, J. M. Materials Evaluation. 1990, 48, 1106.

19. Tabib-Azar, M.; Shoemaker, N. S.; Harris, S. Meas. Sci. Technol. 1993, 4, 583.

20. Keilmann, F.; van der Weide, D. W.; Eickelkamp, T.; Merz, R.; Stockle, D. Optics Communications. 1996, 129, 15.

21. Vlahacos, C. P.; Black, R. C.; Anlage, S. M.; Amar, A.; Wellstood, F. C. Appl. Phys. Lett. 1996, 69, 3272.

22. Golosovsky, M.; Davidov, D. Appl. Phys. Lett. 1996, 68, 1579.

23. Bae, J.; Okamoto, T.; Fujii, T.; Mizuno, K.; Nozokido, T. Appl. Phys. Lett. 1997, 71, 3581.

24. Grober, R. D.; Schoelkopf, R. J.; Prober, D. E. Appl. Phys. Lett. 1997, 70, 1354.

25. Duewer, F.; Gao, C.; Takeuchi, I.; Xiang, X.-D. Appl. Phys. Lett. 1999, 74, 2696.

26. Gao, C.; Duewer, F.; Xiang, X.-D. Appl. Phys. Lett. 1999, 75, 3005.

27. Duewer, F.; Gao, C.; Xiang, X.-D. Rev. Sci. Instrum. 2000, 71, 2414.

28. Pozar, D.M. Microwave Engineering (Addison-Wesley: New York; 1990.

29. Wang, Z.; Kelly, M. A.; Shen, Z. X.; Wang, G.; Xiang, X.-D.; Wetzel, J. T. J. of Appl. Phys. 2002, 92, 808.

30. Gao, C.; Duewer, F.; Lu, Y.; Xiang, X.-D. Appl. Phys. Lett. 1998, 73, 1146.

31. Chang, X.-D.; Gao, C.; Takeuchi, I.; Yoo, Y.; Wang, J.; Schultz, P. G.; Xiang, X.-D.; Sharma, R. P.; Downes, M.; Venkatesan, T. Appl. Phys. Lett. 1998, 72, 2185.

32. Chang, H.; Takeuchi, I.; Xiang, X.-D. Appl. Phys. Lett. 1999, 74, 1165.

33. Yoo, Y. K.; Duewer, F.; Yang, H.; Dong, Y.; Li, J.-W.; Xiang, X.-D. Nature. 2000, 406, 704.

34. www.ariel-tech.com.

12

Combinatorial Computational Chemistry Approach in the Design of New Catalysts and Functional Materials

Rodion V. Belosludov, Seiichi Takami, Momoji Kubo, and Akira Miyamoto
Tohoku University, Sendai, Japan

I. INTRODUCTION

Three decades ago, J. J. Hanak proposed a new approach for materials exploration that he called the "multisample concept." The goal was to make materials discovery processes faster with higher efficiency as compared to the standard methods (1). This idea of producing many different compositions at once appeared to be an elegant art that introduced diversity and parallel processing in experiments. More recently, researchers have begun to use a new idea commonly known as the *combinatorial* approach. At first, this was applied mostly to chemical synthesis and screening of large numbers of organic compounds for drug discovery processes (2). By using the combinatorial approach, researchers have been able to predict and find an optimal composition with the desired characteristics much faster than before. There are many good examples of applications of the combinatorial method in chemistry, including the development of catalytic antibodies (3) and, more recently, the identification of nonpeptide agonists (activators) for different somatostain receptor subtypes (4).

Inorganic materials have many more potential applications than organic materials because they exhibit a wide range of physical/chemical properties, many of which are strongly dependent on compositions. The combinatorial approach has great potential in investigating important inorganic materials, and its utility has been experimentally demonstrated in a variety of materials systems, including superconductors (5), giant magnetoresistance materials (6), and photoluminescent materials (7–9).

Recent rapid advances in computer technology have given rise to numerous opportunities in materials science research. This involves going from the description of a model system by analytical methods to the representation of realistic molecular structures and the associated physical and chemical properties by numerical modeling. Computational chemistry combines all numerical methods based on quantum chemistry (QC), molecular dynamics (MD), Monte Carlo (MC), and molecular mechanics (MM) simulations for the prediction of the structure and electronic and thermodynamics properties of materials. The basic principle of these simulations is an accurate determination of the total energy of an investigated system. In same ways, computational chemistry can be considered "economical" because so many physical properties are related to the total energies. While just one piece of a theoretical tool is necessary to calculate all the physical properties that are related to the total energies, completely different sets of experimental tools are required to measure each type of physical properties of a material. In principle, this represents an enormous advantage of computational chemistry over experimental measurements (10). Simulations are easy to perform, even for very complex systems: their complexity is often no worse than the complexity of the physical description of the systems themselves. As the capabilities of computers increase, simulations of many-body systems will be able to treat more complex physical systems at higher levels of accuracy. Thus, ultimate impact of computational chemistry is enormous for of an extremely wide range of scientific and engineering applications.

The design of functional materials such as high-performance catalysts and electronics/photonics devices nowadays is too complicated to be accomplished without detailed knowledge of the structure, reactivity and electronic properties of the materials. The understanding and modeling of solid surfaces and interfaces at an atomic level can benefit greatly from computational chemistry. Recently, we introduced the concept of a combinatorial approach to computational chemistry for catalyst design (11). In this study, the adsorption energies of NO and water molecules on ion-exchanged ZSM-5 zeolite with various cations were investigated using both the first-principle calculations and molecular dynamics (MD) methods. Based on these results, it was proposed that some cations, such as Ir^{3+}, are new candidates for deNO$_x$ catalysts with high resistance to water (11). This approach can be applied in a cost-effective way to accelerate the process of designing new functional materials using different simulation techniques.

In the present chapter, we review our recent application of combinatorial computational chemistry as well as the integrated computer simulation methods to solid surfaces and interfaces.

II. COMPUTATIONAL DETAILS

The value of simulation can be evaluated based on its ability to rapidly and accurately predict the properties of novel functional materials in a more cost-effective way than is experimentally possible. In parallel with improvements in computer technology, the benefits of simulation can come from improvements in theory, in algorithms, and in strategy. The combinatorial strategy is an attractive method with high potential in various fields of applications, especially in heterogeneous catalysis. For example, the goal in a study of catalysts is often not to completely replace currently used materials, but rather to increase the performance of a given catalytic system by reducing the cost or improving the activity as well as the selectivity. The computational screening of a large number of different constituent elements in targeted compounds can play an important role for quick identification of new multicomponent materials before experiments. This is the main idea of combinatorial computational chemistry (CCC) approach that has been developed by our group. Instead of synthesizing and testing a large number of potential catalysts, it is possible to use the combinatorial computational chemistry approach to screen a large number of candidates, including very expensive elements. Consequently, experiments need to be performed on only a small number of the most promising candidates. Different computational methods and their combinations can be used within the CCC approach, depending on the properties of interest, from the atomic to the macroscopic level. We select the most favorable simulation methods among the well-validated approaches, namely, from the main families of quantum, molecular, analytical, and statistical methods.

The computational methodologies applied in our studies are mainly divided into two types: (a) molecular dynamics and (b) quantum chemical calculations using density-functional theory.

A. Molecular Dynamics

Classical molecular dynamics (MD) method is one of the most popular techniques in computational studies. The strength of this technique lies in the fact that both static and dynamic information of the system under study can be extracted from the calculation (12,13). The molecules in a system can be considered as rigid bodies as well as a collection of atoms or a composition of chemical functional groups. The interaction between particles may be presented by empirical functions, such as a combination of Lennard–Jones and electrostatic interactions, as well as other short-range, hydrogen-bonding or covalent, interactions that result from two or more atoms being brought close together. The relevant potential parameters of these chosen interactions are obtained either from experimental data or more accurately from quantum chemical ab initio calculations on model systems. The dynamics of systems can be solved for a set of coupled differential equations in which the motions of atoms are governed by Newtonian mechanics. This is the basic principle of the classical MD method (12). According to Newtonian mechanics, the equation of motion is

$$m_i \ddot{\mathbf{r}}_i(t) = \mathbf{F}_i(t) = -\frac{\partial U(\mathbf{r}^N)}{\partial \mathbf{r}_i} \tag{1}$$

where $\mathbf{F}_i$ is the force on a particle i caused by the other particles, m_i is the mass, $\mathbf{r}_i$ is the coordinate of particle i, and the dots indicate total time derivatives. The notation $\mathbf{r}^N$ represents the set of vectors that locate the atomic centers of mass, $\mathbf{r}^N = \{\mathbf{r}_1, \mathbf{r}_2, \mathbf{r}_3, \cdots, \mathbf{r}_N\}$ and defines the configuration of a system. U is the potential energy, which can include contributions not only from two-body interactions but also from three-body, four-body, and high-order terms:

$$U(r^N) = \sum_{i<j} U_2(\mathbf{r}_i, \mathbf{r}_j) + \sum_{i<k<j} U_3(\mathbf{r}_i, \mathbf{r}_j, \mathbf{r}_k) + \sum_{i<j<k<j} U_4(\mathbf{r}_i, \mathbf{r}_j, \mathbf{r}_k, \mathbf{r}_l) + \cdots, \tag{2}$$

The equation of motion can be integrated by specifying an appropriate time step Δt, and then the trajectory of each particle can be determined at any time. The choice of Δt is not arbitrary. Since

$$\ddot{\mathbf{r}}_i(t) = \frac{1}{m_i} \mathbf{F}_i(t) \tag{3}$$

the magnitude of Δt can be estimated from

$$\frac{\Delta v}{\Delta t} \approx \frac{F}{m}; \quad \Delta t \approx \frac{m \Delta v}{F} \tag{4}$$

Therefore, the integration time step Δt is strongly dependent on the strength of force. For a very flat potential energy function, a fairly large Δt can be used. However, in most applications, a reasonable Δt is about a few femtoseconds. With the modern computer, the maximum practical number of integration steps is between 10^5 and 10^7. Hence, the total time of a feasible MD simulation is within the picosecond-to-nanosecond range. Thus, in principle, MD calculations are applicable only to the investigation of dynamical phenomena within this time range (12).

One of the most important and useful quantities that can be evaluated from MD calculations is the time correlation function, C_{AB}, which correlates two time-dependent quantities A and B (12,13):

$$C_{AB} = \langle A(t)B(t + t_0) \rangle \tag{5}$$

This important function provides detailed information on the dynamics of a system. The integral of this function is related directly to the macroscopic transport coefficient, and the Fourier transform is related to relevant experimental spectra (12). For example, the single-particle velocity autocorrelation function $v(t)$,

$$v(t) = \langle \mathbf{v}(t).\mathbf{v}(t+t_0) \rangle \tag{6}$$

is related to the macroscopic diffusion coefficient D by

$$D = \frac{1}{3} \int_0^\infty v(t)dt \tag{7}$$

and its Fourier transform gives the spectral density of states, $F(\omega)$. This can be compared with the vibrational density of states obtained from a neutron incoherent inelastic scattering experiment (12) and given by

$$F(\omega) = \int_0^\infty e^{-i\omega t} v(t)dt \tag{8}$$

Many other useful quantities, such as thermal conductivity and viscosity, can be obtained in a similar manner employing relevant Fourier transformations (14).

There are two numerical schemes that are widely used in MD calculations. The Verlet algorithm (15) is derived from the simplest second-order difference equation for the second derivative. It gives the coordinates of all particles (atoms or molecules) at each time step. This algorithm introduces an error of the order of Δt^4 into the integration of the equations of motion. The Coulomb interaction is long-ranged, and it is extremely difficult to compute the Coulomb energy of an ionic system using a direct real-space summation. The Ewald method (13) is a rapidly convergent method for performing Coulomb summations. In this method, the infinite Coulomb summation has been replaced by two summations, one over lattice vectors and the other over reciprocal-lattice vectors, held to each other by the arbitrary constant of the order of unity. The two summations become rapidly convergent in their respective spaces by choosing an appropriate value of this constant.

An epitaxial growth process has been investigated using a new MD code developed by our group (16). This is a modified version of the MD program developed by K. Kawamura (17). Our main modifications deal with the total number of species in the system, which is not fixed but increases with time. Figure 1 shows the model system of the MD simulations for investigation of the epitaxial growth process. It consists of two components: a substrate and a source that emits the molecules. The number of molecules deposited over the substrate surface is increased one by one. The horizontal positions of the emerging mole-cules over the substrates are randomly selected. The molecules are shot to the surface at regular time intervals with definite velocity. Since various kinds of depositing molecules and substrates can be simulated in our MD code, the effect of a particular depositing species and a substrate on the structure of the constructed materials and the interface structures of various heterojunctions can be studied. The temperature of the substrate, the distance between the substrate and the source of emitted molecules, the number of the molecules impinging on the surface, the incident rates of the molecules, and the incident time interval are the variable parameters. Various experimental conditions can be optimized via the new MD program. The temperature is controlled by means of scaling the atom velocities under periodic boundary conditions. The substrate temperature is such that the energy released from deposited molecules is correctly reproduced from the deposition event.

The calculations have been performed for 50,000–80,000 steps, with a time step of 2.0×10^{-15} s at 300–1000 K. The two-body, central force interatomic potential, as shown in Eq. (9), was used for all calculations. In Eq. (9), the first and second terms refer to Coulomb and exchange repulsion interactions, respectively:

$$u(r_{ij}) = \frac{Z_i Z_j e^2}{r_{ij}} + f_0(b_i + b_j) \times \exp\left(\frac{a_i + a_j - r_{ij}}{b_i + bj}\right) \tag{9}$$

Z_i is the atomic charge, e is the elementary electric charge, r_{ij} is the interatomic distance, and f_0 is a constant. The parameters a_i and b_i in Eq. (9) represent the size and stiffness of atom i, respectively, in the exchange repulsion interaction.

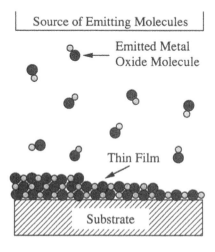

Figure 1   Simulation model for an epitaxial growth process.

B. Quantum Chemistry Calculations Using Density-Functional Theory

Many calculations presented in this chapter were performed in the framework of density-functional theory (DFT) (18). This method is a nonempirical approach and is an alternative to Hartree–Fock–based theories. It is widely used to tackle chemistry problems because of its ability to include a significant part of the electron correlation energy at a relatively low computational cost. And DFT is a theory of electronic ground-state structure, couched in terms of the electronic density distribution $n(\mathbf{r})$. Since the early 1970s, it has been effective in understanding and calculating the ground-state density, $n(\mathbf{r})$, and energy, E, of molecules, clusters, and solids with or without applied static perturbations (19). It is a complementary approach to the traditional methods of quantum chemistry couched in terms of the many-electron wave function $\Psi(\mathbf{r}_1, \mathbf{r}_2, \cdots, \mathbf{r}_N)$, which depends on the position of N electrons ($\mathbf{r}_i$, $i = 1, \cdots, N$). Both Thomas–Fermi (20) and Hartree–Fock–Slater (21) methods can be regarded as ancestors of the modern DFT. In these theories, the statistical consideration is used to approximate the distribution of electrons in an atom, and the effective potential field is determined by its distribution and the nuclear charge of atoms (20,21). In contrast to those intrinsic approximations, the present DFT theory is, in principal, exact to the ground state. The Hamiltonian of systems of N nonrelativistic, interacting electrons is presented in atomic units as

$$H = T + V + U \tag{10}$$

where T is the kinetic energy of electrons, V is the sum of the *arbitrary* external potential $v(\mathbf{r}_j)$, and U is the Coulomb potential (19).

The starting point of DFT is a rigorous lemma of Hohenberg and Kohn (22): ''The specification of the ground-state density $n(\mathbf{r})$ determines the external potential $v(\mathbf{r}_j)$ uniquely (to within an additive constant C).'' In other words,

$$n(\mathbf{r}) \to v(\mathbf{r}) \tag{11}$$

Since $n(\mathbf{r})$ also determines N by integration, it specifies the full Hamiltonian H and therefore all properties determined by H. Examples are the full N-particle ground-state wave function $\Psi(\mathbf{r}_1, \mathbf{r}_2, \cdots, \mathbf{r}_N)$, electrical polarizability, vibrational constants, and energy surfaces for chemical reactions (19).

By using this lemma, a minimal principle for the energy as a functional of $n(\mathbf{r})$ can be derived (22). For a given $v(\mathbf{r}_j)$, one defines the following energy functional of $n(\mathbf{r})$:

$$E_{v(\mathbf{r})}[n(\mathbf{r})] \equiv \int v(\mathbf{r}) n(\mathbf{r}) dr + F[n(\mathbf{r})] \tag{12}$$

where

$$F[n(r)] \equiv (\Psi[n(\mathbf{r})], (T+U)\Psi[n(\mathbf{r})]) \tag{13}$$

$F[n(\mathbf{r})]$ is a functional of $n(\mathbf{r})$, because Ψ is an antisymmetric wave function. The minimal principle holds:

$$E_{v(r)}[n(\mathbf{r}) \geq E_{v(r)}[n_0(\mathbf{r})] \equiv E \tag{14}$$

where $n_0(\mathbf{r})$ and E are the density and energy of the ground state, respectively.

$F[n]$ can be written in terms of its largest and elementary contributions (19) as

$$F[n(\mathbf{r})] = T_s[n(\mathbf{r})] + \frac{1}{2} \int \frac{n(\mathbf{r})n(\mathbf{r}')}{|\mathbf{r}-\mathbf{r}'|} \, dr \, dr' + E_{xc}[n(\mathbf{r})] \tag{15}$$

where $T_s[n(\mathbf{r})]$ is the kinetic energy of a nointeracting system with density $n(\mathbf{r})$. The next term is the classical expression for the interaction energy, and $E_{xc}[n(\mathbf{r})]$ is the exchange correlation energy. The physical content of DFT theory becomes identical to that of the Hartree approximation when E_{xc} is neglected. From the minimal principle, Eq. (14), the Euler–Lagrange equation,

$$\frac{\delta T_s[n(\mathbf{r})]}{\delta n(\mathbf{r})} + v(\mathbf{r}) + \int \frac{n(\mathbf{r}')}{|\mathbf{r}-\mathbf{r}'|} dr' + v_{xc}(\mathbf{r}) - \mu = 0 \tag{16}$$

associated with the stationary of $E[n]$, can be transformed into a new set of self-consistent (or Kohn–Sham) equations:

$$\left(-\frac{1}{2}\nabla^2 + v(\mathbf{r}) + \int \frac{n(\mathbf{r}')}{|\mathbf{r}-\mathbf{r}'|} dr' + v_{xc}(\mathbf{r}) - \varepsilon_j \right) \varphi_j(\mathbf{r}) = 0 \tag{17}$$

$$n(\mathbf{r}) = \sum_{j=1}^{N} |\varphi_j(\mathbf{r})|^2 \tag{18}$$

$$v_{xc}(\mathbf{r}) = \delta E_{xc}[n(r)] / \delta n(\mathbf{r}) \tag{19}$$

which differ from the Hartree equation only by inclusion of the exchange correlation potential $v_{xc}(\mathbf{r})$. Equation (17) must be solved to calculate $v_{xc}(\mathbf{r})$ in each cycle from Eq. (18), with a selected approximation for $E_{xc}[n(\mathbf{r})]$. The additional calculation time, compared to a Hartree calculation, is very small. However, in spite of the appearance of simple, single-particle orbitals, the Kohn–Sham equations are in principle exact, provided that the exact $E_v[n(\mathbf{r})]$ is used in Eq. (19), and the only error in the theory is due to approximations of $E_v[n(\mathbf{r})]$ (19). The ground-state energy is given by

$$E = \sum_{j=1}^{v} \varepsilon_j - \frac{1}{2} \int \frac{n(\mathbf{r})n(\mathbf{r}')}{|\mathbf{r}-\mathbf{r}'|} dr\, dr' - \int v_{xc}(\mathbf{r})n(\mathbf{r})dr + E_{xc}[n(\mathbf{r})] \qquad (20)$$

where ε_j and n are self-consistent quantities.

For practical use, good, simple approximations of $E_{xc}[n(\mathbf{r})]$ are needed. This is the local density approximation (LDA) (18):

$$E_{xc}^{LDA}[n(\mathbf{r})] = \int \varepsilon_{xc}(n(\mathbf{r}))n(\mathbf{r})dr \qquad (21)$$

where $\varepsilon_{xc}(n)$ is the exchange-correlation energy per particle of a uniform interacting electron gas of density n. The LDA becomes exact when the length scale l over which $n(\mathbf{r})$ varies is large, namely,

$$l >> d, (da_0)^{1/2} \qquad (22)$$

where d is the mean particle spacing ($\approx n^{-1/3}$) and a_0 is the hydrogen radius. This approximation quantity is known to vary with a very high accuracy ($\sim 0.1\%$). Moreover, for fairly well-understood reasons, it also gives useful results for most physical and chemical applications, in which Eq. (22) generally is not satisfied (19). The local spin-density approximation (LSDA) provides the spin-dependent generalization of Eqs. (17)–(19), and it is the next logical step in density-functional theory. The LSDA employs the fact that the exchange-correlation energy of a homogeneous electron gas depends on the density and on the magnetization parallel to an applied uniform magnetic field H, which stabilizes the spin-polarized state (23,24).

The next important approximation is the so-called generalized gradient approximation (GGA),

$$E_{xc}^{GGA} = \int f\left(n(\mathbf{r}), |\nabla n(\mathbf{r})|\right)dr \qquad (23)$$

in which $f(n, |\nabla n|)$ is a suitably chosen function of its two variables. The typical GGA exchange energy looks like

$$E_x = E_x^{LSDA} - b\sum_{\sigma} \int n_{\sigma}^{4/3} \frac{\chi_{\sigma}^2}{\left(1 + 6b\chi_{\sigma}\sinh^{-1}\chi_{\sigma}\right)} dr \qquad (24)$$

where

$$E_x^{LSDA} = -\frac{3}{2}\left(\frac{3}{4\pi}\right)^{1/3}\sum_{\sigma}\int n_{\sigma}^{4/3}\, dr \qquad (25)$$

and

$$\chi_{\sigma} = |\nabla n_{\sigma}|/n_{\sigma}^{4/3} \qquad (26)$$

where σ is a spin label and $b = 0.0042$ a.u. is a semiempirical fitted parameter. It should be noted that the exchange energy of an atomic system with this parameter is remarkably accurate and that this functional is effectively an interpolational formula between the small- and large-χ limits of the exchange energy density (19). The most useful exchange functional is Becke's 1988 functional, which includes the Slater exchange along with corrections according to the gradient of the density (25).

Although the chemical consequences of gradient corrections for correlation are relatively small compared to their exchange counterparts (26), accurate estimation of correlation energy in GGA has also received considerable attention. Of the many possible exchange-correlation combinations, several are currently used. There days, the most popular correlation functionals are the Lee, Yang, and Parr (LYP) functional (27), the Perdew 1986 (P86) functional (28), and the Perdew and Wang 1991 (PW91) functional (29).

The important feature of DFT is that the computing time, for a many-atom system with no geometrical symmetries, grows roughly as N_{at}^2 or N_{at}^3 (where N_{at} is the number of atoms). This is much better than the traditional methods, where computing time grows as $e_{at}^{\alpha N}$ ($\alpha \approx 1$) (19). Since the DFT method can be applied for the accurate estimation of the electronic structure for molecules, solids, and surfaces containing any element in the periodic table with better quality/cost ratio than that for traditional ab initio methods, it will be useful as a computational method for the CCC approach.

Thus, in our studies DFT was applied to investigate the electronic properties of bulk and surface structures, adsorption phenomena, etc. We employed the CASTEP (30) and DSolid (31) programs for the periodic density-functional calculations. The density-functional calculations for cluster models were carried out using the Amsterdam density-functional (ADF) program package developed by Baerends, Ellis, and Ros (32).

III. RESULTS AND DISCUSSIONS

A. Investigation of the deNO$_x$ Process on Various Ion–Exchanged Zeolites (11,33–40)

The environmental pollution caused by nitric oxides contained in exhaust gases from diesel or lean-burn engines of mobile sources is a serious problem of global significance that urgently needs to be solved (41). Much effort has been spent since the early 1990s on investigating the removal of nitrogen oxides (NO$_x$) from exhaust gases using different catalysts. Among the catalytic systems, cation-exchange zeolites have been widely studied, due to their valuable catalytic properties in a variety of industrial processes. It has been reported that selective catalytic reduction of the NO$_x$ species by hydrocarbons can be performed by various types of ion–exchanged ZSM-5 (42–50). ZSM-5, developed by Mobil Oil, is an aluminosilicate zeolite with a high silica and low aluminum content. Its structure is based on channels with inserting tunnels. The catalytic activity of ZSM-5 is due to its high acidity. We have previously applied computational chemistry to clarify the mechanism of the $deNO_x$ reaction on Co-, Cu-, Ga-, Ag-, In-exchanged ZSM-5 and other compounds (33–38). It is experimentally found that the water or SO_x molecules are highly poisonous for almost all ion–exchanged ZSM-5 (41,42). Also, various exchange cations present in ZSM-5 seem to exhibit different activity, selectivity, and durability, but one cannot compare those results directly because the experimental conditions are not the same.

We devised combinatorial computational chemistry to investigate adsorption of NO and NO_2 for the purpose of designing new $deNO_x$ catalysts. Resistance to poison is evaluated by the difference in the adsorption energies of the NO_x species and poisonous compounds. We estimated the adsorption energies and adsorption configurations of NO and NO_2 on various ion–exchanged ZSM-5 and examined the difference in the adsorption energies between NO_2 and water or between NO_2 and SO_x molecules to evaluate their resistance to water and SO_x poisoning. We have looked at numerous ion–exchanged ZSM-5 materials using the CCC approach. We have investigated the electronic and energetic properties of various ion–exchanged ZSM-5 materials by screening large number of different cations and have proposed novel effective exchange cations in ZSM-5 for the $deNO_x$ reaction.

The crystal structure of a zeolite-type silicalite with the same topology as ZSM-5 but without Al incorporation has been obtained from x-ray diffraction (XRD) data (51). However, the XRD observation of ZSM-5 does not inform us about the position of the Al atoms in the crystal structure of ZSM-5, because one cannot distinguish Al atoms from Si atoms in ZSM-5. When Al atoms are incorporated into ZSM-5, the structure of the ZSM-5 becomes different from that of a silicalite. In order to determine the structure of ZSM-5 framework with the Al incorporation, one needs to perform the calculations for a ZSM-5 crystal containing more than 288 atoms in the unit cell under three-dimensional periodic boundary conditions. Because DFT calculations for such a large unit cell of ZSM-5 are not available, MD calculations have been carried out to determine the structure of the ZSM-5 framework. Figure 2 shows the ZSM-5 structure used in an MD calculation. The 12 tetrahedral sites can be used for substitution of Al for Si. Earlier quantum chemical studies have reported that one of these sites (called T12) is energetically favorable for the incorporation of Al (34); therefore, in our study this site has been considered for an Al substitution.

Various cations have been selected as exchange cations. We have used K^+, Cu^+, Ag^+, and Au^+ as monovalent ions, Fe^{2+}, Co^{2+}, Ni^{2+}, Cu^{2+}, Zn^{2+}, Pd^{2+}, and Pt^{2+} as divalent ions, and Al^{3+}, Sc^{3+}, Cr^{3+}, Fe^{3+}, Co^{3+}, Ga^{3+}, In^{3+}, Ir^{3+}, and Tl^{3+} as trivalent cations. Based on experimental results (52–54), we have

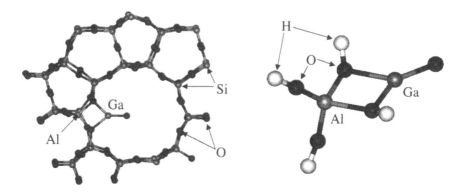

Figure 2 The framework structure and a cluster model of Ga^{3+}-ZSM-5.

assumed that monovalent cations exist as M^+ (M = metal), divalent cations exist as $M^{2+}OH^-$, and trivalent cations exist as $M^{3+}O^{2-}$. For MD calculations we have used a ZSM-5 unit cell (a = 20.06 Å, b = 19.80 Å, c = 13.36 Å) containing 95 Si atoms, 1 Al atom, 192 O atoms, and 1 M atom. An OH or an O has been added to the unit cell when M is divalent or trivalent, respectively. In our recent report (39), MD calculations have been performed on M^+-ZSM-5, where M^+ = H^+, Li^+, Na^+, K^+, or Cs^+. During these calculations, the exchange cations have been attached to the tetrahedral Al site. It has been found that a different nature of the exchange cations in M^+-ZSM-5 has very little influence on the structure of ZSM-5. Taking this into account, we have performed MD calculations on the structure of M^+-ZSM-5, where M^+ = Cu^+. From the optimized crystal structure of Cu^+-ZSM-5, a Cu^+-AlO_4 cluster has been extracted. Four hydrogen atoms have been added to saturate the dangling bonds, and the cluster has a Cu^+-$Al(OH)_4$ configuration. The cluster models for M^{2+}-ZSM-5 and M^{3+}-ZSM-5 have been constructed using the same procedure. In Figure 2, the M^{3+}-ZSM-5 (where M = Ga) cluster model is also shown. This system has been used as a model for an active site in our density-functional theory (DFT) calculations, which are performed using the Amsterdam density-functional (ADF) program (32). In this package, the DFT framework and linear combinations of Slater-type atomic orbitals are used in the Kohn–Sham formulation (23). The Vosko–Wilk–Nusair (VWN) parameterization (24) with Becke 88 (25) and Perdew 86 (28) gradient-corrected functionals have been utilized for an accurate estimation of the exchange-correlation energy and potential. The triple-ζ quality basis sets extended by polarization function are used. Triple-ζ quality means that the each valence electron wave function is presented as a linear combination of three Slater-type orbitals with different coefficients. This is necessary for accurate description of the electronic structure of various atomic and molecular systems. The position of each distinct exchanged cation has been optimized within the DFT calculation. During the DFT calculations, the $Al(OH)_4$ fragment has been fixed at the geometry of the structure of the ZSM-5 framework.

1. Adsorption of NO and NO_2 Molecules on Various Ion–Exchanged ZSM-5 Materials

We have calculated the adsorption states of NO and NO_2 Molecules on various ion–exchanged ZSM-5 materials. Here, the adsorption energy (E_{ads}) has been defined by the following equation:

$$E_{ads} = E_{(adsorption\ system)} - [E_{(ZSM-5)} + E_{(molecule)}] \tag{27}$$

A large negative value of E_{ads} indicates that the molecule is strongly adsorbed on ZSM-5. The effect of more than 20 different exchanged cations in ZSM-5 on the adsorption of NO and NO_2 molecules has been estimated. In the case of NO, we considered two different models of NO adsorption on the active sites of the ion–exchanged ZSM-5: (a) the interaction of the N atom of the NO molecule with the exchanged metal cation (N-end) and (b) the interaction of the O atom of the NO molecule with the exchange metal cation (O-end). Table 1 shows the adsorption energies of a NO molecule on various ion–exchanged ZSM-5 materials with the N-end and O-end configurations, respectively. The most stable adsorption energy for each exchange cation is presented in Figure 3. It has been found that NO is adsorbed strongly on Fe^{2+}, Co^{3+}, In^{3+}, Ir^{3+}, and Tl^{3+} ion–exchanged ZSM-5 but not adsorbed on K^+, Cu^{2+}, and Zn^{2+} ion–exchanged ZSM-5. Figure 4 shows the adsorption energies of a NO_2 molecule on various ion–exchanged ZSM-5 catalysts. A NO_2 molecule is strongly adsorbed on Cu^+, Au^+, Fe^{2+}, Co^{2+}, Pd^{2+}, Pt^{2+}, Cr^{3+}, Fe^{3+}, Co^{3+}, In^{3+}, Ir^{3+} and Tl^{3+} ion–exchanged ZSM-5. It is interesting to note that the NO_2 molecule is adsorbed weakly on Cu^{2+} and Zn^{2+} ion–exchanged ZSM-5. When we compared the adsorption energies of NO_x on Cu^+ and Cu^{2+} ion–exchanged ZSM-5, a significant difference was observed. The results show that ($+1$) positive charge is the most favorable electronic state for the Cu ion–exchanged ZSM-5 in the NO and NO_2 adsorption. Hence, we have confirmed that controlling the electronic state of an exchange cation is an important issue in the deNO$_x$ catalytic process.

The change in the N—O bond length due to the adsorption process is one of the parameters for the activation of NO_x molecules. It has been seen that for the NO molecule, the N—O bond is significantly elongated in the cases of trivalent exchanged cations, especially on Al^{3+}, Sc^{3+}, Co^{3+}, Ga^{3+}, In^{3+}, and Tl^{3+} cations (Figure 5). Figure 6 shows the elongation of the N—O bond length of the adsorbed NO_2 molecule on various ion–exchanged ZSM-5 materials as compared to the NO_2 molecule in the gas phase. In this figure, the sum of elongation of

Table 1 Adsorption Energies of NO on Various Ion-Exchanged ZSM-5 Materials

	N-end (kcal/mol)	O-end (kcal/mol)
Monovalent		
K	4.41	−3.99
Cu	−36.02	−7.51
Ag	−22.73	10.33
Au	−35.96	−14.31
Divalent		
Fe	−44.30	−14.35
Co	−43.50	−9.22
Ni	−35.06	−12.02
Cu	3.20	24.36
Zn	−5.61	−5.03
Pd	−30.00	1.23
Pt	−25.51	−15.42
Trivalent		
Al	−1.21	−25.72
Sc	−5.09	−19.90
Cr	−50.21	−36.14
Fe	−40.41	−35.99
Co	−35.36	−44.01
Ga	−9.25	−33.02
In	−7.39	−51.84
Ir	−42.84	−27.18
Tl	−6.62	−54.22

the two N—O bonds in NO_2 is shown. After NO_2 adsorbs on the active site in all ion–exchanged ZSM-5, the N—O bond is elongated and the NO_2 molecule is activated. We find that a NO_2 molecule is greatly activated on Cr^{3+} and Fe^{3+} ion–exchanged ZSM-5. In order to analyze the effect of exchanged cations on the NO_x activation, we have estimated the charge transfer from ZSM-5 to NO_x molecules for various ion–exchanged cations. It has been found that the elongation of the NO bond is linearly correlated to the charge transfer. Since the semioccupied orbital of the NO and NO_2 molecules is antibonding, the charge transfer from ZSM-5 to adsorption molecules leads to the activation of the latter.

Table 2 shows the partial charges on various exchange cations before the adsorption. It is clear that in the $M=O$ fragment, the O atoms have a large negative charge (-0.55 to -0.85), while the M atoms have a larger positive charge ($+0.84$ to $+1.36$) in M^{3+}-ZSM-5 as compared to those of the monovalent

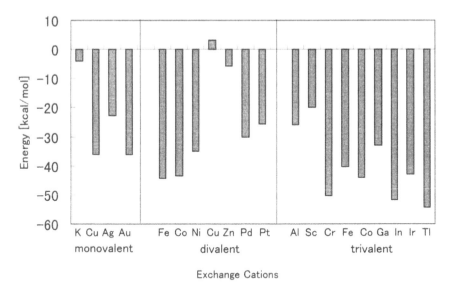

Figure 3 Adsorption energies of NO on various ion–exchanged ZSM-5 materials.

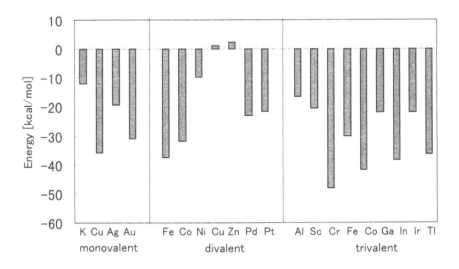

Figure 4 Adsorption energies of a NO_2 molecule on various ion–exchanged ZSM-5 materials.

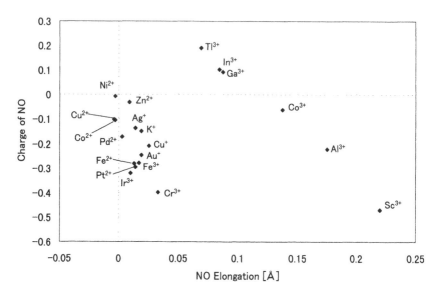

Figure 5   Correlation between the elongation of the bond length of NO and charge transfer from ZSM-5 to NO on various cation-exchanged ZSM-5 materials.

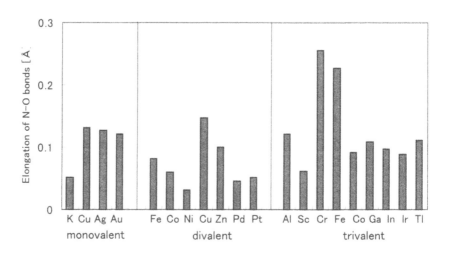

Figure 6   Elongation of the N—O bond length of the adsorbed NO_2 molecule on various ion−exchanged ZSM-5 materials as compared to the NO_2 molecule in the gas phase.

M and the divalent MOH in M^+-ZSM-5 and M^{2+}-ZSM-5, respectively. In Al^{3+}, Sc^{3+}, Co^{3+}, Ga^{3+}, In^{3+}, and Tl^{3+} ion–exchanged ZSM-5, the O atom of the NO molecule is coordinated to the M atom of the $M=O$ group, whereas the N atom of NO is coordinated to the O atom of the $M=O$ moiety. Hence, the reason why the NO molecule is activated strongly on these ion–exchanged ZSM-5 materials is that the N atom of the NO molecule receives a large negative charge from the O atom of the $M=O$ group as the exchange cation in M^{3+}-ZSM-5.

Figure 7 shows the charge transfer from ZSM-5 to the NO_2 molecule for various ion–exchanged ZSM-5 materials. The charge transfer varies from a large negative value to a positive value. NO_2 receives the highest negative charge in Cr^{3+}-ZSM-5 among all ion–exchanged ZSM-5 materials. We conclude that the charge transfer is one of the factors in the activation of the NO_2 molecule on ion–exchanged ZSM-5.

Table 2 Partial Charges on Various Exchange Cations

	Metal	OH or O
Monovalent		
K	0.86	
Cu	0.52	
Ag	0.51	
Au	0.56	
Divalent		
Fe	0.90	−0.44
Co	0.80	−0.40
Ni	0.76	−0.38
Cu	0.72	−0.40
Zn	0.95	−0.48
Pd	0.75	−0.40
Pt	0.82	−0.39
Trivalent		
Al	1.14	−0.83
Sc	1.36	−0.76
Cr	1.17	−0.70
Fe	1.07	−0.69
Co	1.03	−0.64
Ga	1.17	−0.85
In	1.22	−0.79
Ir	0.84	−0.55
Tl	1.26	−0.77

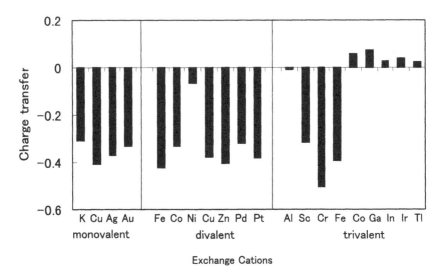

Figure 7 Charge transfer from ZSM-5 to the NO_2 molecule for various ion–exchanged ZSM-5 materials.

2. Design of Novel deNO$_x$ Catalysts with High Resistance to Water

Many highly active catalysts cannot be industrialized because of their short life-times. One of the well-known deactivation processes of catalysts is the poisoning by coexistent gases, such as water and SO_x. Hence, investigation on the adsorption of poisonous gases can provide important information in designing catalysts with high resistance to poisons.

Experimentally, it is well known that ion–exchanged ZSM-5 can easily be deactivated in the presence of water molecules. In order to design deNO$_x$ catalysts with high resistance to water, we have investigated adsorption of a water molecule on various ion–exchanged ZSM-5 materials (11). The adsorption energies of a water molecule on various ion–exchanged ZSM-5 materials are shown in Table 3. It has been observed that a water molecule is strongly adsorbed on Al^{3+}, Sc^{3+}, Co^{3+}, Ga^{3+}, In^{3+}, and Tl^{3+} ion–exchanged ZSM-5.

As mentioned before, the effect of exchanged cations on the adsorption process can be analyzed by the charge transfer from ZSM-5 to the adsorption molecule. Table 4 shows the partial charges on water in various ion–exchanged ZSM-5 materials. The charges on the O atom of the water in all ion–exchanged ZSM-5 except Ir^{3+}-ZSM-5 tend to be more negative (-0.02 to -0.22) than the charge of the isolated water molecule before adsorption. The charges on the H

Table 3 Adsorption Energies of Water, SO_2, and SO_3 on Various Ion-Exchanged ZSM-5 Materials

	H_2O (kcal/mol)	SO_2 (kcal/mol)	SO_3 (kcal/mol)
Monovalent			
K	−8.39	−9.29	−1.28
Cu	−17.86	−19.41	−15.29
Ag	−12.69	−15.35	−9.82
Au	−21.92	−27.61	−23.09
Divalent			
Fe	−20.12	−11.80	−19.08
Co	−26.05	−10.45	−23.55
Ni	−19.65	−7.16	−29.91
Cu	−0.42	−9.98	−14.16
Zn	−20.59	−16.47	−23.51
Pd	−13.70	−2.64	−40.19
Pt	−13.97	−23.77	−37.41
Trivalent			
Al	−79.28	−64.58	−76.64
Sc	−54.41	−49.89	−61.69
Cr	−20.59	−30.41	−43.51
Fe	−16.03	−33.77	−41.94
Co	−37.19	−38.49	−54.78
Ga	−57.89	−44.53	−55.57
In	−50.49	−41.50	−48.73
Ir	−4.43	−16.01	−31.46
Tl	−41.90	−43.17	−37.11

atoms of the water in all ion–exchanged ZSM-5, except Sc^{3+}-ZSM-5, tend to be more positive (0.10 to 0.18) than the charge of the isolated water molecule. This is especially the case for the water dissociated on Al^{3+}, Sc^{3+}, Co^{3+}, Ga^{3+}, In^{3+}, and Tl^{3+} ion–exchanged ZSM-5. The charges on the O atom of water in the cases of these ion–exchanged ZSM-5 materials, except Tl^{3+}-ZSM-5, become negative (-0.03 to -0.21). The reason why the water molecules are dissociated on these ion–exchanged ZSM-5 materials is that the M atom of the M=O group has a large positive charge, while the O atom has a large negative charge (see Table 2). The M—O moiety has a strong polarization effect on the O—H bond of the water, which is the main driving force in facilitating the breakage of the O—H bond in the water molecule.

In order to evaluate the ability of different exchanged cations to exhibit resistance to water poisoning, the difference in the adsorption energies of NO_2

Table 4   Partial Charges on Water in Various Ion-Exchanged ZSM-5 Materials

	O atom	2H atoms	Total
Monovalent			
K	−0.70	0.74	0.04
Cu	−0.63	0.75	0.12
Ag	−0.62	0.75	0.13
Au	−0.65	0.73	0.08
Divalent			
Fe	−0.68	0.74	0.06
Co	−0.65	0.73	0.08
Ni	−0.62	0.74	0.12
Cu	−0.64	0.73	0.09
Zn	−0.66	0.74	0.07
Pd	−0.68	0.78	0.10
Pt	−0.67	0.74	0.07
Trivalent			
Al	−0.79	0.70	−0.09
Sc	−0.80	0.59	−0.21
Cr	−0.66	0.76	0.10
Fe	−0.63	0.72	0.09
Co	−0.69	0.65	−0.04
Ga	−0.79	0.75	−0.04
In	−0.82	0.79	−0.03
Ir	−0.59	0.71	0.12
Tl	−0.78	0.79	0.01
Water	−0.60	0.60	0.00

and water molecules on various ion–exchanged ZSM-5 materials have been calculated and are plotted in Figure 8. The positive value indicates that the water molecule is more strongly adsorbed on the exchanged cation as compared to the NO_2 molecule. Water molecules can readily deactivate ion–exchanged ZSM-5 that has a positive value. On the contrary, the negative value indicates that the NO_2 molecule is more strongly adsorbed than the water molecule. Hence, the larger negative value is favorable for the deNO$_x$ catalysts in the presence of water molecules. From Figure 8, Cu^+, Fe^{2+}, Pd^{2+}, Cr^{3+}, Fe^{3+}, and Ir^{3+} ion–exchanged ZSM-5 materials are found to exhibit high resistance to water poisoning.

Experimentally, it has been shown that Ga-ZSM-5 is easily deactivated in the presence of water as compared to In-ZSM-5 (43). This tendency is in good agreement with our simulation results. Recently, it has been reported from experiments that Fe- and Pd-ZSM-5 have high resistance to water (44,46). This is also

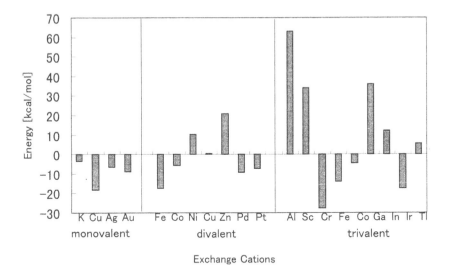

Figure 8 Difference in the adsorption energies between NO_2 and H_2O molecules on various ion–exchanged ZSM-5 materials.

in good agreement with our simulation results. Based on these results, we have proposed Cr^{3+} and Ir^{3+} ion–exchanged ZSM-5 as new candidates for deNO$_x$ catalysts with high resistance to water poisoning.

3. Design of Novel deNO$_x$ Catalyst with High Resistance to SO$_x$

We have also investigated the resistance to SO$_x$ poisoning using an approach similar to that in the case of water. The adsorption energies of SO_2 and SO_3 molecules have been calculated (40); the results are shown in Table 3. The adsorption energy of SO_2 molecules at distinct monovalent cation–exchanged ZSM-5 is larger than that of SO_3 molecules. In contrast, SO_3 molecules are more strongly adsorbed on distinct divalent and trivalent cation–exchanged ZSM-5 than are SO_2 molecules.

Figure 9 shows the difference in the adsorption energies between NO_2 and SO_2 and between NO_2 and SO_3 molecules on various ion–exchanged ZSM-5 materials. It can be seen that Cu^+, Fe^{2+}, Co^{2+}, Pd^{2+}, Cr^{3+}, and Ir^{3+} ion–exchanged ZSM-5 materials exhibit high resistance to SO_2 poisoning and that K^+, Cu^+, Ag^+, Au^+, Fe^{2+}, Co^{2+}, and Cr^{3+} ion–exchanged ZSM-5 materials exhibit high resistance to SO_3 poisoning. Recently, experimental studies have shown that Fe- and Co-ZSM-5 have high resistance to SO_2 (41,44). This is in good agreement

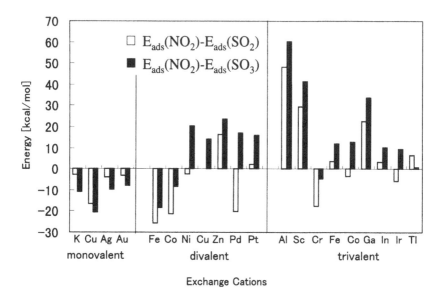

Figure 9   Difference in the adsorption energies between NO_2 and SO_x ($x = 2, 3$) molecules on various ion–exchanged ZSM-5 materials.

with our simulation results. The present results have proposed Cr^{3+}-ZSM-5 as
a new candidate for $deNO_x$ catalysts with high resistance to SO_x.

From the calculation results, Cu^+, Fe^{2+}, Pd^{2+}, Cr^{3+}, Fe^{3+}, and Ir^{3+}
ion–exchanged ZSM-5 materials are found to have high resistance to water poi-
soning. Cu^+, Fe^{2+}, Co^{2+}, Pd^{2+}, Cr^{3+}, and Ir^{3+} ion–exchanged ZSM-5 materials
are found to have high resistance to SO_2 poisoning, while K^+, Cu^+, Ag^+, Au^+,
Fe^{2+}, Co^{2+}, and Cr^{3+} ion–exchanged ZSM-5 materials are found to have high
resistance to SO_3 poisoning. Our calculations predict that Ir^{3+}-ZSM-5 can serve
as a good catalyst for $deNO_x$ reactions with high resistance to water and SO_2.
Moreover, Cr^{3+}-ZSM-5 should serve as a good catalyst for $deNO_x$ reactions with
high resistance to water, SO_2, and SO_3. It is significant that these candidates
have not been reported experimentally yet. Thus, we have demonstrated that
combinatorial computational chemistry can provide a novel approach for catalyst
design.

B.   Electronic and Structural Properties of ZnO and TiO_2: The Role of Metal Dopants (55,56)

Wide-band semiconductor materials are of considerable interest for their use in
blue light–emitting diodes and short-wavelength laser diodes. Initial research

efforts have been focused mainly on ZnSe (57), but lately substantial progress has been made in the field of GaN-based technologies (58). With its large, direct bandgap (3.2 eV) and a wurtzite crystal structure, ZnO is similar to GaN. Because of its relatively close match in lattice constants, ZnO may be used as a suitable substrate for group III nitride compounds (59). Zinc oxide is also of potential interest as a suitable material for high-temperature, high-power electronics devices. In 1997, Kawasaki et al. (60) reported fabrication of ZnO films on α-$Al_2O_3(0001)$ substrates by laser molecular beam epitaxy techniques and observed an ultraviolet laser emission when they were pumped with a YAG laser. Another important subject in the field of semiconductors is the fabrication of a double heterostructure (DH) composed of a thin well layer sandwiched between two barrier layers, which has been utilized in laser diodes to facilitate radiative recombination by carrier confinement (61). Bandgap engineering is one of the challenging problems in the fabrication of a DH laser diode using a ZnO active layer. Modulating the bandgap while keeping the lattice constants similar is critical for this propose.

Titanium dioxide is an important compound with a wide range of current and future applications, including catalysis, oxygen-gas sensors, and coatings. Another attractive use of wide-bandgap oxide semiconductors is the conversion of visible light into electricity (62). The design of titanium-based photocatalysis with high catalytic activities at visible light is also of great interest. The drawback in the requirement of high-energy light for creating electron-hole pairs is overcome by sensitizing these materials to visible light with appropriate dyes (63). Recently, it has been found that the photoreactivities of TiO_2 appear to be a complex function of the dopant concentration, the energy level of dopants within the TiO_2 lattice, their d electronic configuration, the distribution of dopants, and the light intensity (64).

In this section, we discuss the application of the CCC approach to investigating structural and electronic properties of metal oxide electronics materials. The main focus of this work is a systematic study of metal ion doping of zinc oxide (ZnO) and titanium oxide (TiO_2) structures, which have attracted much attention due to their potential applications in many industrial fields, such as fabrication of ultraviolet laser-emitting devices, double heterostructures, and different catalysis processes.

The calculations presented in this section have been performed in the DFT framework; the details of this theory were explained in Section II.B. We recall that electron correlation is essential for a correct description of systems involving transition metal ions. The periodic density-functional calculations have been carried out using the CASTEP code (30). This program is found to be a useful tool for accurate estimation of physical/chemical properties of metals and metal oxides at a relatively low computational cost. In this program, the plane-wave basis and pseudopotential method combined with DFT has provided a simple framework, in which the calculation of the physical forces on ions is greatly simplified so that an extensive geometry optimization is possible. For minimization of the Kohn–Sham energy functional, the conjugated gradient technique is performed. This technique provides a simple and effective procedure for direct minimization of any function. The initial direction of the gradient is taken to be negative at the starting point. A subsequent conjugate direction is then constructed from a linear combination of the new gradient and the previous direction that minimized the Kohn–Sham energy functional (10). In practice, a small number of iterations is required to locate the minimum. This increases the computational speed, and the calculations do not have a large memory requirement. Using pseudopotentials (65–68) to describe the electron–ion interaction allows us to provide very accurate spin-polarized ab initio electron structure calculations and to considerably reduce the calculation time as compared with all-electron calculations, especially for metal oxide systems. Once a consistently accurate level of theory has been selected, we have examined the important properties of the bulk structures of selected metal oxides, depending on a number of dopants. We also focus on obtaining data that are potentially useful in designing future experiments aimed at improving essential features of various electronics devices based on metal oxides.

1.   Bandgap Modulation of ZnO

To obtain the complete band structure of ZnO, calculations have been carried out including unoccupied orbitals. The bandgap modulation of ZnO has been achieved by doping it with various metals, and their structural properties have been calculated. The values of angles of the unit cell (α, β and γ) are fixed, and the other cell parameters have been optimized. The details of these calculations have been described elsewhere (55). The ZnO supercell is represented as $Zn_{1-y}X_yO$ (Figure 10), where X = Zn, Mg, V, Cr, Mn, Fe, Co, Cd, Pd, or Pt. We have used 50% ($y = 0.5$) as the percentage of doping in our calculations.

The bandgaps and optimized structures of ZnO doped with various metals are presented in Table 5. The experimental bandgap values of ZnO and Mg-doped ZnO are reported to be 3.2 eV and 3.8 eV, respectively (69). It is well known that using the LDA energies to estimate the gap leads to underestimated values as compared to the experimental gap (70). In LDA calculations, the Kohn–Sham eigenvalues (band energies) do not rigorously represent excitation energies. Furthermore, there is a fundamental difference between the occupied and unoccupied states, which are treated in the same way by LDA theory. However, the structural and other electronic properties, such as bandwidth or energy differences, are predicted with reasonable accuracy by LDA theory. Moreover, this theory can also correctly predict the trend on the bandgap modulation of semiconductor materials when they are doped with various metals.

From Table 5, it is observed that the bandgap of ZnO increases when it is doped with Mg, V, Mn, Fe, Pd, and Pt, whereas it decreases when it is doped

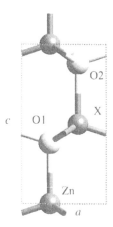

Figure 10   Model of the ZnO supercell $(Zn_{1-y}X_yO, y = 0.5)$; X is a dopant.

with Co and Cd. The largest increase in the bandgap of ZnO is found to occur in the cases of Mg, V, and Fe. Among these three metals, only Mg has an ionic radius similar to that of ZnO. Thus, no significant changes in the structural parameters of ZnO are observed when it is doped with Mg (Table 5). Decreasing the bandgap of ZnO by doping with Co and Cd makes it possible to control the modulation of the bandgap. This is important for fabrication of double heterostructures with varied optical properties.

Table 5   Calculated Bandgap and Selected Structural Parameters of $Zn_{1-y}X_yO$ $(y = 0.5)$.

Species	Band Gap (eV)	$\Theta_{o\,1\text{-}x\text{-}o2}$ (deg)	$R_{Zn\text{-}o1}$ (Å)	$R_{o1\text{-}x}$ (Å)	$R_{o2\text{-}x}$ (Å)	Ion radius (Å)
Zn	0.57	107.8	2.014	2.005	2.014	0.74
Mg	1.30	106.3	2.022	1.988	2.000	0.72
V	0.76	107.1	2.054	2.010	2.032	0.79
Cr	0.51	113.4	2.000	2.050	2.100	0.73
Mn	0.63	109.4	2.003	2.009	2.012	0.83
Fe	0.70	110.7	2.020	2.046	2.001	0.61
Co	0.23	108.3	2.013	2.010	2.019	0.65
Cd	0.42	109.8	1.992	2.114	2.175	0.95
Pd	0.65	126.5	1.970	2.110	2.110	0.86
Pt	0.56	121.4	1.960	2.140	2.060	0.80

   The distortion of the crystal structure can be completely described by the lattice parameters a and c, the ratio c/a, and the volume V of the unit cell. The optimized cell parameters and the volume of ZnO when doped with various metals (Mg, V, Cr, Mn, Fe, Co, Cd, Pd, and Pt) are compared and plotted in Figure 11. This figure proves again that the cell parameters and the volume of ZnO are retained only in the case of doping with Mg.

   Next we correlated the change in the c/a value with the electronegativity of the metals. This is presented in Figure 12. The c/a value decreases with an increase in the strength of the ionic bond; in the cases of Mg and V doping, this value is found to be similar to that of ZnO as compared to the other metals. To retain the wurtzite structure, the c/a ratio should be in a specific range. Figure 12 shows that this range is closer to the doped ZnO only in the cases of Mg and V doping and that the other ions have either lower or higher c/a values. Although a wide bandgap is also obtained by doping with Fe and Mn, it may be difficult to form a stable structure because their ionic radii and the c/a values are different from those of ZnO. The most important factor in designing DH is that a doped ZnO have a wide bandgap as well as lower highest occupied molecular orbital (HOMO) and higher lowest unoccupied molecular orbital (LUMO) energy levels than those in undoped ZnO.

   Figure 13 shows the changes in orbital energy for both the HOMO and LUMO of doped ZnO. It is found that decrease in the HOMO level and increase

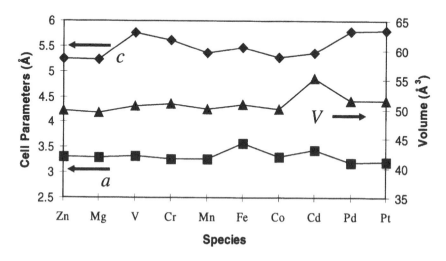

Figure 11   Optimized cell parameters and volume for ZnO doped with various metallic ions.

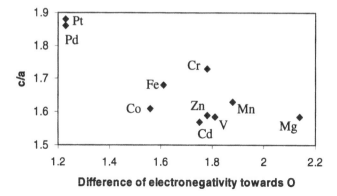

Figure 12 Correlation of *c/a* for various doped ZnO materials with electronegativity of metals.

in the LUMO level are observed only in the case of Mg doping. This clearly suggests that Mg is the best candidate for fabricating DH.

2. The Role of Metal Dopants in Rutile TiO$_2$

The accuracy of the plane-wave pseudopotential DFT technique using ultrasoft pseudopotentials (USPP) (68) has been tested by computing the lattice parameters

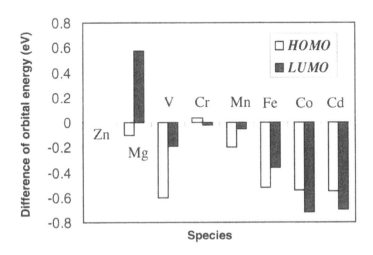

Figure 13 Changes of HOMO and LUMO energy levels of doped ZnO.

of bulk rutile TiO_2. The optimal values of the unit cell volume and the ratio c/a are obtained through minimization of the total energy of the titanium dioxide structure. The calculated structural properties obtained using USPP are in good agreement with the experimental results (71) and with previous calculations (72,73) (Table 6). A direct-forbidden gap of 1.86 eV (indicating that the direct gap at this point is dipole forbidden) has been found at the Γ point. Although this bandgap is lower than the experimental value (74) of 3.0 eV (the reason for such an underestimate has been previously given), the details of the band structure are almost identical to those obtained by Glassford and Chelikowsky (72). The accurate estimate of structural and electronic properties with these pseudopotentials gives us a good reason to believe that accurate calculations can be obtained for TiO_2 doped with transition metals.

Next we investigate the effect of doping on the structural and electronic properties of rutile TiO_2. The TiO_2 supercell is represented as $Ti_{1-y}X_yO_2$ (Figure 14), where X = Si, V, Fe, Sn, or Zr. In the first step, the percentage of doping is fixed at 50% ($y = 0.5$). The calculated bandgap and optimized structures for TiO_2 doped with Si, V, Ge, Sn, and Zr are presented in Table 7. It is clearly seen that the value of the bandgap is strongly dependent on the ionic radius of the dopants. The largest bandgap was found in the case of Zr doping. As mentioned before, the stability of the crystal lattice can be well described by the parameters of lattice geometry, such as a, c/a, and V (Figure 15), as well as the correlation of the change in the c/a value with the electronegativity of the metals. This is given in Figure 16. In the cases of closed-shell ions (X = Si, Ge, Sn, and Zr), the doping of TiO_2 is found to have only a slight effect on the stability of the crystal lattice. However, considerable changes in the values of the bandgap make it possible to control the gap value using various doping elements.

It is known that the photocatalytic activity of TiO_2 also depends on the doping ions. The reaction activity of catalysts increases with shifting of the Fermi level. Since the position of the Fermi level is determined by the positions of the HOMO and LUMO energy levels, it is possible to estimate the catalytic activity using changes in the HOMO energy and the value of the bandgap. The calculated

Table 6 Calculated Values of Bulk Lattice Parameters of TiO_2 Compared with Experiment

Property	Experiment (Ref. 71)	LDA (USPP)	LDA (NCPP) (Ref. 72)	HF (EPC) (Ref. 73)
Cell volume (Å^3)	62.439	60.677	60.210	62.200
A (Å)	4.594	4.556	4.653	4.555
c/a	0.644	0.641	0.637	0.658

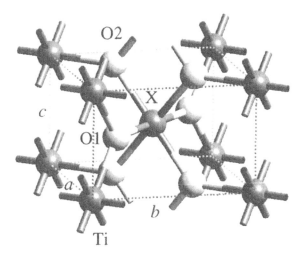

Figure 14 Model of TiO_2 supercell ($Ti_{1-y}X_yO_2$, $y = 0.5$); X is a dopant.

energy shift of the HOMO energy level for different metal ion dopants is shown in Figure 17. It can be seen that the largest shift with practically the same value of the bandgap is observed in the case of the Sn ion. This result suggests that the Sn dopant increases the catalytic activity of TiO_2. This is in agreement with experimental data (75). Although the doping by the open-shell ion V^{4+} (with one unpaired d electron) has little effect on the position of the HOMO energy level, the reduction in the bandgap value (Table 7) is also predicted by the change (of about 0.5 eV) in the position of the Fermi level.

Table 7 Calculated Bandgap and Selected Structural Parameters of $Ti_{1-y}X_yO_2$ ($y = 0.5$)

Species	Bandgap (eV)	R_{Ti-O} (Å)	R_{O1-X} (Å)	R_{X-O2} (Å)	$\Theta_{Ti-O1-X}$ (deg)	Ion radius (Å)
Ti^{4+}	1.86	1.932	1.958	1.932	130.8	0.61
	(3.0)[a]	(1.949)[a]	(1.980)[a]	(1.949)[a]	(130.6)[a]	
Si^{4+}	0.98	1.903	1.788	1.783	133.7	0.40
V^{4+}	0.92	1.923	1.923	1.892	131.9	0.58
Ge^{4+}	1.19	1.933	1.943	1.916	132.2	0.53
Sn^{4+}	1.72	1.952	2.046	2.017	131.7	0.69
Zr^{4+}	2.25	1.990	2.120	2.074	128.8	0.72

[a] Experimental values for TiO_2 (Refs. 71 and 74).

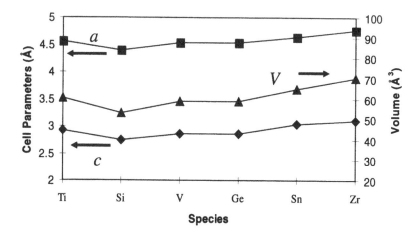

Figure 15 Optimized cell parameters and volume for doped TiO$_2$.

Taking into account the fact that, in the experimental studies, the dopant concentration is significantly lower than 50%, we calculated the Ti$_{1-y}$X$_y$O$_2$ structures, where X = Sn and y = 0.25. The shift of the HOMO energy level is lower than that in the case of 50% dopant concentration. The shifts are -3.05 eV and -4.67 eV for y = 0.25 and y = 0.5, respectively. Moreover, the bandgap value decreases for Sn-doped TiO$_2$ (1.72 eV at 50% and 1.66 eV at 25%). The catalytic activity of doped TiO$_2$ structures is dependent not only on the type of dopant and its concentration but also on the electronic configuration.

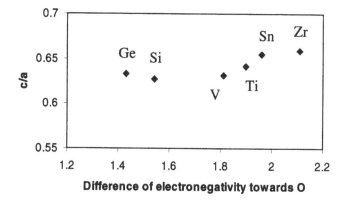

Figure 16 Correlation of c/a for various doped TiO$_2$ with ion tendency.

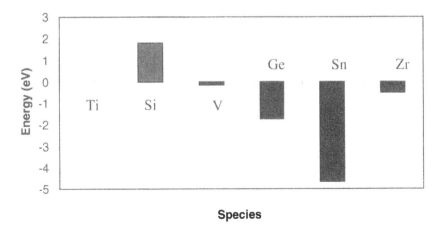

Figure 17 Difference in HOMO energy level for doped TiO_2.

We have found that the CCC approach is an accurate algorithm to predict structures and electronic properties for selected metal oxides. The periodic density-functional calculations combining the plane-wave basis and the pseudopotential method are useful tools to explain available experimental data and to design new functional metal oxides. The analysis of the bandgap modulation of doped ZnO reveals that Mg doping is effective in increasing the bandgap of ZnO while retaining the wurtzite structure. In the case of TiO_2, it is found that it is easier to control the bandgap values by doping with the closed-shell ions, such as Si, Ge, Zr, and Sn, because the stability of the crystal lattice changes slightly, depending on the dopants. The photocatalytic activity appears to be a function not only of the dopant but also of other factors. However, it is possible to qualitatively predict the best dopants for important catalyst systems. The analysis of calculated results confirms the increase in photocatalytic properties of TiO_2 when it is doped with Sn ion.

C. Layer-by-Layer Epitaxial Growth Process on Metal Oxide Surfaces (16,76–81)

Artificial construction of atomically defined metal oxide layers is important in making electronic devices incorporating high-temperature superconducting oxide films, magnetic and optical films, as well as other advanced materials, such as supported metal oxide catalysts (82–84). Hence, good understanding of homo- and heteroepitaxial growth processes on metal oxide surfaces, as well as the formation process of metal clusters on metal oxide substrates, is crucial for realizing atomically controlled structures that exhibit interesting, novel physical properties.

A large amount of information about the epitaxial growth processes on a variety of metal oxide substrates has been experimentally accumulated. In addition, theoretical approaches, such as MD, quantum chemistry, and Monte Carlo simulation, have also been used to understand the mechanisms of the epitaxial growth processes in order to help realize atomically controlled structures. As mentioned in Section II.A, the classical MD method is one of the most popular techniques in computational study (12). However, to date only a small number of MD simulations have been devoted to crystal growth processes. Since it enables us to understand the atomistic mechanism of different crystal growth modes and the effects of various factors, such as substrate species, substrate temperature, and surface defects, on the fabricated materials, the interest in the crystal growth MD simulation has been steadily increasing. Schneider et al. (85) had developed the first MD algorithm to simulate the vapor-ase crystal growth process and applied it to a Lennard–Jones system. Since then, the growth process of silicon films (86–88) and the continuous deposition process of metals on metal (89) and silica (90) surfaces have been actively studied. We have also investigated the deposition process of a single Au or Pd cluster on the MgO(001) surface (77,78).

Previously the deposition process of monoatomic molecules and metal clusters had been simulated, but the processes of continuous deposition of metal oxide molecules and the whole epitaxial growth of metal oxide substrates had not been calculated, to the best of our knowledge. Hence, we have developed a new MD code, which allows us to systematically investigate such processes. The main merit of this MD program is that it can accurately and quickly calculate the different growth mechanisms, taking into account many important effects, such as temperature and the presence of surface defects. Hence, it is an essential and ideal simulator for the CCC approach for systematic study of the epitaxial growth process and the design of atomically defined metal oxide superlattices. We have already provided a brief explanation of this code in Section II.A, and the details of this method have been described elsewhere (16). In the present section, we present the application of our MD code to investigate the homo- and heteroepitaxial growth processes of different metal oxide surfaces.

1. Layer-by-Layer Homoepitaxial Growth Process of MgO(001) Surfaces

Thin-film fabrication of MgO has attracted much attention due to its value as a high-temperature material, a secondary electron emission material, a wide-gap insulator, and a model catalyst and other practical applications. A number of interesting experimental studies on the epitaxial growth of MgO thin films on various substrates, including MgO (91), silicon (92), quartz (93), and sapphire

(94), by various techniques, including molecular beam epitaxy (MBE), chemical vapor deposition (CVD), and sputtering, have been reported. Their properties have been extensively studied by x-ray diffraction (XRD), x-ray photoelectron spectroscopy (XPS), Auger electron spectroscopy (AES), reflection high- and low-energy electron diffraction (RHEED and LEED, respectively), transmission electron microscopy (TEM), atomic force microscopy (AFM), etc. Recently, MgO thin films have been actively studied as a buffer layer for high-T_c superconducting films (95). Since an atomically flat MgO buffer layer is required to enhance the growth of uniform superconducting oxides such as $YBa_2Cu_3O_7$, studies on the atomistic control of the surface, interface, and growth process of MgO have been performed.

In order to understand the entire process of epitaxial growth of the MgO surface, an MD simulation has been applied to the continuous deposition of MgO molecules and to the formation of the metal oxide layer (16) for 50,000 steps with a time step of 2.0×10^{-15} s. A total of 32 MgO molecules have been deposited on the MgO(001) plane, one, by one, at a constant velocity of 900 m/ s. This amount is equal to the number of Mg and O atoms that constitute a single MgO layer in the present MD unit cell. Figure 18 shows the homoepitaxial growth process of the MgO(001) plane at 300 K. The MgO substrate is represented by atomic bonds, while the deposited MgO molecules are represented by spheres. After some MgO molecules are migrated on the Mg(001) plane, two-dimensional epitaxial growth of the MgO thin layer has been observed at 300 K. Moreover, the deposited molecules are seen to retain the NaCl-type structure and the (001) oriented configuration during the MD simulation. The present MD simulation has reproduced the experimental results reported by Yadavalli et al. (96), where they observed the epitaxial growth of MgO(001) plane with a NaCl-type structure over a wide temperature range of 140 K to 1300 K. However, after all of the 32 MgO molecules are deposited on the MgO(001) plane, we have observed that there are some defects in the first constructed MgO layer and that some MgO molecules had already begun to form the second MgO layer. Thus, we conclude that the layer-by-layer homoepitaxial growth of MgO thin films without defects cannot be realized at the low temperature of 300 K.

It is experimentally well known that substrate temperature is one of the important parameters in constructing well-defined MgO thin layers. The homoepitaxial growth process of the MgO(001) surface at the high temperature of 1000 K has been simulated, and the result is shown in Figure 19. Even at this temperature, MgO grows epitaxially, retaining the NaCl-type structure and (001) orientation. Moreover, the formation of a single, two-dimensional, uniform and flat layer of MgO without any defects has been observed at the 50,000 time step. This is significantly different from the result at 300 K and indicates that the epitaxial growth of the MgO(001) plane follows the monolayer overgrowth mode (Frank–van der Merwe mechanism) (97). Thus, a high temperature is found to

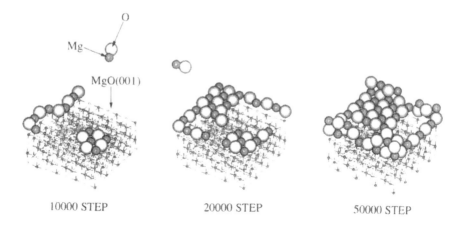

Figure 18   Homoepitaxial growth process of the MgO(001) plane at 300 K.

be favorable for a complete layer-by-layer homoepitaxial growth. This is in agreement with experimental results (91) that revealed that surfaces of MgO films grown at 923 K and 1023 K are smoother than those obtained at a lower temperature of 298 K. Therefore, the effect of substrate temperature on the epitaxial growth process is well reproduced by the present MD code. As mentioned in Section II.A, many thermodynamic properties of the system can also be obtained by the MD method. We have estimated the self-diffusion coefficient (D) responsible for the migration of a single MgO molecule on the MgO surface. The fundamental knowledge of the migration process of deposited molecules on the surface is essential to elucidate the temperature dependence of the homoepitaxial growth process. It has been found that surface diffusivity of deposited MgO molecules is low at 300 K. At this temperature, molecules remain at the position where they land on the surface. This leads to clustering of deposited MgO molecules, and the grown film displays many defects. At a high temperature of 1000 K, the surface diffusivity becomes higher and MgO molecules on the surface can move around to fill up all available sites. Thus, a smooth, flat MgO thin film is constructed, and a layer-by-layer homoepitaxial growth is realized.

2.   Layer-by-Layer Epitaxial Growth Process of SrTiO₃(001) Surfaces

Another substrate of interest from the viewpoint of epitaxial growth of high-T_c superconducting films is $SrTiO_3(001)$. The epitaxial growth process and the atomic surface structure have been extensively studied by various experimental techniques (84,98–100). Interesting theoretical and computational studies have

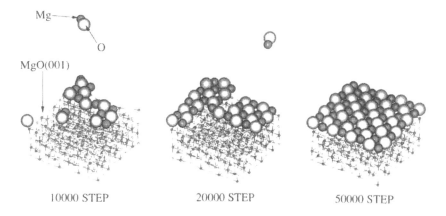

Mg ──── O

MgO(001)

10000 STEP   20000 STEP   50000 STEP

Figure 19   Homoepitaxial growth process of the MgO(001) plane at 1000 K.

been performed to elucidate the surface reconstruction, electronic structure, die-lectric property, ferroelectricity, band structure, and energetics of SrTiO₃ (101–103). However, outside of our work (79–82), there have been no studies devoted to the simulation of the epitaxial growth process of SrTiO₃ surfaces.

The SrTiO₃ crystal has a structure where nonpolar SrO and TiO₂ atomic planes are stacked alternately. Therefore, two different surface terminations, either with SrO or TiO₂, at the (001) top layer are possible. Experimental studies (84,100) have revealed that the SrTiO₃(001) surface is terminated by the TiO₂ atomic plane. In order to make a direct comparison with experimental results, we have employed the SrTiO₃(001) surface model terminated by the TiO₂ atomic plane as a substrate for the MD simulations, and the SrO molecules have been deposited continuously over the SrTiO₃(001) surface to simulate the initial stage of the homoepitaxial growth process.

Prior to the homoepitaxial growth simulations, the surface migration pro-cess of a single SrO molecule on the SrTiO₃(001) surface terminated by the TiO₂ atomic plane has been investigated at various temperatures (81). During the 30,000 time steps, with a time step of 2.0×10^{-15} s, migration of the single SrO molecule has not been observed over a wide temperature range of 300–1000 K. This may be interpreted as a result of the strong interaction between the TiO₂ atomic plane and the SrO molecule.

In the first period of the epitaxial growth, SrO molecules migrate on the TiO₂ atomic plane of the SrTiO₃ surface. During the formation of the SrO plane, they begin to migrate only on the constructed SrO plane and, after some period, search for remaining uncovered TiO₂ positions. The final process is more impor-tant for discussing the temperature dependence of the constructed SrO layer structure, since the flatness of the SrO layer is determined by the migration ability of the SrO molecules on the constructed SrO layer of the SrTiO3(001) surface. Hence, the surface migration process of a single SrO molecule on the $SrTiO_3(001)$ surface terminated by a SrO atomic plane has also been investigated. The results are much different from those in the case when TiO_2 was the topmost plane. Here, even the migration of the SrO molecule is not found at 300 K; the smooth migration of a SrO molecule is observed at 400 K. Moreover, the value of the diffusion coefficient increased drastically with the temperature. This sensitive temperature dependence of the SrO migration ability is likely to affect the quality and flatness of the constructed SrO layer on the $SrTiO_3(001)$ surface.

We have also calculated the adsorption energies of a single SrO molecule on both the $SrTiO_3(001)$ surfaces terminated by the TiO_2 and ones terminated by SrO atomic planes at 700 K during the migration process. The adsorption energy varies mainly from 10 to 50 kcal/mol, and the average adsorbtion energy is 32.4 kcal/mol (81), which is lower than that of a single MgO molecule on the MgO(001) surface (59.2 kcal/mol) (16). The lower adsorption energy of the SrO molecule on the $SrTiO_3(001)$ surface terminated by the SrO atomic plane may lead to a higher migration ability and a lower activation energy of the SrO molecule as compared to the MgO molecule on the MgO(001) surface. This result lends support to the view that the complete epitaxial growth of a SrO layer without defects on the $SrTiO_3(001)$ surface can be achieved at a lower temperature than the temperature required for the homoepitaxial growth of the MgO(001) surface because of the weaker interaction between the SrO atomic plane and the SrO molecule, as compared to that between the MgO surface and MgO molecule. On the other hand, the average adsorption energy of a single SrO molecule on the $SrTiO_3(001)$ surface terminated by the TiO_2 atomic plane is 84.1 kcal/mol, which is much larger than that on the $SrTiO_3(001)$ surface terminated by SrO. Such a strong interaction between the TiO_2 atomic plane and a SrO molecule may be the reason for the low diffusivity of a SrO molecule on the $SrTiO_3(001)$ surface terminated by the TiO_2 atomic plane.

Figure 20 shows the process of deposition of SrO molecules (spheres) on the $SrTiO_3(001)$ surface (atomic bonds) terminated by the TiO_2 atomic plane at 300 K. After the migration of SrO molecules on the $SrTiO_3(001)$ surface, a two-dimensional and epitaxial growth of the SrO thin layer is observed. The deposited SrO molecules retained the perevoskite structure and the (001)-oriented configuration during the MD simulation. After all 32 SrO molecules are deposited on the $SrTiO_3(001)$ surface, some defects are observed in the first SrO layer, and some SrO molecules have begun constructing the second layer. The small diffusivity of the SrO molecule at 300 K leads to the formation of defects in the constructed SrO layer because the deposited molecules are stacked close to the position where they hit the surface and did not reach defect sites at this low temperature.

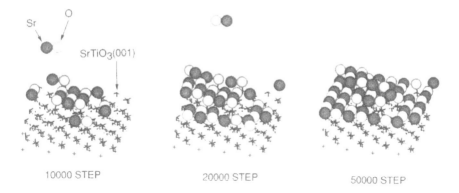

Figure 20   Epitaxial growth process of the SrO layer on a $SrTiO_3(001)$ surface terminated by the TiO_2 atomic layer at 300 K.

However, it is found that high temperature leads to the complete epitaxial growth of the SrO layer on the $SrTiO_3(001)$ surface. The MD simulation at 713 K shows that the SrO grows epitaxially, retaining the perevoskite structure and (001) orientation (Figure 21). Furthermore, the formation of a single, two-dimensional, uniform, flat layer of SrO without any defects has been observed after 50,000 steps. The high value of the diffusion coefficient is obtained at 700–1000 K, and this seems to help fill the defect sites on the surface effectively. Experimentally, Kawasaki and co-workers (84) have achieved epitaxial and two-dimensional growth of a single SrO layer on an atomically flat $SrTiO_3(001)$ surface terminated by the TiO_2 atomic plane at 713 K, which is in good agreement with our simulation results.

The experimental technique for obtaining a $SrTiO_3(001)$ surface terminated by a SrO atomic plane directly from a commercial $SrTiO_3(001)$ substrate without the deposition of the SrO layer is yet to be developed. However, the present results suggest that artificial construction of the $SrTiO_3(001)$ surface completely terminated by the SrO atomic plane is possible. We show that two different atomically flat $SrTiO_3(001)$ surfaces can be obtained. These surfaces are desirable for atomic control of metal oxide interfaces, as in ferroelectric/$SrTiO_3$ (e.g., $BaTiO_3/SrTiO_3$) and superconductor/$SrTiO_3$ (e.g., $YBa_2Cu_3O_{7-x}/SrTiO_3$) multilayers.

$YBa_2Cu_3O_{7-x}$ (YBCO) has a structure similar to that of $SrTiO_3$, and their lattice mismatch is less than 2%. Thus, $SrTiO_3$ is expected to serve as a good insulating barrier and a substrate for fabricating Josephson tunnel junctions based on YBCO, which are yet to be successfully demonstrated. Fabrication of a uniform YBCO/$SrTiO_3$ heterojunction has been extensivly studied using various experi-

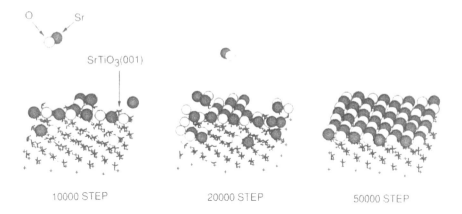

Figure 21 Epitaxial growth process of the SrO layer on a SrTiO₃(001) surface terminated by the TiO₂ atomic layer at 713 K.

mental techniques (104–106). In order to understand such a system, a detailed
investigation of the interface structure of the YBCO/SrTiO₃ substrate at atomic
level using theoretical techniques is necessary. The simulation of the interface
structure of a YBCO/SrTiO₃ bilayer has not been performed explicitly because
of the complexity of the materials system involved. Instead, we have investigated
the formation of a BaO/SrTiO₃(001) heterojunction. This study is useful for the
design and fabrication of YBCO/SrTiO₃ Josephson tunnel junctions, because the
BaO plane can be looked at as one part of the layered YBCO structure, which
has a stacked sequence of BaO/CuO₂/Y/CuO₂/BaO/CuO.

We have applied our crystal growth MD simulation code to the deposition
process of BaO molecules on the SrTiO₃(001) and SrO/SrTiO₃(001) substrates
(82) in order to examine whether atomically uniform BaO/SrTiO₃(001) and BaO/
SrO/SrTiO₃(001) heterojunctions can be fabricated (Figure 22). Our results indicate that during the process of deposition of BaO on the SrTiO₃(001) surface, a
flat two-dimensional BaO layer without any defects can be obtained at 700 K,
as shown in Figure 23. In this case, the deposited molecules are inclined to adhere
directly to the substrate instead of aggregating, due to the strong atomic interaction
between the BaO molecules and the surface. This leads to the formation of a
flawless thin BaO layer on a SrTiO₃(001) surface.

The [010]-direction heterointerface stress due to the lattice mismatch during
the entire heteroepitaxial growth process of BaO/SrTiO₃(001) has been calculated
(Figure 24). The stress gradually increases from the beginning of the emission
of BaO molecules (40 ps) during the heteroepitaxial process and reaches as high as
approximately 1.2 GPa. The linear increase in the stress with time is an interesting

(a)

(b)

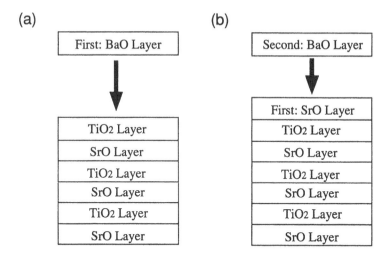

First: BaO Layer

↓

TiO2 Layer
SrO Layer
TiO2 Layer
SrO Layer
TiO2 Layer
SrO Layer

Second: BaO Layer

↓

First: SrO Layer
TiO2 Layer
SrO Layer
TiO2 Layer
SrO Layer
TiO2 Layer
SrO Layer

Figure 22 Models employed for the heteroepitaxial growth process of the BaO layer on (a) $SrTiO_3(001)$ and (b) $SrO/SrTiO_3(001)$ substrates. Here, $SrTiO_3(001)$ is terminated by a TiO_2 atomic plane.

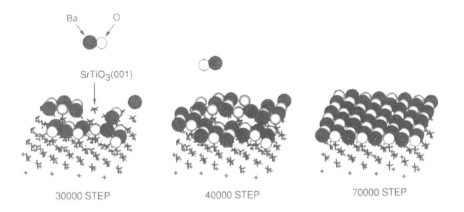

Figure 23 Epitaxial growth process of the BaO layer on $SrTiO_3(001)$ substrate terminated by a TiO_2 atomic plane at 700 K.

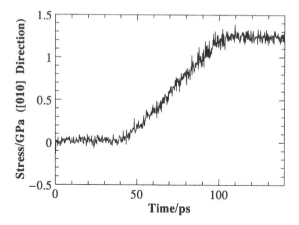

Figure 24 [010]-Direction stress of the BaO/SrTiO$_3$(001) heterojunction due to the lattice mismatch during the heteroepitaxial growth process at 700 K.

finding. We have expected that the large stress above is suddenly induced when the last defect on the constructed BaO layer is filled by a deposited BaO molecule to complete an epitaxial BaO film. Because the constructed BaO layer is attached directly to the TiO$_2$ atomic plane, and BaTiO$_3$ also has a stacked structure with alternating BaO and TiO$_2$ atomic planes, the BaO layer is incorporated as a part of the BaTiO$_3$ crystal (a = 4.012 Å) instead of a BaO crystal ($1/\sqrt{2}a$ = 3.905 Å). Hence, the lattice mismatch between the BaTiO$_3$ (a = 4.012 Å) crystal and the SrTiO$_3$ (a = 4.012 Å) crystal produces the above large stress at the heterointerface. It is surprising that the BaO layers grow epitaxially and uniformly on the SrTiO$_3$(001) substrate, even with such a large heterointerface stress. However, the large stress may disturb the subsequent fabrication of the YBCO/BaO/SrTiO$_3$(001) heterojunction with an atomically smooth surface and interface.

Figure 25 shows the process of continuous deposition of BaO molecules on the SrO/SrTiO$_3$(001) substrate at 700 K. The results indicate that a uniform BaO/SrO/SrTiO$_3$(001) heterojunction with an atomically smooth surface and interface can be fabricated at 700 K, as in the case of BaO deposition on SrTiO$_3$(001) terminated by TiO$_2$.

However, in this case a different trend in the calculated [010] heterointerface stress has been observed (Figure 26) as compared to that in the heteroepitaxial growth process of BaO deposition on SrTiO$_3$(001) terminated by TiO$_2$. Although the stress is slightly increased during the deposition of BaO (40–104 ps), almost no stress is induced at the heterointerface after the epitaxial growth. Here, the topmost single SrO layer of the SrO/SrTiO$_3$(001) substrate inherits the structure

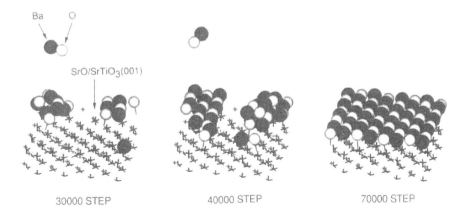

Ba   O

SrO/SrTiO₃(001)

30000 STEP   40000 STEP   70000 STEP

Figure 25   Epitaxial growth process of the BaO layer on a SrO/SrTiO₃(001) substrate at 700 K.

of the SrTiO₃(001) substrate because the SrO layer is attached directly to the TiO₂ plane of the SrTiO₃ substrate. Since the constructed BaO layer is attached directly to the SrO atomic plane, the BaO layer is constructed as a part of a BaO crystal instead of BaTiO₃. Furthermore, the lattice constant of the SrTiO₃ crystal ($a = 3.905$ Å) is the same as that of a BaO crystal ($1/\sqrt{2}a = 3.905$ Å); thus, the BaO/SrO/SrTiO₃(001) heterojunction does not gain stress due to the lattice mismatch.

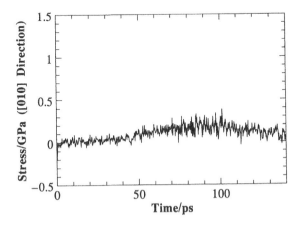

Figure 26   [010]-Direction stress of the BaO/SrO/SrTiO₃(001) heterojunction due to the lattice mismatch during the heteroepitaxial growth process at 700 K.

Kawasaki and co-workers applied the laser MBE technique to the formation of BaO layers on $SrTiO_3(001)$ terminated by a TiO_2 atomic plane and SrO/$SrTiO_3(001)$ at around 700 K. They observed persistent RHEED oscillations during the growth of BaO layers on SrO/$SrTiO_3(001)$ but not on $SrTiO_3(001)$ terminated by TiO_2 (107). Moreover, they confirmed the atomic flatness of the constructed BaO layer on the SrO/$SrTiO_3(001)$ substrate using atomic force microscopy (AFM).

On the basis of these experimental data and the present MD results, we confirm that an atomically uniform and smooth BaO/SrO/$SrTiO_3(001)$ heterojunction can be fabricated and that the BaO layers/single SrO layer is a suitable buffer layer for the YBCO/$SrTiO_3$ heterojunction. In conclusion, the utility of our new MD code has been successfully demonstrated in the clarification of the crystal growth mechanism and in the design and growth of atomically controlled structures on substrates.

D. Catalytic Properties of $V_2O_5(010)$ Surfaces (108–111)

Vanadium pentoxide (V_2O_5) is a very important material that has been extensively used as an n-type semiconductor, a cathode material in lithium batteries, and an industrial catalyst. As a catalyst, it is used in selective oxidation processes of hydrocarbons and sulfur dioxide, selective catalytic reduction (SCR) of nitric oxide by ammonia, and in photocatalytic reactions. The experimental studies show that the $V_2O_5(010)$ surface plays a crucial role in catalysis (112,113) and that the supported (010) surface monolayer has excellent catalytic activity (114,115). It is believed that the catalytic properties of a V_2O_5-based catalyst depend strongly on its ability to provide lattice oxygens for the reactions. The investigation of the V_2O_5 lattice oxygen is crucial for understanding the active sites and reaction mechanisms. Three different kinds of oxygen atoms exist in layered V_2O_5. They are a singly coordinated O_1, which is a vanadyl oxygen (V = O species), a dicoordinated O_2, and a tricoordinated O_3, which bridges two and three vanadium atoms, respectively. However, the role of lattice oxygens present in the vanadium pentoxide surface has not been clarified yet, mainly because of the limitation in experiments and computational techniques. There are many unresolved experimental issues concerning the active site and the oxidation process. A recent AFM study of the $V_2O_5(001)$ surface (116) has shown that the O_2 is more negatively charged than O_3 and has proposed O_2 as an active site. On the other hand, electronic spin resonance (ESR) spectroscopy and infrared (IR) spectroscopy studies have indicated that CO, SO_2, and C_2H_4 adsorb at the O_1 oxygen (117) or at O_3 oxygen centers (118). There have been several theoretical studies using the cluster approach to tackle this problem (119,120). However, this approach often fails to provide a comprehensive picture because of the cluster size limit. In recent years, with the rapid development of computational techniques, it has become feasible to deal with the adsorption systems using periodic boundary first-principle calculations. In the present section, it will be shown that the periodic DFT approach is very reliable for investigating the catalytic proper-ties of the $V_2O_5(001)$ surface. The combinatorial algorithm used here is divided into two steps. First, we select the surface model, which includes all possible active sites of the V_2O_5 surface, and investigate those sites using an accurate computational method. Second, we investigate the adsorption processes of differ-ent molecules using the method and model selected in the first step.

The CASTEP program (described in Section III.B) has been used to opti-mize the geometries of both the stoichiometric (010) surface and adsorption sys-tems at the generalized gradient approximation (GGA) level. After that, the elec-tronic structure has been analyzed using the DSolid program. The surface has been modeled as a periodic slab composed of a single unit-cell (SUC) monolayer, which had previously been successfully employed for band calculations (108) (Figure 27). The shortest distance between atoms belonging to successive slabs is greater than 8.3 Å. The interlayer interaction is not significant (109), and we can investigate the adsorption mechanism of small molecules on the V_2O_5 surface using this model. The adsorption energy (E_{ads}) has been calculated according to the following expression:

$$E_{ads} = E_{(adsorption\,system)} - [E_{(surface)} + E_{(molecule)}] \tag{28}$$

where $E_{(adsorption\,system)}$, $E_{(molecule)}$, and $E_{(surface)}$ are the total energies of the adsorp-tion system, an isolated molecule, and the V_2O_5 isolated surface, respectively. As already mentioned in Section III.A, the negative E_{ads} value corresponds to a thermodynamically stable adsorption system.

1. Reactivity of Lattice Oxygens and H Adsorption and OH Desorption Processes

One of the key steps in both selective oxidation of hydrocarbons and the selective catalytic reduction (SCR) reaction is the evolution of an H atom from a reactant by one of the surface oxygens to form a surface OH species. Therefore, a study of adsorption of the H atom at V_2O_5 lattice oxygens is necessary to understand the active sites and the reaction mechanism. To investigate the effect of both local chemical environment of an adsorption site and lattice relaxation, three different optimization conditions have been applied: (I) optimization of the O—H group in an orientation perpendicular to the surface while the geometry of the other atoms has been fixed; (II) optimization of all three coordinates of the O—H group for examining the influence of local environment of the adsorption site; (III) full optimization of all the constituent atoms of the adsorbate–substrate system. The obtained results show that all three lattice oxygens have high reactiv-ity for H adsorption and that the O_1 site is found to be the most active in an

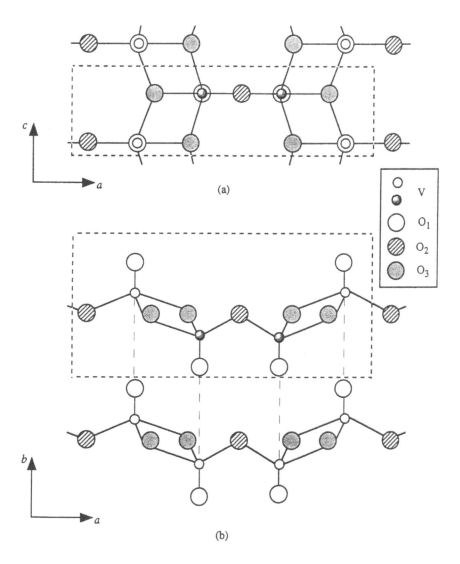

Figure 27 The periodic boundary model of $V_2O_5(010)$ used in the present study is indicated by broken lines. They are projections onto (a) the ac plane and (b) the ab plane. The closed small circles represent the exposed V atoms on the surface, and the opened small ones designate the V atoms beneath the singly coordinated oxygens.

unsupported V_2O_5 catalyst (Table 8), which is consistent with the experimental identification (113,117). Between optimizations I and II, there is no significant difference for H adsorption at the O_1 site as well as at the O_2 site. This indicates that there is almost no effect of the local environment at both O_1 and O_2 sites on their reactivity. However, the O_3 site exhibits a completely different behavior. For optimization condition II, the H atom binding to the O_3 site also interacts with its nearest O_2 site with a hydrogen bond. As a consequence, the adsorption energy is negatively increased by 13.8 kcal/mol. It shows that the local chemical environment of the O_3 site affects both geometry and energy. On the other hand, a remarkable difference is observed after full geometry optimization is performed at both O_1 and O_2 sites. A negative increment in adsorption energy of 15.8 and 21.3 kcal/mol has been obtained, respectively. These comparatively higher-energy differences show a significant contribution of the lattice relaxation on adsorption systems with respect to both the O_1 and O_2 sites. Similarly, full geometry optimization for H adsorption at the O_3 site also leads to a negative increment of adsorption energy by 13.5 kcal/mol where the surface relaxation is the main contributor.

When the full optimization (III) is performed, a large charge redistribution occurs. Surprisingly, charges on both the O_2 and O_3 sites are depleted. However, the O_1 site has accumulated more charge than in the cases of I and II optimizations. In the case of a clean $V_2O_5(010)$ surface, the vanadyl oxygen is the least ionic among these three oxygens, indicating that the O_1 site still has potential to gain extra electrons. Hence, the O_1 site exhibits a different behavior. In the case of H adsorption, the charge on the V atom beneath the O_1 adsorption site is almost identical to that of the clean surface, while the other three V atoms are reduced instead. On the other hand, the two V atoms bonded to the O_2 site and the two V atoms closer to the O_3 site are observed to become substantially reduced in the cases of H adsorption at the O_2 and O_3 sites, respectively.

The calculated vibrational frequencies of the O_1H, O_2H, and O_3H groups are 3792.17, 3658.26, and 3561.13 cm^{-1}, respectively. The least coordinated hydroxyl, O_1H, shows the highest frequency value, and the most coordinated hydroxyl, O_3H, exhibits the lowest one. These values are within the range of the

Table 8 Adsorption Energies of a H Atom at O_1, O_2, and O_3 Oxygen Sites Present in Monolayer SUC of $V_2O_5(010)$ Calculated by Optimization Methods Discussed in Text

Optimization method	I	II	III
E_{ads}, O_1 (kcal/mol)	−47.1	−46.4	−62.2
E_{ads}, O_2 (kcal/mol)	−39.9	−39.0	−60.3
E_{ads}, O_3 (kcal/mol)	−32.0	−45.8	−59.3

observation by Bond et al. (121). In particular, the calculated frequency of the O_1H group is very close to the experimentally obtained value (122).

A series of calculations were performed to evaluate the desorption ability, because the reactivity of surface oxygens depends on both their adsorption and desorption abilities. The OH desorption process was investigated by keeping the surface and OH group fixed at their equilibrium coordinates and varying only the position of the OH species on the surface. We considered the three types of oxygen atoms, and three structures were optimized.

The results show that the O_1H and O_3H species essentially reach their free positions, where they are out of the interaction with the (010) surface when they are 2.0 Å away, while the O_2H species still undergoes significant interaction with the surface at that distance. The removal of O_1H species from the (010) surface is found to require the least energy if one considers the effect contributed by the local environment of the adsorption site. These results indicate that the vanadyl oxygen not only has the strongest binding ability on the H atom but has also the greatest desorption ability of the hydroxyl groups to form the OH^- species and an oxygen vacancy.

2. Adsorption of H_2O

It is well known that the interaction of H_2O with metal oxide surfaces has important consequences for their catalytic behavior. Catalysts are typically exposed under ambient conditions, and water generally exists in both reactants and products. Numerous investigations on the adsorption states of water on titanium dioxides have been performed, but very few studies concerning the adsorption states of water on V_2O_5 have been undertaken, despite the fact that V_2O_5 is often supported on TiO_2 in many catalytic processes. Hence, we have looked at the adsorption mechanism of water on the $V_2O_5(010)$ surface in detail by considering all possible configurations of the adsorbate–surface system (110).

The water molecule has a large dipole moment and lone-pair electrons on the oxygen. Therefore, it is a good electron donor, and hydrogens of water molecules can interact with surface oxygens by hydrogen bonding. The adsorption on the V_2O_5 surface may be considered as a Lewis acid–base reaction. The H (Lewis acid) and the O atoms (Lewis base) of a water molecule are able to interact with the Lewis-base sites (surface oxygens) and Lewis-acid sites (surface vanadiums), respectively. Figure 28 shows the equilibrium structures of the adsorption systems in the case of adsorption of a H_2O molecule on the $V_2O_5(010)$ surface at all possible adsorption sites. It has been found that the molecular adsorption of water at three kinds of surface oxygens and the exposed surface vanadium are energetically stable and that their adsorption abilities decrease in the order of O_1 ($E_{ads} = -23.1$ kcal/mol) > V (-15.5 kcal/mol) > O_2 (-12.9 kcal/mol) > O_3 (-9.69 kcal/mol). This implies that the surface vanadyl oxygen is the most active,

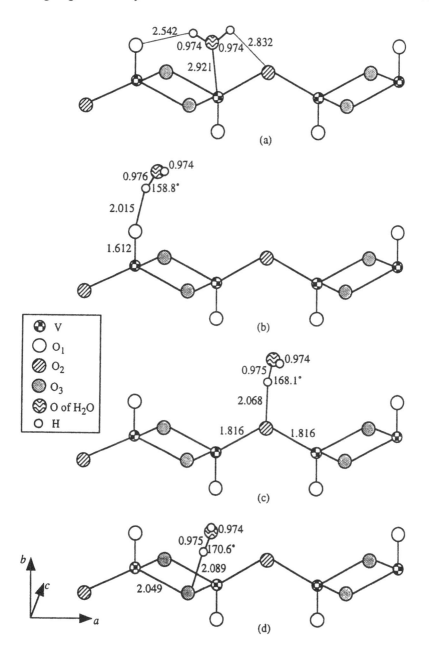

Figure 28 Equilibrium interatomic distances (Å) and hydrogen bond angles (deg) of the adsorbed water molecule at (a) exposed V, (b) O_1 (c) O_2, and (d) O_3 sites on the $V_2O_5(010)$ surface.

as in the case of H adsorption. As shown in Figure 28, the H-bond distance between the proton and each adsorption site increases in the order of $O_1 < O_2 < O_3$, indicating that the interaction of the O—H group of water with the adsorption site decreases accordingly. That is, the most active adsorption site, O_1, forms the strongest H-bond, while the least active O_3 site produces the weakest one. The significance of these phenomena is that the hydrogen bonding plays a key role in the molecular adsorption of water at the surface oxygens of V_2O_5. Because the vanadyl oxygen forms the strongest H-bond, it acts as the preferable adsorption site, rather than the O_2 and O_3 sites. In adsorption, the adsorbate of water donates its electrons to the substrate and becomes positively charged, and the substrate becomes polarized (Table 9). Such a redistribution of charge involving the surface is believed to be a very important factor in the water adsorption. To find a quantitative correlation between the redistribution of the charge and the adsorption ability, we have focused mainly on the difference in charge on the adsorption site before and after adsorption (Δq_{site}) and on the increased charge on the water molecule $[(H_2O)^{x+}]$ due to the adsorption. It is observed that the adsorption abilities of the surface oxygens correlate with the ratio $\Delta q_{site}/(H_2O)^{x+}$, as shown in Figure 29. The curve indicates a linear relationship between the adsorption energies and the ratio of $\Delta q_{site}/(H_2O)^{x+}$. The higher ratio corresponds to the higher adsorption abilities of the oxygen site. This trend has not been observed for the adsorption of H atoms experimentally. The adsorption of H atoms at all different surface oxygens is energetically very stable, and the vanadyl oxygen acts as the preferable adsorption site. This is probably due to the difference in the bonding properties: The interaction of water with the surface oxygen is mainly contributed by the hydrogen bonding, whereas the interaction of H atoms with these oxygens is attributed to a covalent force.

In addition, change in the charges on the vanadium atoms connected to O_1, O_2, and O_3 sites reveals that the donation from the surface oxygen to the water molecule is larger for the O_1 site than for the other two. The net charge transfer is found to increase in the order of $O_1 < O_2 \approx O_3$, while the stability order decreases in the order of $O_1 > O_2 > O_3$. This reverse trend proves that the stability of the adsorption system is decided mainly by the donating ability of the surface oxygen sites rather than the acceptance of electrons from the water molecule.

3.   Adsorption of NH$_3$ on the Brönsted and Lewis Acid Sites of V$_2$O$_5$ (010)

Many different types of reaction mechanisms have been proposed for the SCR of NO by NH_3 using the V_2O_5 catalyst in the presence of oxygen (123–127). The focus of the argument toward the SCR mechanisms is mainly on the adsorption states of both NO and NH_3 molecules, which are related directly to the

Table 9 Atomic Charges on Adsorbate (q_{Hads}, q_H, and q_O) and Adsorption (q_{site}) Sites with Respect to the Molecular Adsorption System.

	Adsorption site			
	V	O_1	O_2	O_3
q_{Hads}	+0.292	+0.297	+0.324	+0.338
q_H	+0.289	+0.330	+0.323	+0.319
q_O	−0.496	−0.489	−0.495	−0.505
$(H_2O)^{x+}$	+0.085	+0.138	+0.152	+0.152
q_{site}	+1.240	−0.456	−0.621	−0.747
	(+1.261)	(−0.316)	(−0.527)	(−0.681)
Δq_{site}	−0.021	−0.140	−0.094	−0.066

The values in the brackets are charges of atoms on the clean surface. Δq_{site} is the charge difference of the adsorption site before and after adsorption, and $(H_2O)^{x+}$ denotes the increased charge on water molecule due to adsorption.

pathway of the reaction. In general, it is accepted that the ammonia strongly adsorbs on the Brönsted site of V_2O_5 to form the NH_4^+ ion, which is followed by the SCR process (124–126). However, different processes where the ammonia molecule is adsorbed on Lewis acid sites of the exposed vanadiums have also been reported (127). Therefore, it is necessary to scrutinize the adsorption states of ammonia on V_2O_5 in order to understand both the SCR reaction and the experimental acidic properties of the V_2O_5 surface.

The hydroxyl V_2O_5 surface, formed by atomic hydrogen adsorption on the three kinds of lattice oxygens, has been used to reproduce the real surface with various Brönsted acid sites (111). It has been observed that an ammonia molecule is strongly adsorbed on these sites (Figure 30). As shown in Table 10, all three adsorption systems are calculated to be very stable energetically. The adsorption ability decreases in the order corresponding to the hydroxyl groups ($O_1H > O_3H > O_2H$). This order indicates that the adsorption of ammonia on the O_1H group is the most favorable, which is in good agreement with an earlier observation (124,125). Moreover, at the O_1H site (Figure 30a) the proton of the newly formed NH_4^+ ion strongly interacts with the vanadyl oxygen adjacent to the O_1H species in the form of a strong H-bond with a distance of 1.58 Å. This is in agreement with experimental data (126). Similarly, one proton of the NH_4^+ ion adsorbed on the O_3—H hydroxyl group also interacts with the vanadyl oxygen through the H-bond (1.86 Å). This bond is weaker than that observed in the case of adsorption at O_1—H.

It is also found that a NH_3 molecule donates its electrons to the surface (Table 10), and the surface becomes polarized. The N atom of an NH_4^+ ion

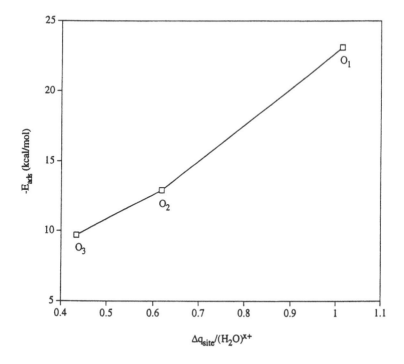

Figure 29 Correlation between the adsorption energy (E_{ads}) and the ratio of $\Delta q_{site}/$ $(H_2O)^{x+}$ ($\times$ 100%). Δq_{site} is the difference in charge on the adsorption site before and after the adsorption. $(H_2O)^{x+}$ is the increased charge on the water molecule due to the adsorption. The corresponding linear regression equation is $y = 0.235\,x - 0.982$. The correlation coefficient is $r = 0.996$.

belonging to the most stable adsorption system with the O_1H site (Figure 30a) possesses the least negative charge, while the N atom of the least stable adsorption system with the O_2H site (Figure 30b) takes the most negative charge (Table 10). This seems to indicate that the charge on the N atom of the NH_4^+ ion is correlated with the corresponding adsorption ability. The length of the newly formed N—H bond in NH_4^+ ions decreases in the same order as in the adsorption ability, as shown in Table 10. This result clearly shows that it is easier for the O_1H to release the proton and subsequently interact with the NH_3 molecule than for the O_2H and O_3H species. In addition, interatomic distances of the O—H species after the adsorption are significantly elongated, and the increase in the O—H bond length is found to correlate with the adsorption ability and length of the newly formed N—H bond. The adsorption ability of NH_3 at hydroxyl species

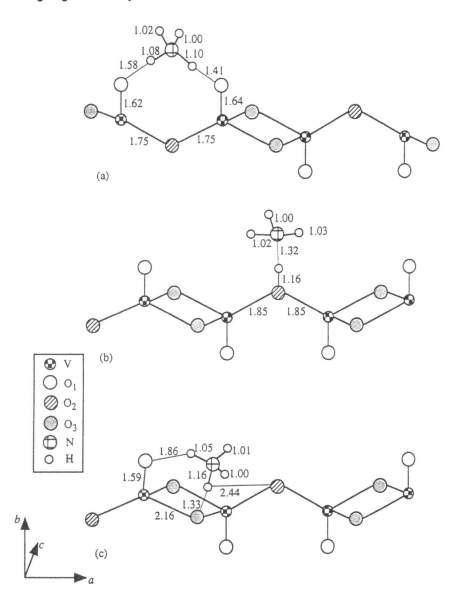

Figure 30 Geometries of ammonia adsorption at the Brönsted acid sites consisting of (a) singly coordinated oxygen, (b) dicoordinated oxygen, and (c) tricoordinated oxygen sites on $V_2O_5(010)$.

Table 10 Selected Parameters of NH_3 Adsorption on Brönsted Acid Sites of $V_2O_5(010)$ Surface (Corresponding Values Before Adsorption Are in Parentheses)

Adsorption site	O_1H	O_2H	O_3H
q_N	$-0.492(-0.577)$	$-0.591(-0.577)$	$-0.493(-0.577)$
q_H	$+0.339(+0.300)$	$+0.480(+0.388)$	$+0.447(+0.421)$
q_O	$-0.658(-0.511)$	$-0.688(-0.567)$	$-0.764(-0.621)$
	$(-0.618)^a$		$(-0.678)^b$
q_V	$+1.267(+1.275)$	$+1.163(+1.180)$	$+1.190(+1.153)$
	$(+1.255)^a$	$+1.163(+1.180)$	$+1.190(+1.153)$
			$+1.297(+1.280)$
$(NH_4)^{x}+$	$+0.726$	$+0.712$	$+0.786$
$d_{N-H}(\text{Å})$	1.10	1.32	1.16
$d_{O-H(\text{Å})}$	1.41(0.97)	1.16(0.98)	1.33(0.99)
			$(2.44)^c$
$\Delta d_{O-H}(\text{Å})$	0.44	0.18	0.34
E (kcal/mol)	-27.4	-15.0	-20.7
	(-44.0)	(-34.3)	(-35.9)

[a] Charges on the nearest atom of V=O group.
[b] Charge on the nearest atoms of O_2 atom.
[c] Distance between the H and the closet O_2 atom.

seems to be determined by the ability of the species to provide lone-pair electrons of the hydroxyl oxygens to the adsorbed NH_3 species rather than the electron-transfer ability from the adsorbate to the surface.

Adsorption of NH_3 on the Lewis acid side depends on the interaction of its highest occupied molecular orbital (HOMO) and lowest unoccupied molecular orbital (LUMO) with the electronic orbitals of vanadium atoms. The exposed V atoms are electron deficient and can receive electrons from the nitrogen lone-pair orbitals of NH_3 through interaction of vanadium and NH_3 orbitals. Due to the difference in the nature of the interaction, the adsorption on the Lewis acid sites is expected to be much weaker than that on the Brönsted acid site. Figure 31 displays the fully relaxed geometry of the two forms of ammonia adsorption. In the first, the electron pair of the ammonia molecule is donated to the vanadium site, which leads to the formation of the V—N bond almost perpendicular to the surface plane (on-top site, labeled T in Table 11). The second form is the configuration where the ammonia binds to both the singly coordinated oxygen and its counterpart vanadium (bridging site, labeled B in Table 11). The calculated adsorption energies (Table 11) of both configurations are very close. The former is more stable than the latter by 0.9 kcal/mol. Due to the weak interaction, as predicted, these energies are found to be much lower than those in the case of

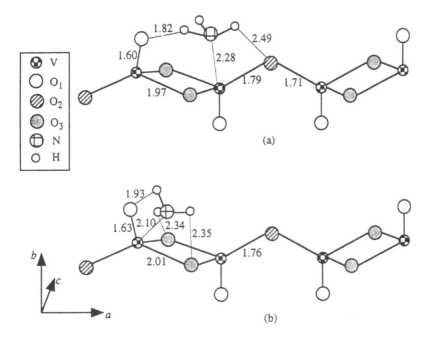

Figure 31 Geometries of ammonia adsorption on the Lewis acid (a) on-top and (b) bridging sites present in the $V_2O_5(010)$ surface.

the Brönsted acid sites. These results reveal that the coordination interaction between the N and the V site plays a very important role in bonding. On the other hand, it has been found that the hydrogen bonding has significant contribution because the H-bond lengths are observed to be 1.82 Å for the on-top site (T) and 1.93 Å for the bridging site (B). The bonding distances indicate that the interaction between N and V atoms for the T site is weaker than that of the B site, but the H-bonding ($O_1 \cdots H$—N) for the former is stronger than that for the latter. This suggests that the similarity in interaction leads to a similar stability order.

At both Brönsted and Lewis acid sites, it is found that coordination interaction and the hydrogen bonding between N—H and the vanadyl oxygen contribute to the bonding. As for the adsorption at the Brönsted site, the H-bonding interaction plays a crucial role, and the ammonium species is formed when NH_3 adsorbs to the O_1H group. This is in agreement with IR observations. Moreover, the hydroxyl group containing the vanadyl oxygen exhibits high reactivity toward ammonia adsorption.

Table 11 Optimized Interatomic Distances and Adsorption
Energies (per NH_3 molecule) with Respect to NH_3 Adsorption
on Lewis Acid Sites of $V_2O_5(010)$ (LDA Values are in
Parentheses)

Adsorption site	On-top (T)	Bridging (B)
d_{V-N}	2.28	2.10
d_{O1-H}	1.82	1.93
d_{O2-H}	2.47	2.96
d_{O3-H}	2.60	2.34;2.35
E_{ads} (kcal/mol)	−2.3	−1.4
	(−30.6)	(−30.3)

Summarizing the results obtained by periodic DFT calculations in this section, we have successfully implemented an algorithm and a model that accurately predict catalytic properties of the V_2O_5 surface. We have demonstrated that the present methodology within the combinatorial strategy can be used to investigate the adsorption processes of different molecules, such as oxides of sulfur, carbon, and nitrogen. Since the properties of catalysts can be modified by various combinations of supports, the role of a large number of promoters on a V_2O_5 catalyst can also be estimated using the CCC approach. Such results will be highly valuable in explaining the available experimental data and in designing new catalysts based on vanadium pentoxide.

E. Permeation Dynamics of Gas Molecules: An Application of Molecular Dynamics to the Design of Inorganic Membranes (128–132)

Porous and dense inorganic membranes have been investigated extensively as possible gas separation media. In particular, porous membranes exhibit high permeability relative to dense membranes and high thermal and chemical stability relative to organic membranes (133). In order to design a high-performance membrane, investigation of the transport mechanism in membranes is needed. The transport mechanism depends on various factors, such as the intrinsic properties of the molecules, the membrane structure, the affinity of the molecule for the membrane, and the physical conditions of separation processes (temperature, pressure, etc.). The separation mechanism in a microporous membrane has not been well established. For example, the experimental studies on permeation of *iso*- and *n*-butane through a zeolite membrane have showed that *n*-butane has a higher permeance than *iso*-butane (134,135). However, other studies (136) have observed that *iso*-butane shows greater permeability than *n*-butane at high temperatures.

In this section, we present our recent results of MD studies on different inorganic membrane systems. The advantage of the present methodology is the possibility of obtaining an accurate estimate of a complex separation mechanism and predicting new functional materials for this process in a relatively short period of time. This feature makes this methodology ideal for combinatorial computational chemistry.

We used a modified MD program for our calculations. Our modifications were related to the temperature control and the many-body potential describing gas molecules. The brief explanation of the MD simulations was given in Section II.A, and the details of the modified MD code are described elsewhere (128,130).

1. Permeation Based on Knudsen Flow Mechanism

For gas transportation in inorganic membranes with a pore size of less than 1 nm, the contribution of viscous flow can be neglected and the Knudsen flow becomes the most important transport mechanism. Calculations have been performed under a condition where the Knudsen diffusion mechanism is obeyed. The permeability of the system governed by the Knudsen flow is given by (137)

$$q = 4/3 \times pr(2RT/\pi M)^{1/2}(n_1/V_1 - n_2/V_2)/d, \tag{29}$$

where d is the membrane thickness, p is the porosity, V is the volume, and n is the number of molecules. The subscripts 1 and 2 refer to the gas and the vacuum phase, respectively. The MD simulations have been performed using an *NVT* ensemble (where number of particles, volume, and temperature are held constant). Helium and argon are the gas atoms used with a Knudsen factor of 0.94 and 0.71, respectively. We have employed a MgO ceramic membrane model with a thickness of 15 Å and with a cylindrical pore placed at the center of the membrane having a diameter of 11 Å. Another poreless MgO membrane (bulkhead membrane) with a thickness of 2 Å is added to the bottom of the unit cell to prevent periodic diffusion. The driving force for the permeation is the pressure difference between the gas and vacuum phases divided by the membrane.

One hundred twenty-eight helium atoms have been placed above the MgO membrane, and the temperature is kept at 773 K. In order to attain the condition of Knudsen flow, a very weak affinity of the gas for the membrane has been maintained. Figure 32 presents the computer graphic images of the He permeation through the membrane after 0.5, 24.0, and 45.5 ps. The helium gas flows directly through the pore without any adsorption on the membrane. The Knudsen flow is reproduced well by our simulation method.

We then investigated the permeation rate for a mixture of He-Ar gas (He = 64 atoms, Ar = 64 atoms) using the same procedure. Since the permeability

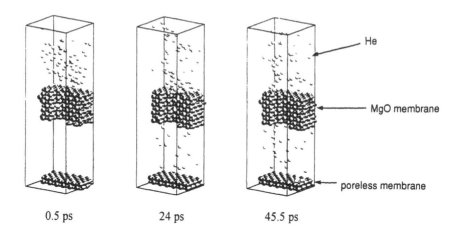

Figure 32   He permeation through a MgO membrane at 773 K.

depends on the molecular weight, the light species are expected to permeate faster. However, the separation factor decreases with increase in the simulation time. This behavior can be explained as follows. In the beginning of simulation, the He atoms permeate faster; consequently, the He concentration decreases in the gas phase. This leads to an increase in the Ar concentration in the gas phase, which in turn decreases the separation factor with time.

2.   Permeation Based on Other Flow Mechanisms

The separation factor in Knudsen flow is defined as the inverse of the square root of the molecular mass. Therefore, for molecules having similar masses, other transport mechanisms are required to achieve a high separation factor. The capillary condensation and the surface diffusion mechanisms are based on the affinity differences of the molecules, where diffusion selectivity favors the molecules with the strongest affinity for the membrane surface. These two mechanisms can be considered as suitable for separating molecules of similar weight and size, such as CO_2 and N_2.

Thus, the affinity of gas molecules for the membrane can be used for the direct separation of CO_2 at a high temperature ($>800°C$) from the exhaust gas. CO_2 is released from almost all stationary sources, such as thermal power plants and fossil fuel combustion in large factories (137). We have employed MD simulations to investigate the effect of affinity for CO_2 separation from CO_2/N_2 mixed gas (131). The affinity strength in this simulation is represented by the following interatomic potentials:

$$U(r_{ij}) = 4u_{ij} \left\{ \left(\frac{r^*_{ij}}{r_{ij}} \right)^6 - \left(\frac{r^*_{ij}}{r_{ij}} \right)^{12} \right\}$$

(30)

where u_{ij} is the affinity strength and r^*_{ij} is the interatomic distance. Different affinity models are applied in order to examine their effect on separation behavior. All calculations have been performed at 1073 K because of our interest in high-temperature separation processes. Figure 33 shows the results obtained from the simulated diffusion of CO_2 and N_2 on a MgO membrane using the weak-affinity model. In this model, the affinity of gas molecules for the wall of a membrane atom is 0.2 kcal/mol (u_{ij}). It has been observed that both CO_2 and N_2 diffuse through the membrane to the vacuum phase in a given simulation time. The change in the number of molecules in the vacuum phase with an increment in time steps is plotted in Figure 34. This is an indication that N_2 has greater diffusion ability than CO_2. Selective CO_2 separation does not seem to take place. The number of permeated molecules in the vacuum phase increased logarithmically in the beginning and saturated at 70 ps. It follows from this result that after 70 ps the driving force of the permeation from the gas phase to the vacuum phase reduces, due to the decrease in pressure difference between the phases, and the phases reach an equilibrium state.

To study the strong-affinity model, we have applied different affinity strengths for CO_2 and N_2 (1.2 and 0.2 kcal/mol, respectively). It is found that both molecules could diffuse to the vacuum phase at the beginning of the simulation (Figure 35). However, with increasing time, CO_2 molecules are captured and

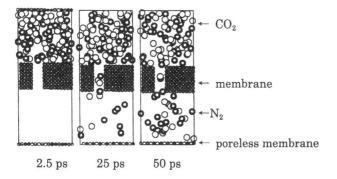

2.5 ps   25 ps   50 ps

Figure 33   The dynamic behavior of the model represented a weak affinity between the CO_2 molecule and the wall of the membrane. The affinity of CO_2 and N_2 for the wall of the membrane was 0.2 kcal/mol. Light sphere molecules represent CO_2, and the dark ones represent N_2.

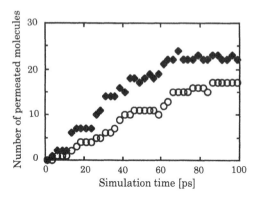

Figure 34 Change in number of permeated molecules as a function of the simulation time in a weak-affinity model. Open circle represents CO_2, and the closed symbols correspond to N_2.

preferentially condensed in the micropore. Because the lengthy time-separation process is closer to the real situation, further dynamic behavior has been investigated. For this purpose, we have performed a simulation of the process in which the membrane pore has already been filled with permeable gas molecules (Figure 36). At any time step, CO_2 molecules are preferentially condensed in the micropore and block the diffusion of N_2 into the micropore. Figure 37 shows the change in the number of CO_2 and N_2 molecules in the vacuum phase. It shows the occurrence of selective CO_2 permeation, while the permeation of N_2 molecules is suppressed. This result shows that the affinity of the membrane is important in the separation of CO_2 from a CO_2/N_2 mixed gas at high temperatures.

Recently, zeolite membranes have gained much attention as novel membrane materials. We have also investigated the separation process of *iso-* and *n*-butanes through a ZSM-5 silicate membrane (132). Figure 38 shows the MD calculations results for *n*-butane single-gas diffusion through the membrane at 373 K. After 40 ps, all allowed spaces for adsorption in the zeolite channels are saturated, but the molecules are not permeated yet. At 86 ps, the permeation through the silicate membrane has begun. Permeation starts only after the saturation of the membrane; the diffusion mechanism is then through capillary condensation or surface diffusion. This is in agreement with an experimental observation, which has reported the condensation of *n*-butane inside the silicalite pores during permeation (135). In the case of *iso-*butane, permeation is not observed even after 250 ps (Figure 39). Only some molecules are found to diffuse into the straight channels. This indicates that the permeance of *iso-*butane is significantly smaller than that of *n*-butane. This result is also consistent with experimental data [134].

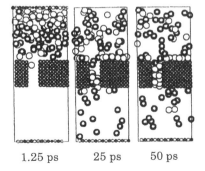

1.25 ps   25 ps   50 ps

Figure 35 The dynamic behavior of the model represented a strong affinity between the CO_2 molecule and the wall of the membrane. The affinity of CO_2 and N_2 for the wall of the membrane was 1.2 and 0.2 kcal/mol, respectively.

Based on the real-time visualization, we have observed the aggregation of *iso*-butane molecules around the entrance of a straight channel. This is because the molecule size is close to the pore diameter. Most of the molecules collide with the pore wall, which acts to decrease their velocity and enhances their condensation around the entrance of the pore.

    In order to analyze the effect of temperature on the adsorption, the number of species adsorbed in the silicate pores per picosecond at 373 and 773 K have been determined. The results are shown in Figure 40. At 373 K, the amount of *n*-butane increases rapidly with time and then reaches a constant value. This indicates that the transport mechanism is governed by the interaction of molecules

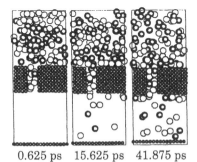

0.625 ps   15.625 ps   41.875 ps

Figure 36 Further dynamic behavior of the model representing a strong affinity. The initial configuration is taken from the result in Figure 35 at 50 ps.

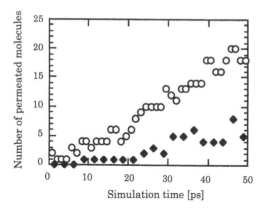

Figure 37 Change in the number of permeated molecules as a function of the simulation time in a strong-affinity model. Open circles represent CO_2, and the closed symbols correspond to N_2.

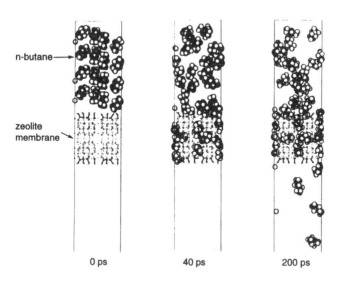

Figure 38 Process of permeation of *n*-butane gas through a silicalite membrane.

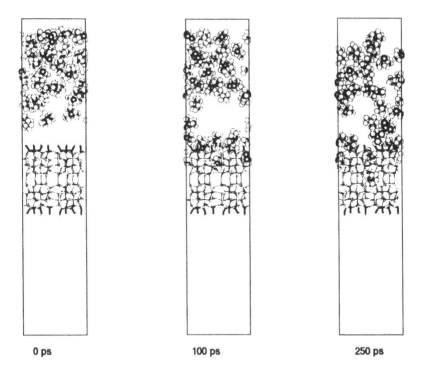

0 ps   100 ps   250 ps

Figure 39   Process of diffusion of process of *iso*-butane through a silicate membrane at 373 K.

with the pore wall. At 773 K, the amount of *n*-butane is smaller than that at 373 K because the contribution of the interaction of molecules with the pore is reduced at high temperatures. In the case of *iso*-butane, its permeation is not observed after 250 ps, even at high temperatures. The amount of adsorbed *iso*-butane (Figure 39) is smaller than that of *n*-butane, and the effect of temperature seems less important. The present calculations show that a ZSM-5 membrane can be used for the separation of *iso*- and *n*-butane isomers.

In summary, our molecular dynamics studies are effective in understanding gas permeation through inorganic membranes. The results are in good agreement with experimental data. We have shown that our molecular dynamics program can contribute substantially to designing novel, effective inorganic membranes.

IV. CONCLUSIONS

Computational chemistry is rapidly becoming an essential tool for the investigation of a variety of physical and chemical properties of materials. Improvements

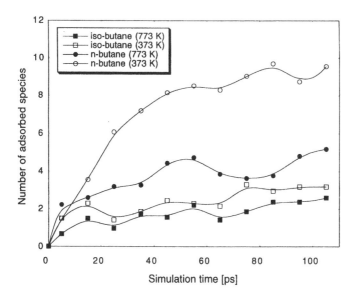

Figure 40   Amounts of adsorbed *iso-* and *n*-butane in silicalite pores at 373 K and 773 K obtained from MD calculations.

in computer technology will continue to increase the possibilities of molecular modeling. However, the benefits of simulation will also come from improvements in theory, in algorithms, and in overall strategy. The CCC approach is an attractive method with high potential in various fields of applications, especially in heterogeneous catalysis. And CCC can play an important role in the development of new multicomponent materials with high-performance and previously unexplored properties by screening a large number of different constituent elements in targeted compounds. The CCC approach can also provide trends and new insights that may be difficult to define experimentally. We continue using this approach for various urgent topics related to catalysis, environmental chemistry, and materials science. Recently, we have successfully applied the CCC approach to design new types of catalysts, which can be used in the Fisher–Tropsch synthesis for the production of ecologically high-quality transportation fuels (138). The obtained results are in agreement with available experimental data.

Currently, we use the most favorable simulation methods among the well-validated approaches in the framework of the combinatorial strategy. Despite significant research results presented by computational chemistry, the problems of computational time and accuracy remain for each of the computational methods. Density-functional theory is frequently used to accurately predict the electronic properties for molecules, solids, and surfaces in ground state, but only a quantitative understanding of the properties in excited states, such as bandgap and reaction barrier, can be obtained. High-level quantum chemical calculations remain prohibitively expensive for large systems, and one is always faced with tradeoffs in terms of system size, quality of basis sets, and level of theory. To make the CCC approach even more attractive, researchers need to continue developing new computational methods that can be applied for investigating large systems with more favorable computational costs and accuracies comparable to traditional first-principle methods.

Recently, first-principle quantum chemical MD simulations have gained much attention in investigating the dynamics of catalytic reaction under different temperature conditions. However, the calculation time for this method is very large, and current calculations are limited by the size of the system (about 100 atoms) as well as the electronic structure of the chemical elements in this system. Thus, it is also not possible to calculate the reaction dynamics on the catalysts with a large number of metal atoms. To overcome these limitations, we have developed a new method that can accurately simulate the dynamic behavior of catalytic reactions for systems containing a huge number of atoms. This method is based on our accelerated quantum chemical molecular dynamics simulation algorithm. In this method, the parameters are fitted to first-principle calculations. Thus, the accuracy of calculation is comparable to that of the first-principle methods, but computation time is significantly reduced as compared with first-principle calculations. Now we are in the process of applying this algorithm to investigating the dynamic behavior of different adsorption species over catalytic surfaces with more than 1000 atoms per unit cell, which is very close to a realistic structure of supported metal catalysts. It is also important to note that such a large system is not possible to calculate by using first-principle molecular dynamics simulations. Our preliminary results show that the electronic and structure configurations for such a large model can be accurately and rapidly calculated. We believe that methods such as these will pave the way for new directions in the CCC approach in the future.

REFERENCES

1. Hanak, J.J. The "multiple-sample concept" in materials research: synthesis, compositional analysis and testing of entire multicomponent systems. J Mater Sci. 1970, 5, 964–971.
2. Bunin, B.A.; Plunkett, M.J.; Ellman, J.A. The combinatorial synthesis and chemical and biological evaluation of a 1,4-benzodiazepine library. Proc Natl Acad Sci. 1994, 91, 4708–4712.

3. Pollack, S.J.; Jacobs, J.W.; Schultz, P.G. Selective chemical catalysis by an antibody. Science. 1986, 234, 1570–1573.

4. Rohrer, S.P.; Birzin, E.T.; Mosley, R.T.; Berk, S.C.; Hutchins, S.M.; Shen, D.M.; Xiong, Y.; Hayes, E.C.; Parmar, R.M.; Foor, F.; Mitra, S.W.; Degrado, S.J.; Shu, M.; Klopp, J.M.; Cai, S.-J.; Blake, A.; Chan, W.W.S.; Pasternak, A.; Yang, L.; Patchett, A.A.; Smith, R.G.; Chapman, K.T.; Schaeffer, J.M. Rapid identification of subtype-selective agonists of the somatostatin receptor through combinatorial chemistry. Science. 1998, 282, 737–740.

5. Xiang, X.D.; Sun, X.; Briceño, G.; Lou, Y.; Wang, K.A.; Chang, H.; Wallace-Freedman, W.G.; Chen, S.W.; Schultz, P.G. A combinatorial approach to materials discovery. Science. 1995, 268, 1738–1740.

6. Briceño, G.; Chang, H.; Sun, X.; Schultz, P.G.; Xiang, X.D. A class of cobalt oxide magnetoresistance materials discovered with combinatorial synthesis. Science. 1995, 270, 273–275.

7. Wang, J.; Yoo, Y.; Gao, C.; Takeuchi, I.; Sun, X.; Chang, H.; Xiang, X.D.; Schultz, P. Identification of a blue photoluminescent composite material from a combinatorial library. Science. 1998, 279, 1712–1714.

8. Danielson, E.; Golden, J.H.; Mcfarland, E.W.; Reaves, C.M.; Weinberg, W.H.; Wu, X.D. A combinatorial approach to the discovery and optimization of luminescent materials. Nature. 1998, 389, 944–948.

9. Matsumoto, Y.; Murakami, M.; Jin, Z.; Ohtomo, A.; Lippmaa, M.; Kawasaki, M.; Koinuma, H. Combinatorial laser molecular beam epitaxy (MBE) growth of Mg-Zn-O alloy for bandgap engineering. Jpn J Appl Phys. 1999, 38, L603–L605.

10. Payne, M.C.; Teter, M.P.; Allan, D.C.; Arias, T.A.; Joannopoulos, J.D. Iterative minimization techniques for ab initio total-energy calculations: molecular dynamics and conjugate gradients. Rev Mod Phys. 1992, 64, 1045–1097.

11. Yajima, K.; Ueda, Y.; Tsuruya, H.; Kanougi, T.; Oumi, Y.; Ammal, S.S.C.; Takami, S.; Kubo, M.; Miyamoto, A. Combinatorial computational chemistry approach to the design of $deNO_x$ catalysts. Appl Catal A. 2000, 194–195, 183–191.

12. Tse, J.S. Molecular Modelling and Related Computational Techniques. In: Comprehensive Supramolecular Chemistry; Atwood, J.L., Davies, J.E.D., MacNicol, D.D., Vögtle, F., Lehn, J.M., eds.; Elsevier: Amsterdam, 1996; Vol. 8, 593–616.

13. Allen, M.P.; Tildesley, D.J. Computer Simulation of Liquids; Oxford University Press: Oxford, 1987, 1–385.

14. McQuarrie, D. Statistical Mechanics; Harper & Row: New York, 1976, 1–641.

15. Verlet, L. Computer "experiments" on classical fluids. I. Thermodynamical properties of Lennard–Jones molecules. Phys Rev. 1967, 159, 98–103.

16. Kubo, M.; Oumi, Y.; Miura, R.; Fahmi, A.; Stirling, A.; Miyamoto, A.; Kawasaki, M.; Yoshimoto, M.; Koinuma, H. Layer-by-layer homoepitaxial growth process of MgO (001) as investigated by molecular dynamics, density functional theory, and computer graphics. J Chem Phys. 1997, 107, 4416–4422.

17. Kawamura, K. Interatomic potential model for molecular dynamics simulations of multicomponent oxides. In: Molecular Dynamics Simulations; Yonezawa, F., ed.; Springer: Berlin, 1992, 88–97.

18. Parr, R.G.; Yang, W. Density-Functional Theory of Atoms and Molecules; Oxford University Press: New York, 1989, 1–333.

19. Kohn, W.; Becke, A.D.; Parr, R.G. Density-functional theory of electronic structure. J Phys Chem. 1996, 100, 12974–12980.

20. Fermi, E. A statistical method for the determination of some properties of atom. II Application to the periodic system of elements. Z Phys. 1928, 48, 73–79.

21. Slater, J.C. A simplification of the Hartree–Fock method. Phys Rev. 1951, 81, 385–390.

22. Hohenberg, P.; Kohn, W. Inhomogeneous electron gas. Phys Rev B. 1964, 136, 864–871.

23. Kohn, W.; Sham, L.J. Self-consistent equations including exchange and correlation effects. Phys Rev A. 1965, 140, 1133–1138.

24. Vosko, S.H.; Wilk, L.; Nusair, M. Accurate spin-dependent electron liquid correlation energies for local spin density calculations: a critical analysis. Can J Phys. 1980, 58, 1200–1211.

25. Becke, A.D. Density-functional exchange-energy approximation with correct asymptotic behavior. Phys Rev A. 1988, 38, 3098–3100.

26. Becke, A.D. Density-functional thermochemistry. I. The effect of the exchange-only gradient correction. J Chem Phys. 1992, 96, 2155–2160.

27. Lee, C.; Yang, W.; Parr, R.C. Development of the Colle–Salvetti correlation-energy formula into a functional of the electron density. Phys Rev B. 1988, 37, 785–789.

28. Perdew, J.P. Density-functional approximation for the correlation energy of the inhomogeneous electron gas. Phys Rev B. 1986, 33, 8822–8824.

29. Perdew, J.P.; Wang, Y. Accurate and simple analytic representation of the electron-gas correlation energy. Phys Rev B. 1992, 45, 13244–13249.

30. CASTEP User Guide; MSI: San Diego, 1997.

31. DSolid User Guide; Biosym/MSI: San Diego, 1996.

32. Baerends, E.J.; Ellis, D.E.; Ros, P. Self-consistent molecular Hartree–Fock–Slater calculations. I. Computational procedure. Chem Phys. 1973, 2, 41–51.

33. Miyamoto, A.; Himei, H.; Oka, Y.; Maruya, E.; Katagiri, M.; Vetrivel, R.; Kubo, M. Computer-aided design of active catalysts for the removal of nitric oxide. Catal Today. 1994, 22, 87–96.

34. Himei, H.; Yamadaya, M.; Kubo, M.; Vetrivel, R.; Broclawik, E.; Miyamoto, A. Study of the activity of Ga-ZSM-5 in the deNO$_x$ process by a combination of quantum chemistry, molecular dynamics, and computer graphics methods. J Phys Chem. 1995, 99, 12461–12465.

35. Yamadaya, M.; Himei, H.; Kanougi, T.; Oumi, Y.; Kubo, M.; Stirling, A.; Vetrivel, R.; Broclawik, E.; Miyamoto, A. Quantum chemical investigation of reactants in selective reduction of NO$_x$ on ion–exchanged ZSM-5. Stud Surf Sci Catal. 1997, 105, 1485–1492.

36. Kanougi, T.; Furukawa, K.; Yamadaya, M.; Oumi, Y.; Kubo, M.; Stirling, A.; Fahmi, A.; Miyamoto, A. NO$_2$ adsorption on ion–exchanged ZSM-5: a density functional study. Appl Surf Sci. 1997, 119, 103–106.

37. Ueda, Y.; Tsuruya, H.; Kanougi, T.; Oumi, Y.; Kubo, M.; Chatterjee, A.; Teraishi, K.; Broclawik, E.; Miyamoto, A. Density-functional study on the transition state of methane activation over ion–exchanged ZSM-5. In: Transition State Modeling for Catalysis; Truhlar, D.G., Morokuma, K., eds.; American Chemical Society: Washington, DC, 1999, 321–332.

38. Kanougi, T.; Tsuruya, H.; Oumi, Y.; Chatterjee, A.; Fahmi, A.; Kubo, M.; Miyamoto, A. Density-functional calculation on the adsorption of nitrogen oxides and water on ion–exchanged ZSM-5. Appl Surf Sci. 1998, 561, 130–132.

39. Miyamoto, A.; Matsuba, K.; Kubo, M.; Kawamura, K.; Inui, T. Structures and dynamics of alkali ion–exchanged ZSM-5 as investigated by molecular dynamics and computer graphics. Chem Lett. 1991, 2055–2058.

40. Yajima, K.; Ueda, Y.; Tsuruya, H.; Kanougi, T.; Oumi, Y.; Ammal, S.S.C.; Takami, S.; Kubo, M.; Miyamoto, A. Computer-aided design of novel heterogeneous catalysts—a combinatorial computational chemistry approach. Stud Surf Sci Catal, 2000, 130, 401–408.

41. Armor, J.N. Catalytic reduction of nitrogen oxides with methane in the presence of excess oxygen: a review. Catal Today. 1995, 26, 147–158.

42. Iwamoto, M.; Yahiro, H. Novel catalytic decomposition and reduction of NO. Catal Today. 1994, 22, 5–18.

43. Tabata, T.; Kokitsu, M.; Okada, O. Relationship between methane adsorption and selective catalytic reduction of nitrogen oxide by methane on gallium and indium ion–exchanged ZSM-5. Appl Catal B. 1995, 6, 225–236.

44. Feng, X.; Hall, W.K. FeZSM-5: a durable SCR catalyst for NO_x removal from combustion streams. J Catal. 1997, 166, 368–376.

45. Li, Y.; Battavio, P.J.; Armor, J.N. Effect of water vapor on the selective reduction of NO by met over cobalt-exchanged ZSM-5. J Catal. 1993, 142, 561–571.

46. Ogura, M.; Hayashi, M.; Kikuchi, E. Role of zeolite structure on reduction of NO_x with methane over In- and Pd-based catalysts. Catal Today. 1998, 45, 139–145.

47. Shin, H.K.; Hirabayashi, H.; Yahiro, H.; Watanabe, M.; Iwamoto, M. Selective catalytic reduction of NO by ethane in excess oxygen over platinum ion–exchanged MFI zeolites. Catal. Today. 1995, 26, 13–21.

48. Yokoyama, C.; Misono, M. Catalytic reduction of nitrogen oxides by propene in the presence of oxygen over cerium ion–exchanged zeolites. J Catal. 1994, 150, 9–17.

49. Iwamoto, M.; Hernandez, A.M.; Zengyo, T. Highly selective reduction of NO in excess oxygen through the intermediate addition of reductant between oxidation and reduction catalysts. Res Chem Intermed. 1998, 24, 115–122.

50. Kikuchi, E.; Ogura, M.; Aratani, N.; Sugiura, Y.; Hiromoto, S.; Yogo, K. Promotive effect of additives to In/H-ZSM-5 catalyst for selective reduction of nitric oxide with methane in the presence of water vapor. Catal Today. 1996, 27, 35–40.

51. Flanigen, E.M.; Bennett, J.M.; Grose, R.W.; Cohen, J.P.; Patton, R.L.; Kirchner, R.M.; Smith, J.V. Silicalite, a new hydrophobic crystalline silica molecular sieve. Nature. 1978, 271, 512–516.

52. Ward, J.W. A spectroscopic study of the surface of zeolite Y. II. Infrared spectra of structual hydroxyl groups and adsorbed water on alkali, alkaline earth, and rare earth ion–exchanged zeolites. J Phys Chem. 1968, 72, 4211–4223.

53. Meitzner, G.D.; Iglesia, E.; Baumgartner, J.E.; Huang, E.S. The chemical state of gallium in working alkane dehydrocyclodimerization catalysts. In situ gallium K-edge x-ray absorption spectroscopy. J Catal. 1993, 140, 209–225.

54. Zhou, X.; Xu, Z.; Zhang, T.; Lin, L. The chemical status of indium impregnated HZSM-5 catalysts for the SCR of NO with CH_4. J Mol Catal A. 1997, 122, 125–129.

55. Oumi, Y.; Takaba, H.; Ammal, S.S.C.; Kubo, M.; Teraishi, K.; Miyamoto, A.; Kawasaki, M.; Yoshimoto, M.; Koinuma, H. Periodic boundary quantum chemical study on ZnO ultra violet laser emitting materials. Jpn J Appl Phys. 1999, 38, 2603–2605.

56. Belosludov, R.V.; Ammal, S.S.C.; Inaba, Y.; Oumi, Y.; Takami, S.; Kubo, M.; Miyamoto, A.; Kawasaki, M.; Yoshimoto, M.; Koinuma, H. Combinatorial computational chemistry approach to the design of metal oxide electronics materials. Proc SPIE. 2000, 3941, 2–10.

57. Taniguchi, S.; Hino, T.; Itoh, S.; Nakano, K.; Nakayama, N.; Ishibashi, A.; Ikeda, M. 100h II–VI blue-green laser diode. Electron Lett. 1996, 32, 552–553.

58. Nakamura, S.; Mukai, T.; Senoh, M. Candala-class high-brightness InGaN/AlGaN double-heterostructure blue-light-emitting diodes. Appl Phys Lett. 1994, 64, 1687–1689.

59. Hamdani, F.; Botchkarev, A.; Kim, W.; Morkoç, H.; Yeadon, M.; Gibson, J.M.; Tsen, S.C.Y.; Smith, D.J.; Reynolds, D.C.; Look, D.C.; Evans, K.; Litton, C.W.; Mitchel, W.C.; Hemenger, P. Optical properties of GaN grown on ZnO by reactive molecular beam epitaxy. Appl Phys Lett. 1997, 70, 467–469.

60. Zu, P.; Tang, Z.K.; Wong, G.K.L.; Kawasaki, M.; Ohtomo, A.; Koinuma, H.; Segawa, Y. Ultraviolet spontaneous and stimulated emissions from ZnO microcrystallite thin films at room temperature. Solid State Commun. 1997, 103, 459–463.

61. Hayashi, I.; Panish, M.B.; Foy, P.W.; Sumski, S. Junction lasers which operate continuously at room temperature. Appl Phys Lett. 1970, 17, 109–111.

62. Morrison, S.R. Electrochemistry of Semiconductor and Oxides Metal Electrodes; Plenum Press: New York, 1980, 1–401.

63. Heimer, T.A.; Bignozzi, C.A.; Meyer, G.J. Molecular level photovoltaics: the electro-optical properties of metal cyanide complexes anchored to titanium dioxide. J Phys Chem. 1993, 97, 11987–11994.

64. Choi, W.; Termin, A.; Hoffmann, M.R. The role of metal dopands in quantum-sized TiO_2: correlation between photoreactivity and charge carrier recombination dynamics. J Phys Chem. 1994, 98, 13669–13679.

65. Yin, M.T.; Cohen, M.L. Theory of ab inito pseudopotential calculations. Phys Rev B. 1982, 25, 7403–7412.

66. Kleinman, L.; Bylander, D.M. Efficacious form for model pseudopotentials. Phys Rev Lett. 1982, 48, 1425–1428.

67. Vanderbilt, D. Soft self-consistent pseudopotentials in a generalized eigenvalue formalism. Phys Rev B. 1990, 41, 7892–7895.

68. Moroni, E.G.; Kresse, G.; Hafner, J.; Furthmüller, J. Ultrasoft pseudopotentials applied to magnetic Fe, Co, and Ni: from atoms to solids. Phys Rev B. 1997, 56, 15629–15646.

69. Ohtomo, A.; Kawasaki, M.; Koida, T.; Masubuchi, K.; Koinuma, H.; Sakurai, Y.; Yoshida, Y.; Yasuda, T.; Segawa, Y. $Mg_xZn_{1-x}O$ as a II–VI widegap semiconductor alloy. Appl Phys Lett. 1998, 72, 2466–2468.

70. Trickey, S.B.; Green, F.R.; Averill, F.W. One-electron theory of the bulk properties of crystalline Ar, Kr, and Xe. Phys Rev B. 1973, 8, 4822–4832.

71. Abrahams, S.; Bernstein, J.L. Rutile: normal probability plot analysis and accurate measurement of crystal structure. J Chem Phys. 1971, 55, 3206–3211.

72. Glassford, K.M.; Chelikowsky, J.R. Structural and electronic properties of titanium dioxide. Phys Rev B. 1992, 46, 1284–1298.

73. Silvi, B.; Fourati, N.; Nada, R.; Catlow, C.R.A. Pseudopotential periodic Hartree–Fock study of rutile titanium dioxide. J Phys Chem Solids. 1991, 52, 1005–1009.

74. Pascual, J.; Camassel, J.; Mathieu, H. Resolved quadrupolar transition in TiO_2. Phys Rev Lett. 1977, 39, 1490–1493.

75. Lin, J.; Yu, J.C.; Lo, D.; Lam, S.K. Photocatalytic activity of rutile $Ti_{1-x}Sn_xO_2$ solid solutions. J Catal. 1999, 183, 368–372.

76. Miyamoto, A.; Hattori, T.; Inui, T. Molecular dynamics simulation of deposition process of ultrafine metal particles on MgO(100) surface. Appl Surf Sci. 1992, 60/61, 660–666.

77. Miyamoto, A.; Yamauchi, R.; Kubo, M. Atomic processes in the deposition and sintering of ultrafine metal particles on MgO(001) as investigated by molecular dynamics and computer graphics. Appl Surf Sci. 1994, 75, 51–57.

78. Miyamoto, A.; Takeichi, K.; Hattori, T.; Kubo, M.; Inui, T. Mechanism of layer-by-layer homoepitaxial growth of $SrTiO_3(100)$ as investigated by molecular dynamics and computer graphics. Jpn J Appl Phys. 1992, 31, 4463–4464.

79. Kubo, M.; Oumi, Y.; Miura, R.; Stirling, A.; Miyamoto, A.; Kawasaki, M.; Yoshimoto, M.; Koinuma, H. Atomic control of layer-by-layer epitaxial growth on $SrTiO_3(001)$: molecular dynamics simulations. Phys Rev B. 1997, 56, 13535–13542.

80. Kubo, M.; Oumi, Y.; Miura, R.; Stirling, A.; Miyamoto, A.; Kawasaki, M.; Yoshimoto, M.; Koinuma, H. Molecular dynamics simulation on a layer-by-layer homoepitaxial growth process of $SrTiO_3(001)$. J Chem Phys. 1998, 109, 8601–8606.

81. Kubo, M.; Oumi, Y.; Miura, R.; Stirling, A.; Miyamoto, A.; Kawasaki, M.; Yoshimoto, M.; Koinuma, H. Layer-by-layer heteroepitaxial growth process of a BaO layer on $SrTiO_3(001)$ as investigated by molecular dynamics. J Chem Phys. 1998, 109, 9148–9153.

82. Yoshimoto, M.; Maeda, T.; Ohnishi, T.; Koinuma, H.; Ishiyama, O.; Shinohara, M.; Kubo, M.; Miura, R.; Miyamoto, A. Atomic-scale formation of ultrasmooth surfaces on sapphire substrates for high-quality thin-film fabrication. Appl Phys Lett. 1995, 67, 2615–2617.

83. Tsukada, M.; Kawazu, A. Eds. Atomically Controlled Surfaces and Interfaces; North-Holland: Amsterdam, 1994.

84. Kawasaki, M.; Takahashi, K.; Maeda, T.; Tsuchiya, R.; Shinohara, M.; Ishiyama, O.; Yonezawa, T.; Yoshimoto, M.; Koinuma, H. Atomic control of the $SrTiO_3$ crystal surface. Science. 1994, 266, 1540–1542.

85. Schneider, M.; Rahman, A.; Schuller, I.K. Role of relaxation in epitaxial growth: a molecular dynamics study. Phys Rev Lett. 1985, 55, 604–606.

86. Schneider, M.; Schuller, I.K.; Rahman, A. Epitaxial growth of silicon: a molecular dynamics simulation. Phys Rev B. 1987, 36, 1340.

87. Gawlinski, E.T.; Gunton, J.D. Molecular dynamics simulation of molecular-beam epitaxial growth of the silicon(100) surfaces. Phys Rev B. 1987, 36, 4774–4781.

88. Biswas, R.; Grest, G.S.; Soukoulis, C.M. Molecular-dynamics simulation of cluster and atom deposition on silicon(111). Phys Rev B. 1988, 38, 8154–8162.

89. Luedtke, W.D.; Landman, U. Metal-on-metal thin-film growth: Au/Ni(001) and Ni/Au(001). Phys Rev B. 1991, 44, 5970–5972.

90. Athanasopoulos, D.C.; Garofalini, S.H. Molecular dynamics simulations of the effect of adsorption on SiO_2 surfaces. J Chem Phys. 1992, 97, 3775–3780.

91. Chambers, S.A.; Tran, T.T.; Hileman, T.A. Molecular beam homoepitaxial growth of MgO(001). J Mater Res. 1994, 9, 2944–2952.

92. Huang, R.; Kitai, A.H. Temperature dependence of the growth orientation of atomic layer growth MgO. Appl Phys Lett. 1992, 61, 1450–1452.

93. Kwak, B.S.; Boyd, E.P.; Zhang, K.; Erbil, A.; Wilkins, B. Metalorganic chemical vapor deposition of [100]-textured MgO thin films. Appl Phys Lett. 1989, 54, 2542–2544.

94. DeSisto, W.J.; Henry, R.L. Deposition of (100)-oriented MgO thin films on sapphire by a spray pyrolysis method. Appl Phys Lett. 1990, 56, 2522–2523.

95. Tseng, M.Z.; Hu, S.Y.; Chang, Y.L.; Jiang, W.N.; Hu, E.L. Minimum thickness MgO buffer layers for $YBa_2Cu_3O_{7-x}$/GaAs structures: assessment using photoluminescence of multiple-quantum-well structures. Appl Phys Lett. 1993, 63, 987–989.

96. Yadavalli, S.; Yang, M.H.; Flynn, C.P. Low-temperature growth of MgO by molecular-beam epitaxy. Phys Rev B. 1990, 41, 7961–7963.

97. Frank, F.C.; van der Merwe, J.H. One-dimensional dislocation. I. Static theory. Proc R Soc London Ser A. 1949, 198, 205–216.

98. Koinuma, H.; Nagata, H.; Tsukahara, T.; Gonda, S.; Yoshimoto, M. Ceramic layer epitaxy by pulsed laser deposition in an ultrahigh vacuum system. Appl Phys Lett. 1991, 58, 2027–2029.

99. Lang, Y.; Bonnell, D.A. Atomic structures of reduced $SrTiO_3$(001) surface. Surf Sci. 1993, 285, L510–L516.

100. Yoshimoto, M.; Maeda, T.; Shimozono, K.; Koinuma, H.; Shinohara, M.; Ishiyama, O.; Ohtani, F. Topmost surface analysis of $SrTiO_3$(001) by coaxial impact-collision ion scattering spectroscopy. Appl Phys Lett. 1994, 65, 3197–3199.

101. Cherry, M.; Islam, M.S.; Gale, J.D.; Catlow, C.R.A. Computational studies of protons in perovskite-structured oxides. J Phys Chem. 1995, 99, 14614–14618.

102. Ravikumar, V.; Wolf, D.; Dravid, V.P. Ferroelectric-monolayer reconstruction of the $SrTiO_3$(100) surface. Phys Rev Lett. 1995, 74, 960–963.

103. Zhong, W.; Vanderbilt, D. Competing structural instabilities in cubic perovskites. Phys Rev Lett. 1995, 74, 2587–2590.

104. Koinuma, H. Ed. Crystal Engineering of High Tc-Related Oxide Films. MSR Bulletin; 1994; Vol. 14.

105. Terashima, T.; Iijima, K.; Yamamoto, K.; Hirata, K.; Bando, Y.; Takada, T. In situ reflection high-energy electron diffraction observation during growth of $YBa_2Cu_3O_{7-x}$ thin films by activated reactive evaporation. Jpn J Appl Phys. 1989, 28, L987–L990.

106. Terashima, T.; Bando, Y.; Iijima, K.; Yamamoto, K.; Hirata, K.; Hayashi, K.; Kamigaki, K.; Terauchi, H. High-energy electron diffraction oscillations during epitaxial growth of high-temperature superconducting oxides. Phys Rev Lett. 1990, 65, 2684–2687.

107. Kawasaki, M.; Yashimoto, M.; Koinuma, H., et al. (personal communication, 1998).

108. Yin, X.; Fahmi, A.; Endou, A.; Miura, R.; Gunji, I.; Yamauchi, R.; Kubo, M.; Chatterjee, A.; Miyamoto, A. Periodic density study on V$_2$O$_5$ bulk and (001) surface. Appl Surf Sci. 1998, 130–132, 539–544.
109. Yin, X.; Han, H.; Endou, A.; Kubo, M.; Teraishi, K.; Chatterjee, A.; Miyamoto, A. Reactivity of lattice oxygens present in V$_2$O$_5$(010): a periodic first-principle investigation. J Phys Chem B. 1999, 103, 1263–1269.
110. Yin, X.; Fahmi, A.; Han, H.; Endou, A.; Ammal, S.S.C.; Kubo, M.; Teraishi, K.; Miyamoto, A. Adsorption of H$_2$O on the V$_2$O$_5$(010) surface studied by periodic density-functional calculations. J Phys Chem B. 1999, 103, 3218–3224.
111. Yin, X.; Han, H.; Gunji, I.; Endou, A.; Ammal, S.S.C.; Kubo, M.; Miyamoto, A. NH$_3$ Adsorption on the Brönsted and Lewis acid sites of V$_2$O$_5$(010): a periodic density-functional study. J Phys Chem B. 1999, 103, 4701–4706.
112. Miyamoto, A.; Yamazaki, Y.; Inomata, M.; Murakami, Y. Determination of the number of V $=$ O species on the surface of vanadium oxide catalysts. 1. Unsupported V$_2$O$_5$ and V$_2$O$_5$/TiO$_2$ treated with an ammoniacal solution. J Phys Chem. 1981, 85, 2366–2372.
113. Inomata, M.; Miyamoto, A.; Murakami, Y. Determination of the number of V $=$ O species on the surface of vanadium oxide catalysts 2. V$_2$O$_5$/TiO$_2$ catalysts. J Phys Chem. 1981, 85, 2372–2377.
114. Saleh, R.Y.; Wachs, I.E.; Chan, S.S.; Chersich, C.C. The investigation of V$_2$O$_5$ with TiO$_2$(anatase): Catalyst evolution with calcination temperature and O-xylene oxidation. J Catal. 1986, 98, 102–114.
115. Bond, G.C. Preparation and properties of vanadia/titania monolayer catalysts. Appl Catal A: General. 1997, 157, 91–103.
116. Sayle, D.C.; Catlow, C.R.A.; Perrin, M.A.; Nortier, P. Computer modeling of the V$_2$O$_5$/TiO$_2$ interface. J Phys Chem. 1996, 100, 8940–8945.
117. Tarama, K.; Yoshida, S.; Ishida, S.; Kakioka, H. Spectroscopic studies of catalysis by vanadium pentoxide. Bull Chem Soc Jpn. 1968, 41, 2840–2845.
118. Ramirez, R.; Casal, B.; Utrera, L.; Ruiz-Hitzky, E. Oxygen reactivity in vanadium pentoxide: electronic structure and infrared spectroscopy studies. J Phys Chem. 1990, 94, 8960–9865.
119. Zhanpeisov, N.U.; Bredow, T.; Jug, K. Quantum chemical SINDO1 study of vanadium pentoxide. Catal Lett. 1996, 39, 111–118.
120. Miyamoto, A.; Inomata, M.; Hattori, A.; Ui, T.; Murakami, Y. A molecular orbital investigation of the mechanism of the NO–NH$_3$ reaction on vanadium oxide catalyst. J Mol Catal. 1982, 16, 315–333.
121. Bond, G.C.áá; Parfitt, G.D. The vanadium pentoxide–titanium dioxide system: structural investigation and activity for the oxidation of butadiene. J Catal. 1979, 57, 476–493.
122. Busca, G.; Ramis, G.; Lorenzolli, V. FT-IR study of the surface properties of polycrystalline vanadia. J Mol Catal. 1989, 50, 231–240.
123. Miyamoto, A.; Kobayashi, K.; Inomata, M.; Murakami, Y. Nitrogen-15 tracer investigation of the mechanism of the reaction of NO with NH$_3$ on vanadium oxide catalysts. J Phys Chem. 1982, 86, 2945–2950.
124. Dumesic, J.A.; Topsøe, N.Y.; Topsøe, H.; Chen, Y.; Slabiak, T. Kinetics of selective catalytic reduction of nitric oxide by ammonia over vanadia/titania. J Catal. 1996, 163, 409–417.

125. Topsøe, N.Y.; Topsøe, H.; Dumesic, J.A. Vanadia/Titania catalysts for selective catalytic reduction of nitric oxide by ammonia. J Catal. 1995, 151, 241–252.
126. Takagi, M.; Kawai, T.; Soma, M.; Onishi, T.; Tamaru, K. Mechanism of catalytic reaction between NO and NH$_3$ on V$_2$O$_5$ in the presence of oxygen. J Phys Chem. 1976, 80, 430.
127. Lietti, L.; Svachula, J.; Forzatti, P.; Busca, G.; Ramis, G.; Bregani, F. Surface and catalytic properties of vanadia-titania and tungstem oxide–titania systems in the selective catalytic reduction of nitrogen oxides. Catal Today. 1993, 17, 131–139.
128. Takaba, H.; Mizukami, K.; Kubo, M.; Fahmi, A.; Miyamoto, A. Permeation dynamics of small molecules through silica membranes: molecular dynamics study. AIChE J. 1998, 44, 1335–1343.
129. Takaba, H.; Mizukami, K.; Oumi, Y.; Kubo, M.; Chatterjee, A.; Fahmi, A.; Miyamoto, A. Application of integrated computational chemistry system to the design of inorganic membranes. Catal Today. 1999, 50, 651–660.
130. Mizukami, K.; Takaba, H.; Ito, N.; Kubo, M.; Fahmi, A.; Miyamoto, A. Permeability of Ar and He through an inorganic membrane: a molecular dynamics study. Appl Surf Sci. 1997, 119, 330–334.
131. Takaba, H.; Mizukami, K.; Kubo, M.; Stirling, A.; Miyamoto, A. The effect of gas molecule affinities on CO$_2$ separation from the CO$_2$/$_2$ gas mixture using inorganic membranes as investigated by molecular dynamics simulation. J Membr Sci. 1996, 121, 251–259.
132. Takaba, H.; Kosita, R.; Mizukami, K.; Oumi, Y.; Ito, N.; Kubo, M.; Fahmi, A.; Miyamoto, A. Molecular dynamics simulation of *iso*- and *n*-butane permeations through a ZSM-5–type silicate membrane. J Membr Sci. 1997, 134, 127–139.
133. Hsien, H.P. Inorganic Membranes for Separation and Reaction; Elsevier: Amsterdam, 1996.
134. Kusakabe, K.; Yoneshige, S.; Murata, A.; Morooka, S. Morphology and gas permeance of ZSM-5-type zeolite membrane formed on a porous-alumina support tube. J Membr Sci. 1996, 116, 39–46.
135. Vroon, Z.A.E.P.; Keizer, K.; Gilde, M.J.; Verweij, H.; Burggraaf, A.J. Transport properties of alkanes through ceramic thin zeolite MFI membranes. J Membr Sci. 1996, 113, 293–300.
136. Bai, C.; Jia, M.D.; Falconer, J.L.; Noble, R.D. Preparation and separation properties of silicalite composite membranes. J Membr Sci. 1996, 105, 79–87.
137. Suda, T.; Fuji, M.; Yoshida, K.; Iijima, M.; Seto, T.; Mitsuoka, S. Development of flue gas carbon dioxide recovery technology. Energy Convers Mgmt. 1992, 33, 317–324.
138. Belosludov, R.V.; Sakahara, S.; Yajima, K.; Takami, S.; Kubo, M.; Miyamoto, A. Combinatorial computational chemistry approach as a promising method for design of Fischer–Tropsch catalysts based on Fe and Co. Appl Surf Sci, 2002, 189, 245–252.

13
Computational Informatics: Guided Discovery for Combinatorial Experiments

Krishna Rajan
Rensselaer Polytechnic Institute, Troy, New York, U.S.A.

I. INTRODUCTION

Combinatorial sciences offer an attractive paradigm to search for new materials. The challenge is to expand this paradigm from one of high-throughput *screening* to high-throughput *knowledge*. In order to accomplish this, there is a need to integrate the knowledge of the physics of the materials process into the statistically based strategies governing the combinatorial synthesis. In this chapter, we outline some of the challenges and approaches that exist to developing such integration. Materials discovery is still a process governed by empiricism and accidental discoveries (high-temperature ceramic superconductors and carbon fullerenes, to mention some recent examples). While incremental progress is made in specific technological areas of interest, we need to have a means of exploring vast combinations of structure–property relationships. If new, significant advances in materials science are to be made, we need to have search tools that can accelerate the discovery process.

The development, improvement, or adaptation of materials is usually built on prior data. This is based on knowledge derived through experience, experiments, and theory. However, for each new application and engineering demand, the testing and modeling process has to be repeated or modified to account for the new conditions, and the knowledge-building process repeats in an iterative manner. Because many of the paradigms in materials science and engineering are in fact often derived through empirical correlations between observed behaviors;

theoretical frameworks of these observations have often been established or developed afterwards in order to achieve an explanation of existing data. The advent of high-speed computing has opened up of simulation studies, which have effectively provided another source of data generation besides traditional experiments (1–2). However, the process of sequentially building on experiment, theory, and simulation to iterate a search for a solution to a particular engineering problem has not changed. In this chapter we outline some of the requirements for the establishment of a "toolkit" that significantly challenges this paradigm by developing a methodology/information infrastructure that builds on the use of data to generate knowledge.

The concept of *computational informatics* that we are introducing here refers to the collective integration of phenomenological and experimentally based data sets with computationally derived data based on theoretical formulations with statistical tools that help to seek new patterns of description or prediction. To make a combinatorial experimental strategy successful for materials discovery, one needs to build upon prior data and knowledge and to use those data in a way that permits one to associate and anticipate structure–property correlations before conducting further experiments (either real or "virtual"). In this manner we can develop a methodology that will take advantage of prior knowledge as "stored" in massive databases to search for potential links in structure–processing–property relationships. Integrating information from different types of databases using the appropriate data-mining techniques permits one to link information across length and time scales in a useful manner.

II. ESTABLISHING QUANTITATIVE STRUCTURE–ACTIVITY RELATIONSHIPS FOR MATERIALS SCIENCE

The concept of structure–property relationships in materials science is usually taken for granted as the foundation on which materials design is based. Yet it is interesting to note that other disciplines, especially in biology/pharmacology and drug design areas, have established a formalism termed *quantitative structure–activity relationships* (QSARs) or *quantitative structure–property relationships* (QSPRs). The exact context for use of the terms QSAR and QSPR can vary among the organic-based sciences community, but for the purposes of our discussion, we shall view the two terms as equivalent. The basic functional form for QSARs is

$$\Phi = f(\xi\,\xi\,\xi) \tag{1}$$

where Φ represents the "property" of interest and ξ represents the numerous parameters or "descriptors" associated with that material that appear to have an influence on that property. The QSAR may in fact involve a number of such functions, which collectively describe a set of structure–activity relationships (3).

Many approaches in the engineering design of materials of course take into account the numerous variables that influence the process, resulting in some type of equation very similar to those used in the bioinformatics/organic chemistry field. Such equations are some sort of continuous function, which maps the variability of the functionality in terms of parameters such as composition. Some classical examples where the materials science community has in fact developed such formulations in the ferrous metallurgical research community include:

Chemical reactivity as a function of composition in multicomponent slags
Hardenability in steels
Martensite start temperature as a function of composition in multicomponent alloys

Sometimes the relationships are represented not by some form of continuous function but rather by mapping regions of association of materials descriptors with some functionality. A good example of this is the "bandgap engineering" diagram, which is used to help select and design materials combinations for tuning bandgaps by controlling the combinatorial parameters of lattice matching and alloy composition (Figure 1).

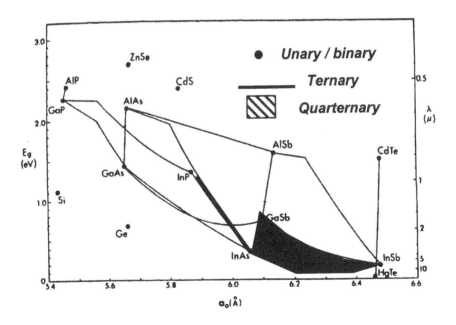

Figure 1 A "QSAR" type of diagram for structure (lattice parameter)–activity (bandgap) relationships. (From Ref. 3.)

Other forms of functionality maps derived from sets of QSAR-type relationships include the Ashby type of deformation maps(4) and Pourbaix and Kroger–Vink(5) diagrams mapping out electrochemical-based property descriptors in fluid/solid interactions as well as defect chemistry in solid-state reactions.

The challenges in materials science are numerous, not least of which is that the functionality of the material is dependent not only on composition but on a wide array of descriptors covering a wide range of length and time scales. For instance, the concept of fracture toughness in ceramics is one where there are innumerable parameters that have been both empirically as well as theoretically shown to have an influence on this property. These include from bonding influences of small amounts of chemical additives, microstructure, sintering conditions, and many other combinations of process variables. If one maps the vast array of combinations of parameters or descriptors affecting this property, the challenge is immense.

Hence to directly develop a singular relationship to quantitatively predict fracture toughness to take into account all these variables is not realistic. Hence there is a need to formulate other strategies for designing combinatorial experiments when dealing with such multivariate data sets. What kinds of physically based computational strategies can provide a guide for the selection of parameters for combinatorial experiments? While the field of computational materials science is of course vast, in this section we will provide some representative examples where computational strategies have been employed to search for new materials or at least identify key parameters or descriptors, which can serve to guide future experiments. Of course, as noted earlier, the examples that we shall identify demonstrate some of the key length scales that one needs to develop. A key aspect of developing an "informatics" approach to materials discovery is the need to establish the critical array of descriptors of materials attributes that may be subsequently input to a database. Having physically meaningful descriptors is key if one is to develop and search for associations between apparently disparate or disjointed data sets. This in turn of course provides possible insights into the mechanisms that govern properties in these new classes of materials. It is worthwhile to note that the need for examining and searching for appropriate descriptors to build on prior materials discovery is getting further attention in the literature. For instance:

> Lacorre et al. (6) have suggested a stereochemical model for designing oxide ion conductors. Their approach involves the use of developing generic descriptors involving substitution of elements based on their oxidation state. The concept here is based on the idea of finding features or descriptors among iso-structural compounds that may provide a clue for developing new materials with similar but enhanced properties. By discovering that $La_2Mo_2O_9$ is a fast oxide ion conductor that is structur-

ally similar to β-SnWO$_4$ (both compounds crystallize in the space group, with identical cationic positions), these workers suggested a structural model for the origin of fast ion conduction. Based on such considerations of the possible mechanisms associated with fast ion conduction, a combinatorial array of elements with oxygen was proposed to produce an array of oxide compositions.

In a similar fashion, MacGlashan et al. (7) have proposed that the structures based on the structure LiAsF$_6$ may provide a basis for developing appropriate compositions/molecular structures of new conducting polymers. As in the case of ceramic systems, the incorporation of appropriate elemental species that alter the local oxidation state within a given coordination complex is suggested as a means of developing strategies for synthesizing new materials/molecular arrangements based on ionic conduction in polymeric systems. The molecular structure was expressed not as crystallographic coordinates but as internal stereochemical descriptors: bond lengths, bond angles, and torsion angles. The characterization of structure provided clues as to potential properties and, more specifically, mechanisms (e.g., ionic conduction). This in turn established guidelines for combinatorial experiments for designing new molecular chemistries for polymeric electrolytes.

The searches for structural descriptors can be more involved in crystalline systems involving network-type structures, such as zeolites, molecular cages, and other structural chemistry problems. For example, Friedrichs et al. (8) developed an enumeration scheme based on tiling theory to examine the combinatorial possibilities of topological arrangements to narrow down the possible choices of molecular arrangements that can exist. By using a combination of group theory, topology, and combinatorics, they explored all the possible permutations of topological arrangements in these network structures and then searched for structural uniqueness among them. This helped significantly reduce the vast array of combinatorial possibilities in predicting possible structures for a variety of structural chemistry applications.

At the chemical-length scale, for instance, there have been many attempts to assess combinatorial parameters, which map out trends in the functionality of materials based on elemental chemistry. Historically this has produced a variety of forms of ''structure maps'' that attempt to identify some critical metric associated with the chemistry at the crystal structure level (which may be viewed as a materials ''gene''). Using crystal chemistry descriptors, for instance, a number of structure maps have been proposed, including (9) (see Figure 2):

Mooser–Pearson plots: empirical plots showing relationships between structure type (classification according to space group) and the effects of bonding and electronegativity

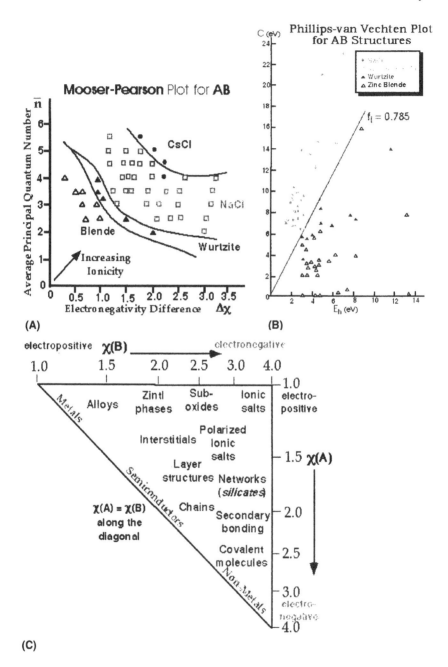

Figure 2 Examples of structure maps associating electronic and atomistic information to crystal chemistry. (From Ref. 10.)

Philips–van Schetken plots: similar to Mooser–Pearson plots but based on
theoretically derived bonding-size maps
Kettlars' triangle: which finds different structure types associated with
binary compounds

These structure maps essentially serve as an informational tool for indicating regimes of stability and existence. These plots attempt to seek associations
between functionality and structure based on crystal chemistry considerations.

As an illustration of this logic as applied to a materials database problem,
we show the example of the utilization of "structure maps" based on searching for
crystallographic descriptors for association of crystal stoichiometry with structure
type. A recent example is that proposed by Pettifor (11) using descriptors based
on a sequencing pattern of the Mendeleev number. As an example of the association of vastly different length and time scales, Figure 3 shows a mapping of the
association of structure type with an engineering-scale metric describing the quality of welding or joining dissimilar materials. Clearly there is no ab initio theory
to correlate such vastly different types of data sets, but this provides an example
of the disparate forms of data sets that exist in materials science. However, when
these data are coupled with data on the formation of compounds with the stoichiometry of AB_2 for all A-B combinations of elements, we find that the best welds
between dissimilar materials appears to form when there is an *absence* of compound formation. Hence a potential materials design screening metric has been
developed here for forming good-quality welds between dissimilar materials,
based on known data sets. This association of disparate data sets demonstrated
graphically here can be accomplished for multidimensional and massive data sets
using state-of-the-art computational algorithms. This will aid in developing a
screening tool for materials selection that is unconstrained in terms of the number
of parameters.

The example provided in Figure 3 provides a combinatorial mapping of
phenomenological structure–property relationships. There are also other situations where the physical foundations for combinatorial design are founded on
principles that have a fundamental theoretical foundation. One such case, which
is not particularly taken advantage of in combinatorial synthesis, is the design
of multifunctional materials.

The concept of functionality of materials can be broadly distinguished according to equilibrium and nonequilibrium properties. The former are founded on
thermodynamic principles as defined by Maxwell's equations (12). The concept of
a relating structure to property and more specifically to multifunctional attributes
of material is in fact well established in materials science. This involves relating
thermal, electrical, and mechanical properties of crystals in terms of their tensorial
representations. Such representations provide useful quantitative descriptors of

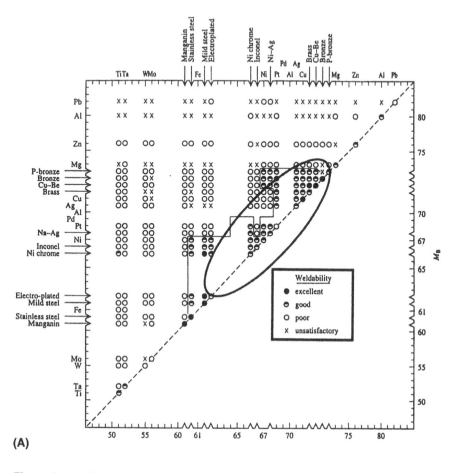

Figure 3 Pettifor structure maps showing association between joining characteristics between a combinatorial array of dissimilar metals and the nature of AB_2-type compound formation. The circled region in the second diagram (B) indicates the parameter space covered in the first diagram (A). (From Ref. 11.)

equilibrium crystal properties in terms of the crystallographic unit-cell symmetry (Figure 4).

III. THERMODYNAMICS AND KINETIC COMPUTATIONS FOR COMBINATORIAL LIBRARIES

A major approach in combinatorial library studies, especially in the fabrication of multicomponent thin-film libraries, is to utilize a variety of geometrical strate-

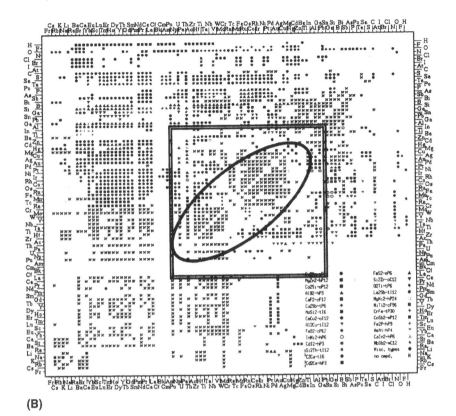

(B)

Figure 3

gies in developing binary, pseudobinary ternary, and quarternary "phase diagrams." Yet the intrinsic assumption in all these studies is that the final composition of the material deposited is in fact a simple linear superposition of elemental compositions based on the source of deposition. Hence a "materials discovery" associated with a predetermined screening metric is less a discovery of a new material and more a discovery of an apparent compositional mixture that appears to satisfy the requirements of the screening process. The natural tendency of course is to associate the combinatorial library with an equilibrium-phase diagram, but, as noted earlier, this is a tenuous assumption. It is here that thermochemical databases need to be integrated into the development of combinatorial library production, especially in the field of inorganic synthesis.

Thermochemical database calculations, while traditionally being used to identify regions of stability in response to external variables such as pressure and

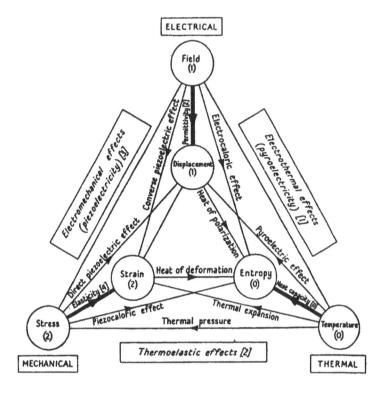

Figure 4 Multifunctional relationships between properties and variables. The tensor rank of a variable is in parentheses and that of a property in brackets. (From Ref. 12.)

temperature, can also be used as a tool for computationally estimating the existence of new stoichiometries. However, despite the well-established nature of these calculations, they are not widely used to *deliberately* search for new materials or compositions. However, there are examples of work that have not been taken advantage of in the combinatorial synthesis community. The process in most thin-film libraries tends to assume a simple rule of mixtures, yet what is needed to be recognized is that in multicomponent systems the prediction of phase relations is a complex nonlinear process (13). For the calculation of phase equilibria in a multicomponent system, it is necessary to minimize the total Gibbs energy, G, of all the phases that take part in this equilibrium. For multicomponent systems, it has proven useful to distinguish three contributions from the concentration dependence on the Gibbs energy of a phase:

1. The Gibbs energy of a mechanical mixture of the constituents of the phase
2. The contribution from the entropy of mixing for an ideal solution
3. The so-called excess free energy

The process of accounting for all these contributions results in multiple nonlinear equations (often in the form of what are termed Redlich–Kister polynomials), which can be robustly solved using a variety of numerical techniques (see Figures 5 and 6).

It should of course be pointed out that this represents only one approach to computational phase stability studies. Others, such as ab initio and cluster variational methods, are also used; and this in fact raises the important issue of the need to integrate these techniques in a manner that provides the maximum level of accurate information for combinatorial experiments.

Integral to thermochemistry-based calculations regarding multicomponent systems is the role of diffusion in the formulation of combinatorial libraries. The establishment of these stoichiometric libraries in multicomponent systems needs to be guided by modeling of the kinetic pathways associated with the resulting stoichiometries. The literature on multicomponent diffusion is extensive and will

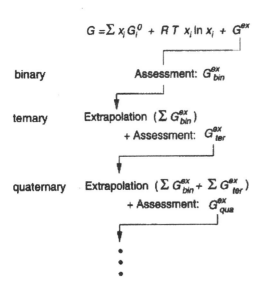

$$G = \Sigma\, x_i\, G_i^0 + RT\, x_i \ln x_i + G^{ex}$$

binary   Assessment: G_{bin}^{ex}

ternary   Extrapolation $(\Sigma\, G_{bin}^{ex})$
              + Assessment: G_{ter}^{ex}

quaternary   Extrapolation $(\Sigma\, G_{bin}^{ex} + \Sigma\, G_{ter}^{ex})$
               + Assessment: G_{qua}^{ex}

Figure 5 The computational strategy used by the CALPHAD system, where the assessment of the excess free energies of the constituent subsystems are for extrapolation to a higher multicomponent system. (From Ref. 13.)

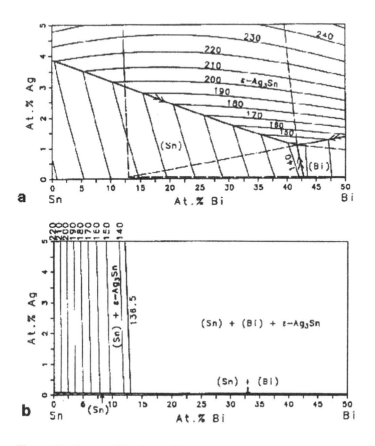

Figure 6   Computed isotherms for the Sn-Bi-Ag system. (From Ref. 13.)

not be reviewed in detail here; however, the main point to be made here is that the use of binary diffusion models in the interpretation of combinatorial libraries can be misleading (14). While Ficks laws still apply, each component flux depends on the gradients of all the components. Solutions to the multicomponent diffusion equation can be obtained using techniques such as the *square root diffusivity method* (Figure 7), which provides an accurate and systematic approach to solving multicomponent diffusion problems in one spatial dimension.

The value of such a method includes the fact that it can capture information on:

The rate constants for interdiffusion
The qualitative classification of penetration curves

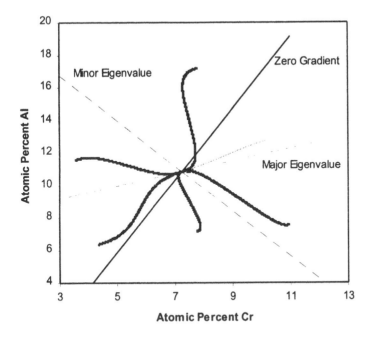

Figure 7 Multicomponent diffusion paths in a Ni–Cr–Al ternary system calculated using the square root diffusivity method.

The classification of diffusion paths
The occurrence or nonoccurrence of zero-flux planes

IV. INFORMATICS STRATEGIES FOR FUNCTIONALITY

The discussion in the previous section clearly alludes to the need to take advantage of existing informatics tools, such as computational thermochemical data sets, as a source for designing combinatorial experiments in materials science. Despite the fact that such databases are well established, their use in combinatorial materials science is still to be explored. Part of the reason for the slow incorporation of such ideas into the framework of combinatorial design of materials is that the "success" of combinatorial experiments is defined by the apparent discovery of a property attribute and not the material itself. Only afterwards is the combinatorial choice of elements identified. Understandably, materials properties provide the engineering motivation for most combinatorial studies, and hence pre-existing data on materials properties are rarely used in a systematic manner. Part of the reason for this state of affairs in combinatorial materials science is that databases in materials science applications tend to be phenomenological in nature. In other words, they are built around taxonomy of specific classes of properties and materials characteristics. In order for databases to serve as less of a "search and retrieve" infrastructure and more for a tool for "knowledge discovery," they need to have functional capabilities and to be able to guide not just property identification but also *structure–chemistry–property relationships*. The recent advances in genomics and proteomics, for instance, provide a good example of the development of such "functional" databases. A first step to achieve this is to develop descriptors of materials properties that can be sorted and classified using appropriate data-mining algorithms.

Such a compilation of chemistry–property–structure information in one class of systems is rarely found in materials science. The sorting of that information to seek potentially new relationships is the purpose of data mining. The main challenge, however, is that there are numerous correlations across length and time scales to seek for a viable combinatorial experiment. The multivariate nature of structure–property relationships is the major stumbling block in using data-mining tools in the materials sciences (see Figure 8).

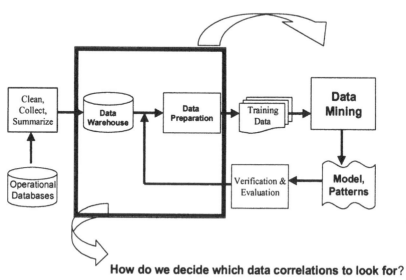

Figure 8   Schematic of infrastructure needed to apply data-mining techniques. (From Ref. 17.)

The challenge in linking length scales in materials science is that we do not necessarily have theories linking every aspect of materials characteristics in a unified manner. Much of materials design is based on phenomenological paradigms, which provide guidelines for materials selection. One needs to know how to integrate data at different length scales in such a way as to detect patterns of behavior (using statistical techniques), which could lead to (or suggest) new data or information (validated by experiments and theoretical formulations). The data-preparation sequence outlined earlier has built into it a further level of detail that needs to be considered (Figure 9).

A statistical evaluation to search for each descriptor is computationally expensive and most possibly ineffective. One example of the numerous data-mining tools is principal component analysis (PCA), which serves as a technique to reduce the information dimensionality that is often needed from the vast arrays of data as obtained from a combinatorial experiment, in a way that minimizes the loss of information (15). (see Figure 10.) It relies on the fact that most of the descriptors are intercorrelated and that these correlations in some instances are high. From a set of N correlated descriptors, we can derive a set of N uncorrelated descriptors (the principal components). Each principal component (PC) is a suitable linear combination of all the original descriptors. The first principal component accounts for the maximum variance (eigenvalue) in the original data set. The second principal component is orthogonal to (uncorrelated with) the first and accounts for most of the remaining variance. Thus the mth PC is orthogonal to all others and has the mth largest variance in the set of PCs. Once the N PCs have been calculated using eigenvalue/eigenvector matrix operations, only PCs

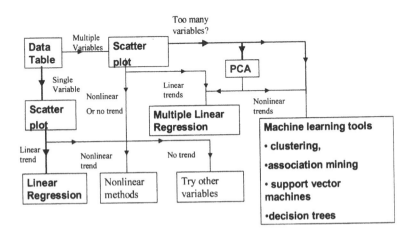

Figure 9   Data-processing tools for different types of data sets. (From Ref. 17.)

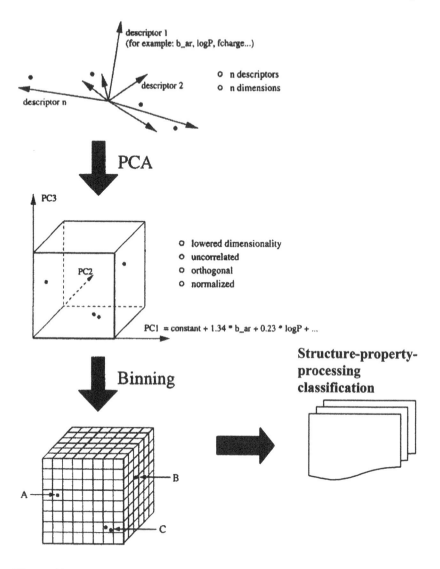

descriptor 1
(for example: b_ar, logP, fcharge...)

descriptor 2

descriptor n

○ n descriptors
○ n dimensions

PCA

PC3

PC2

○ lowered dimensionality
○ uncorrelated
○ orthogonal
○ normalized

PC1 = constant + 1.34 * b_ar + 0.23 * logP + ...

Binning

A

B

C

Structure-property-processing classification

Figure 10   Schematic of PCA procedures. (Adapted from Ref. 16.)

with variances above a critical level are retained. The M-dimensional principal component space has retained most of the information from the initial N-dimensional descriptor space, by projecting it onto orthogonal axes of high variance. The complex tasks of prediction or classification are made easier in this compressed space.

Mathematically we can describe this as follows. Consider p random variables X_1, X_2, $\cdots$, X_p. The original system can be rotated and a new coordinate system obtained, with the new axes representing the directions with maximum variability. The new axes, which are linear combinations of the original axes, are the principal components.

Let Σ be the covariance matrix associated with the random vector $\mathbf{X}' = [X_1, X_2, \cdots, X_p]$. The corresponding eigenvalue–eigenvector pairs are $(\gamma_1, \mathbf{e}_1)$, $(\gamma_2, \mathbf{e}_2)$, $\cdots$, $(\gamma_p, \mathbf{e}_p)$, where $\gamma_1\ \gamma_2\ \cdots\ \gamma_p\ 0$. Then the ith principal component is given by

$$Y_i = \mathbf{e}'_i\, \mathbf{X} = e_{i1}X_1 + e_{i2}X_2 + \ldots + e_{ip}X_p \quad i = 1, 2, \ldots, p \tag{2}$$

Then

$$\text{var}(Y_i) = \mathbf{e}'_i \sum \mathbf{e}_i = \lambda_i \quad i = 1, 2, \ldots, p \tag{3}$$

and

$$\text{cov}(Y_i, Y_k) = \mathbf{e}'_i \Sigma \mathbf{e}_k = 0 \quad i\ k \tag{4}$$

Thus the principal components are uncorrelated and have var variances equal to the eigenvalues of Σ. Another property of principal components is

$$\sigma_{11} + \sigma_{22} + \ldots + \sigma_{pp} = \text{var}(X_1) + \text{var}(X_2) + \ldots + \text{var}(X_p)$$
$$= \lambda_1 + \lambda_2 + \ldots + \lambda_p = \text{var}(Y_1) + \text{var}(Y_2) + \ldots + \text{var}(Y_p) \tag{5}$$

Then the proportion of total population variance due to the kth principal component is

$$\lambda_k / (\lambda_1 + \lambda_2 + \ldots + \lambda_p) \quad k = 1, 2, \ldots, p \tag{6}$$

Consequently, if most of the total population variance , for large p, can be attributed to the first two or three components, them these can replace the original variables with minimal loss of information.

When the variables have different ranges and are measured on different scales (as is the case with most materials problems), they are standardized:

$$Z_1 = (X_1 - \mu_1)/\sigma_{11}$$
$$Z_2 = (X_2 - \mu_2)/\sigma_{22}$$
$$\vdots$$
$$Z_p = (X_p - \mu_p)/\sigma_{pp} \tag{7}$$

Then cov $(\mathbf{Z}) = \mathbf{\rho}$ and var $(Z_i) = 1$, and the principal components of $\mathbf{Z}$ are obtained from the eigenvectors of the correlation matrix $\mathbf{\rho}$ of $\mathbf{X}$. The corresponding eigenvalue–eigenvector pairs for $\mathbf{\rho}$ are $(\gamma_1, \mathbf{e}_1), (\gamma_2, \mathbf{e}_2), \cdots, (\gamma_p, \mathbf{e}_p)$, where $\gamma_1 \ \gamma_2 \ \cdots \ \gamma_p \ 0$. Then the ith principal component is given by

$$Y_i = \mathbf{e'}_i \, \mathbf{Z} = e_{i1} Z_1 + e_{i2} Z_2 + \ldots + e_{ip} Z_p \quad i = 1, 2, \ldots, p \tag{8}$$

Then the proportion of total population variance due to the kth principal component is γ_k/p, where $k = 1, 2, \cdots, p$ and the γ_k are the eigenvalues of $\mathbf{\rho}$.

The complex tasks of prediction or classification are made easier in this compressed space. The PCA technique reduces the redundancy contained within the data by creating a new series of components in which the axes of the new coordinate systems point in the direction of decreasing variance. The resulting components are often more interpretable than the original data set. For instance, Rajan et al. (3,17–18) have used such an approach to reduce the dimensionality of the multivariate problem in developing descriptors for high-temperature-compound superconductors. For the purposes of this discussion, we shall focus on situations where there exists a vast array of variables associated with a single set of compounds or chemistry. MgB_2, a well-known compound, was recently and accidentally found to possess superconducting characteristics (19). We explored a wide array of descriptors based on "legacy" data of other inorganic compounds (intermetallic and ceramic systems) possessing high-temperature superconducting behavior, including: average number of valence electrons, electronegativity difference, radii difference, elemental concentration/ mole fraction stoichiometry, cohesive energy, and ionization energy. The choice of these "descriptors" was based initially on prior studies, which attempted to search for correlations between crystal chemistry and crystal structure and high-T_c properties (20). However, these studies attempted to look for correlations only between "raw" data, and much information can be lost in that manner.

An organized data set that has been "warehoused" (e.g., Ref. 17) can be synthesized into multiple scatter plots, which is shown in Figure 11. We like to refer to this format of data representation as a "combinatorial response map" because it maps out the vast arrays of combinations of materials response to a variety of descriptors. This also serves to graphically represent the challenge and need for techniques, which can condense this information in a statistical manner to seek which combinations of descriptors appear to have the most influence on the response function of interest (in this case, high-temperature superconductivity).

An illustrative plot showing how correlations associated with numerous variables can be reduced to three dimensions, which aids in visualization is shown in Figure 12. By combining six levels of descriptors of materials known to be superconductors, the information is compressed into three dimensions, as shown in the Figure.

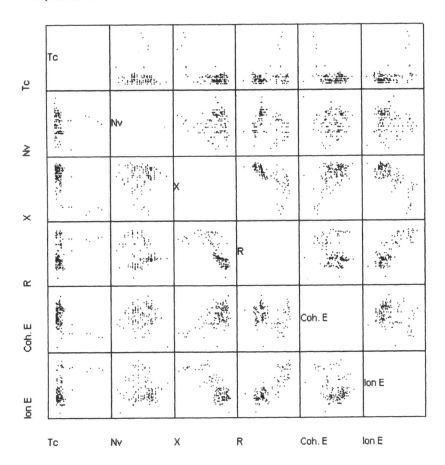

Figure 11 A "combinatorial response map" for high-temperature superconductivity data

The results of the PCA analysis shows that MgB_2 appears to be clustered among other similar iso-stoichiometry compounds. However, if we had relied solely on structure maps, it would have been considered an outlier. This simple example serves to emphasize the value of using multivariate data-analysis techniques.

It should be kept in mind that the examples provided here are only a small part of the arsenal available for data mining. A key for multiscale data mining is to combine the strengths of complementary data-mining tools such as association rules (AR), principal component analysis (PCA), and support vector machines

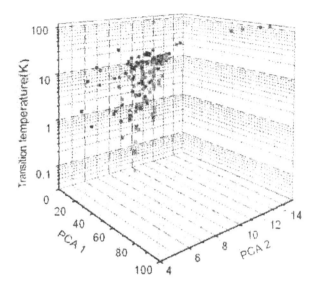

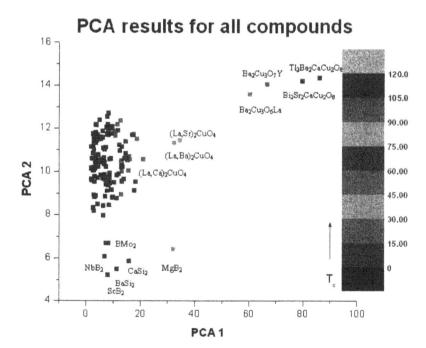

Figure 12   Three-dimensional and two-dimensional PCA plot derived from the multivariate data shown in Figure 10.

(SVMs). The AR method is well suited to descriptive knowledge-discovery tasks, PCA (and variations such as partial least squares) is good for visualization and linear models in high-dimensional spaces, and SVM are well suited for the predictive tasks (21–23). The AR method enable users to find interesting patterns and trends in the data through the creation of relatively simple interpretable rules. Key benefits of the method are discovery, scalability, feature selection, robustness to noise and missing data, suitability for categorical attributes, and method flexibility. The AR method generates interpretable concepts and hypotheses that have not proven well suited to predictive tasks such as classification and regression. The PCA technique can be used to visualize high-dimensional data but is limited to modeling linear relationships. In contrast, SVM methods are a set of highly accurate predictive modeling methods based on statistical learning theory and mathematical programming that work very well in high-dimensional attribute space for well-defined problems such as classification regression and novelty detection. The SVMs perform many of the same functions as neural networks but have overcome many of the limitations of neural network (NN) approaches. The SVM approach is systematic, reproducible, and properly motivated by statistical learning theory. Training involves optimization of a convex mathematical model with few algorithmic or model parameters, whereas in NN one must choose many model parameters, such as the number of hidden units and the connectivity of the hidden units: Also, unlike NN, there are no false local minima to complicate the learning process.

V. CONCLUSIONS: INTEGRATING COMPUTATION WITH COMBINATORIAL EXPERIMENTS

Combinatorial experiments can rapidly generate large amounts of data. But how do we rapidly generate knowledge? The discussion in the previous sections has outlined some of the specific approaches to how one may develop physically based models that are applicable to multiscale problems. The next question is how we can rapidly, efficiently, and accurately search for patterns of behavior that may provide us further insight into choosing multifunctional materials that we have not yet "discovered". Data mining is envisaged as a tool to exploit the masses of available data to accelerate the discovery of these relationships and possible new associations. Data mining acts as a descriptive tool for hypothesizing relationships between structures and materials that are interpretable by the materials scientist. However, we would also like data-mining tools with high predictive accuracy, in order to identify materials likely to possess desirable properties from massive combinatorial libraries of materials. Thus materials science presents a challenging testing ground for the development of new algorithmic and mathematical foundations for integrating discovery and prediction in data mining.

Presently the two types of computation are done almost independent of each other, and the latter is still in its infancy in materials science.

Use computational materials science coupled to combinatorial experimentation to discover what attributes (or combination of attributes) in a material may govern specific properties. This objective addresses the role of data mining as a tool for *description*. Since no single ab initio scientific theory can address all the relevant properties of a catalyst, we need a data-mining tool that helps suggest whch attributes of a given material over many length scales (e.g., crystal structure, chemical stoichiometry, grain size, processing route) may govern macroscopic behavior (chemical characteristics, photonic properties, mechanical properties etc.). The experimental aspects of combinatorial materials sciences need to be intelligently designed so as to generate a data warehouse that can have value beyond short-term goals. Organizing and developing relevant data sets (both experimental as well as "virtual") forms the foundations of data warehousing. The classification of these structural descriptors has to be coupled, to varying degrees, with other attributes regarding materials chemistry, materials properties, and processing parameters. Integrating this information into computational databases allows one to calculate a multitude of chemical and physical properties of the given structure.

Use data mining to select from a combinatorial or any other type of library those compounds most likely to have desired properties. This objective addresses the role of data mining as a tool for *prediction*. To meet the challenge of the nature of the data sets, a strategy is needed that combines complementary data-mining tools. For instance, advanced techniques such as association mining are well suited to descriptive knowledge discovery tasks, while others, such as support vector machines, are well suited to the predictive tasks. Despite having highly flexible underlying models, no one data-mining technique is sufficient to overcome the data difficulties described earlier or to accomplish both the descriptive and predictive tasks. Having an accurate predictive capability is critical for materials science applications and can and will be used to help search for materials with desired properties from both experimentally based libraries as well as "virtual" libraries, based on simulation techniques. In fact, data mining can be used to guide simulation experiments as well as combinatorial experimental libraries.

The full scientific potential of combinatorial experiments can only be achieved by integrating computational materials science techniques with the advanced tools of data mining. In this context, "computational informatics" represents a new and emerging field that can significantly revolutionize materials science research.

ACKNOWLEDGMENTS

The author gratefully acknowledges the work of his students who have contributed primarily to the work described here, C. Suh, A. Rajagopalan, and X. Li. The author also wishes to acknowledge the support of the National Science Foundation through the International Materials Institute for Materials Informatics and Combinatorial Materials Science (DMR-0231291).

REFERENCES

1. Car, R.; Parrinello, M. Unified approach for molecular dynamics and density functional theory. Phys. Rev. Lett. 1985, 55, 2471.
2. Kohn, W.; Sham, L.J. Self-consistent equations including exchange and correlation effects. Phys. Rev. 1965, 140A, 1133.
3. Rajan, K.; Suh, C.; Rajagopalan, A.; Li, X. Quantitative structure–activity relationships (QSARs) for materials science. In: Artificial Intelligence and Combinatorial Materials Science; Takeuchi, I., ed.; MRS: Pittsburgh, 2002, In press.
4. Ashby, M.F. Materials Selection in Mechanical Design; Pergamon Press: Oxford, 1972.
5. Chiang, Y-M; Birnie, D., III; Kingery, W.D. Physical Ceramics; Wiley: New York, 1997.
6. Lacorre, P.; Goutenoire, F.; Bohnke, O.; Retoux, R.; Laligant, Y. Designing fast ion conductors based on $La_2Mo_2O_9$. Nature. 2000, 404, 856–858.
7. MacGlashan, G.S.; Andreev, Y.G.; Bruce, P.R. Structure of the polymer electrolyte poly(ethylene oxide)$_6$: $LiAsF_6$. Nature. 1999, 398, 792–794.
8. Friedrichs, O.D.; Dress, A.W.M.; Hudson, D.H.; Klinowski, J.; Mackay, A.L. Systematic enumeration of crystalline networks. Nature. 1999, 400, 644–647.
9. Pearson, W.B. The Crystal Chemistry and Physics of Metals and Alloys; Wiley Interscience: New York, 1972.
10. Heyes, S.J. Structures of simple inorganic solids. http://www.chem.ox.ac.uk/icl/heyes/structure_of_solids/Strucsol.html; 1999.
11. Pettifor, D.G. Structure maps for pseudobinary and ternary phases. Mat. Sci. Technol. 1988, 4, 675–691.
12. Nye, J.F. Physical Properties of Crystals; Oxford University Press: Oxford, 1957.
13. Kattner, U.R. The theromodynamic modeling of multicomponent phase equilibria. J. Metals. 1997, 40, 14–19.
14. Glicksman, M.E. Diffusion in Solids: Field Theory, Solid-State Principles, and Applications; Wiley: New York, 1999.

15. Preisendorfer, R.W. Principal Component Analysis in Meterology and Oceanography; Elsevier: Amsterdam, 1988.
16. Godden, J.; Stahura, F.L.; Bajorath, J. Variability of molecular descriptors in compound databases revealed by Shannon entropy calculations. J. Chem. Inf. Comput. Sci. 2000, 40, 796–800.
17. Rajan, K.; Rajagopalan, A.; Suh, C. Data mining and multivariate analysis in materials science. In: Molten Salts—Fundamentals to Applications; Gaune-Escard, M., ed.; Kluwer Academic: Dordrecht, 2002, 241–248.
18. Suh, C.; Rajagopalan, A.; Li, X.; Rajan, K. Applications of principal component analysis in materials science. Data Sci. J. 2002, 1, 19–26.
19. Nagamatusu, J.; Nakagawa, N.; Muranaka, T.; Zenitani, Y.; Akimitsu, J. Superconductivity at 39K in magnesium boride. Nature. 2001, 410, 53–65.
20. Villiars, P.; Phillips, J.C. Quantum structure diagrams and high-temperature T_c superconductivity. Phys. Rev. B. 1988, 37, 2345–2348.
21. Zaki, M.; Rajan, K. Data mining: a tool for materials discovery. Proceedings of 17th CODATA meeting; Baveno: Italy, 2002.
22. Bennett, K.; Michaels, A.; Momma, M.; Suh, C.; Rajan, K. Classification of superconductors via support vector machines. In: Artificial Neural Networks in Engineering; 2002, In press.
23. Fayyad, U.; Piatetsky-Shapiro, G.; Smyth, P. From data mining to knowledge discovery: an overview. In: Advances in Knowledge Discovery and Data Mining; Fayyad, U., Piatetsky-Shapiro, G., Smyth, P., Uthurusamy, R., eds.; MIT Press: Cambridge, MA, 1996, 1–36.

Index